I0051726

Molecular Analyses

Medical Genomics and Proteomics

Series Editor: Scott O. Rogers

Science Publisher: Charles R. Crumly, CRC Press/Taylor & Francis Group

Published Titles

Molecular Analyses
Edited by Scott O. Rogers

For more information about this series, please visit: https://www.routledge.com/Medical-Genomics-and-Proteomics/book-series/CRCMOLGENPRO

Molecular Analyses

Edited by

Scott O. Rogers

CRC Press

Taylor & Francis Group

Boca Raton London New York

CRC Press is an imprint of the
Taylor & Francis Group, an **informa** business

First edition published 2022
by CRC Press
6000 Broken Sound Parkway NW, Suite 300, Boca Raton, FL 33487-2742

and by CRC Press
4 Park Square, Milton Park, Abingdon, Oxon, OX14 4RN

CRC Press is an imprint of Taylor & Francis Group, LLC

© 2022 Taylor & Francis Group, LLC

Reasonable efforts have been made to publish reliable data and information, but the author and publisher cannot assume responsibility for the validity of all materials or the consequences of their use. The authors and publishers have attempted to trace the copyright holders of all material reproduced in this publication and apologize to copyright holders if permission to publish in this form has not been obtained. If any copyright material has not been acknowledged please write and let us know so we may rectify in any future reprint.

With the exception of Chapter 5, no part of this book may be reprinted or reproduced or utilised in any form or by any electronic, mechanical, or other means, now known or hereafter invented, including photocopying and recording, or in any information storage or retrieval system, without permission in writing from the publishers.

Chapter 5 of this book is available for free in PDF format as Open Access at www.taylorfrancis.com. It has been made available under a Creative Commons Attribution-NonCommercial-ShareAlike (CC-BY-NC-SA) 4.0 International license.

Except as permitted under U.S. Copyright Law, no part of this book may be reprinted, reproduced, transmitted, or utilized in any form by any electronic, mechanical, or other means, now known or hereafter invented, including photocopying, microfilming, and recording, or in any information storage or retrieval system, without written permission from the publishers.

For permission to photocopy or use material electronically from this work, access www.copyright.com or contact the Copyright Clearance Center, Inc. (CCC), 222 Rosewood Drive, Danvers, MA 01923, 978-750-8400. For works that are not available on CCC please contact mpkbookspermissions@tandf.co.uk

Trademark notice: Product or corporate names may be trademarks or registered trademarks and are used only for identification and explanation without intent to infringe.

Library of Congress Cataloging-in-Publication Data

Names: Rogers, Scott O., 1953- editor.
Title: Molecular analyses / edited by Scott O. Rogers.
Description: First edition. | Boca Raton : CRC Press, 2022. |
Series: Medical genomics and proteomics | Includes bibliographical references and index.
Identifiers: LCCN 2021056926 (print) | LCCN 2021056927 (ebook) |
ISBN 9780367903718 (hardback) | ISBN 9781032161907 (paperback) |
ISBN 9781003247432 (ebook)
Subjects: LCSH: Nucleotide sequence.
Classification: LCC QP625.N89 M65 2022 (print) | LCC QP625.N89 (ebook) |
DDC 572.8/633--dc23/eng/20211209
LC record available at https://lccn.loc.gov/2021056926
LC ebook record available at https://lccn.loc.gov/2021056927

ISBN: 978-0-367-90371-8 (hbk)
ISBN: 978-1-032-16190-7 (pbk)
ISBN: 978-1-003-24743-2 (ebk)

DOI: 10.1201/9781003247432

The Open Access version of chapter 5 was funded by Dr. Mark Benecke.

Contents

Preface

In 2005, the Encyclopedia of Medical Genomics and Proteomics (MGP) was published. It consisted of two large volumes of analytical methods and equipment that were applicable to medical and other studies of genomics and proteomics. While it was a comprehensive collection at the time, many variations, improvements, and new technological developments have occurred during the past 17 years. Also, the alphabetical arrangements of the chapters made searching by topic somewhat cumbersome. Finally, the Encyclopedia was originally available in print version only. All of these factors necessitated an extensive revision in content, format, and organization, with the new title, Series in MGP. Currently, at least one dozen volumes are planned, including those in methods of fluorescence in situ hybridization (FISH), analyses for cancer detection, bacterial pathogens, and others.

This is the inaugural volume in the Series in MGP, entitled Molecular Analyses. The volume and the series are roughly based on the Encyclopedia of MGP, published in 2005. However, there has been significant updating and reorganization of the chapters. This volume is divided into five sections including 35 chapters. The first section includes chapters on the extraction and preservation of nucleic acids. This is followed by sections on the detection of nucleic acids, delivery of nucleic acids using both viral and nonviral delivery systems, methods of nucleic acid sequencing and analyses, and a variety of other methods for analyzing nucleic acids as well as proteins. Some of the original chapters have been included because they are still relevant, while others are included because they contain information of historical interest. That is, they show some of the historical developmental stages of methods and analyses currently being employed. Many of these may be important to the development of new technological directions. Additionally, several chapters have been updated to include new developments in specific areas. Several new chapters also have been added that describe some of the latest methods and analyses used in molecular research.

Additional volumes are currently in active development.

Scott Orland Rogers is a professor emeritus of molecular biology and evolution at Bowling Green State University, Bowling Green, Ohio. He received his PhD in plant molecular biology from the University of Washington, Seattle. He was an assistant professor and associate professor at the State University of New York College of Environmental Science and Forestry before moving to BGSU. He has taught courses in biology, botany, cell physiology, molecular biology, molecular genetics, bioinformatics, and molecular evolution. Research in his lab includes studies of microbes and nucleic acids preserved in ice, life in extreme environments, group I introns, molecular microbial phylogenetics, microbial metagenomics/metatranscriptomics, ancient DNA, and plant development.

Contributors

Farid E. Ahmed
East Carolina University
Greenville, North Carolina, U.S.A.

Philip L. Beales
Institute of Child Health (UCL)
London, U.K.

Mark Benecke
International Forensic Research &
    Consulting
Cologne, Germany

Martin Bengtsson
TATAA Biocenter AB
Göteborg, Sweden

Paula M. Cannon
Childrens Hospital Los Angeles
University of Southern California Keck
    School of Medicine
Los Angeles, California, U.S.A.

Stuart C. Clarke
University of Southampton
Southampton, U.K.

Yvonne E. Cruickshank
University of Strathclyde
Glasgow, U.K.

Stephen Day
Third Wave Technologies
Madison, Wisconsin, U.S.A.

Dino A. De Angelis
Memorial Sloan-Kettering Cancer
    Center
New York, New York, U.S.A.

Luciana Cresta de Barros Dolinsky
Universidade Federal do Rio de Janeiro
    (UFRJ) and Universidade do Grande
    Rio (UNIGRANRIO)
Rio de Janeiro, Brazil

Hermann Einsele
Eberhard-Karls-Universität Tuebingen
Tuebingen, Germany

Falko Fend
Technical University Munich
Munich, Germany

Janin Germer
Ludwig-Maximilians-Universität
Munich, Germany

Sadanand Gite
First Light Diagnostics
Chelmsford, MA U.S.A.

John Gray
University of Toledo
Toledo, Ohio U.S.A.

Thomas Hankeln
Johannes Gutenberg Universität Mainz
Mainz, Germany

Holger Hebart
Eberhard-Karls-Universität Tuebingen
Tuebingen, Germany

James A. Higgins
United States Department of
    Agriculture–Agricultural Research
    Service
Beltsville, Maryland, U.S.A.

Martin Hofmann
Institute of Virology and Immunology
Mittelhaeusern, Switzterland

Christiane Honisch
SEQUENOM Inc.
San Diego, California, U.S.A.

Agnes Hotz-Wagenblatt
Deutsches Krebsforschungszentrum
Heidelberg, Germany

Tim J. Inglis
PathCentre
Perth, Western Australia, Australia

Alexandre Izmailov
Visible Genetics, Inc.
Toronto, Ontario, Canada

Christian Jurinke
SEQUENOM Inc.
San Diego, California, U.S.A.

Jonas Karlsson
TATAA Biocenter AB
Göteborg, Sweden

Kagan Kerman
Ege University
Bornova, Izmir, Turkey;
Japan Advanced Institute of Science
    and Technology
Ishikawa, Japan

C. William Kilpatrick
University of Vermont
Burlington, VT, U.S.A.

Mikael Kubista
TATAA Biocenter AB
Göteborg, Sweden

Juergen Loeffler
Eberhard-Karls-Universität Tuebingen
Tuebingen, Germany

Mikael Leijon
LightUp Technologies AB
Huddinge, Sweden

Mark Lim
Ambergen, Inc.,
Chelmsford, MA, U.S.A.

Adrian Linacre
University of Strathclyde
Glasgow, U.K.

Chunyu Liu
University of Chicago
Chicago, Illinois, U.S.A.

Andrea Mast
Third Wave Technologies
Madison, Wisconsin, U.S.A.

Philip Ng
Baylor College of Medicine
Houston, Texas, U.S.A.

Mehmet Ozsoz
Ege University
Bornova, Izmir, Turkey

Janos Posfai
New England Biolabs, Inc.
Beverly, Massachusetts, U.S.A.

Lluís Quintana-Murci
Institut Pasteur
Paris, France

R. J. Roberts
New England Biolabs, Inc.
Beverly, Massachusetts, U.S.A.

Scott O. Rogers
Bowling Green State University
Bowling Green, OH, U.S.A.

Kenneth Rothschild
Molecular Biophysics Laboratory,
    Photonics Center, Department of
    Physics, Boston University
Boston, MA, U.S.A.

Yayoi Sato
Chiba University
Chiba, Japan

Kathrin D. Schmidt
Eberhard-Karls-Universität Tuebingen
Tuebingen, Germany

Nancy Smyth Templeton
Baylor College of Medicine
Houston, Texas, U.S.A.

Robert J. Trumbly
University of Toledo
Toledo, Ohio, U.S.A.

Dirk van den Boom
SEQUENOM Inc.
San Diego, California, U.S.A.

Tamas Vincze
New England Biolabs, Inc.
Beverly, Massachusetts, U.S.A.

Ernst Wagner
Ludwig-Maximilians-Universität
Munich, Germany

Erin L. Weber
Childrens Hospital Los Angeles,
    University of Southern California
    Keck School of Medicine
Los Angeles, California, U.S.A.

Gunar Westman
TATAA Biocenter AB
Göteborg, Sweden

Neven Zoric
TATAA Biocenter AB
Göteborg, Sweden

1 Nucleic Acid Extraction from Diverse Samples

Scott O. Rogers
Bowling Green State University, Bowling Green,
Ohio, U.S.A.

CONTENTS

INTRODUCTION

DNA and RNA can be used in clinical, forensic, research, and environmental set-tings to identify, analyze, or investigate organisms of interest, including potential pathogens.[1–11] Often the results can be more rapid than traditional methods, such as culturing.[8] In many cases, organisms are difficult or impossible to culture or iden-tify using non-molecular methods.[12] However, nucleic acid sequencing and other molecular biology methods often can yield detailed information leading to identifi-cation at the species, strain, variety, and/or individual levels.[13–16]

There are confounders that can thwart or delay correct identification of disease organisms. In some cases, this can delay accurate treatment of patients with disease symptoms. In the case of forensic investigations, some of these confounders can lead to inadmissibility of evidence. In the worst case, the extraction of nucleic acids can completely fail. The first confounder is that in any biopsy, sputum, blood, stool, envi-ronmental, or other samples, there will be cells from many species included. Cells of various types will be those from a patient/suspect, and these will be of primary interest in analyses of tumors and other diseases specifically involving patient cells. Also present will be other microbial cells that are normally present as part of the biome of the individual, as well as any disease-causing organisms and cells from any surrounding environmental sources (including from other persons). Similarly, for a forensic or environmental sample, many different species will be present. Therefore, for any analysis of the samples, the various species must be separated from one another by means of sorting (e.g., selective culturing, microbial physiology testing, polymerase chain reaction methods, selective probes, and/or sequence analyses).

The second confounder is that nucleic acid extraction methods must be chosen that are effective for the organisms of interest. Extraction from human cells is fairly

DOI: 10.1201/9781003247432-1

routine, and commercial kits are readily available.[17] Similarly, for most viruses and bacteria, commercial kits are available that can be used to obtain purified RNA and/ or DNA. However, organisms that are encapsulated or form spores and archaeal cells that have tough cell walls and cell membranes may present challenges to obtaining adequate amounts of usable nucleic acids. Fungi present additional challenges, due to the presence of tough cell walls consisting of chitin and other biomolecules, the presence of chemicals that inhibit enzymes used in molecular biological procedures, and small genomes (leading to low DNA yields). Many protists also have tough cell walls, as well as inhibiting compounds, although many have relatively large genomes. Plants, many of which are toxic or carry pathogenic organisms, have tough cell walls and many produce compounds that inhibit efforts to extract and utilize DNA. Therefore, extraction methods that include breakage of cell walls, membranes, and removal of inhibiting compounds must be included for many types of organisms.

The third confounder is that a large majority of organisms have never been cultured, and many are known only from their DNA sequences. Therefore, the only methods that will currently lead to accurate identification of the organism are based on molecular biological methods leading to the determination of characteristic sequences. Another potential problem is that sequences are unavailable for many organisms, because they have yet to be sequenced. In this case, organisms that are closely related may provide sufficient information to determine a genus and/or species.

The fourth confounder is that the condition of the tissue is critical to obtaining accurate conclusions of taxonomic determinations.[2,3,16] As soon as cells begin to die and break apart, degradative enzymes and invading organisms begin to fragment the nucleic acids. DNA and especially RNA must be extracted soon after selection of the sample, or the sample should be immediately frozen in liquid nitrogen or dry ice. For forensic samples, this is crucial, because the initiation of nucleic acid degradation has likely already progressed to some extent. For environmental samples, if they are in an area that is warm and moist, the nucleic acids will degrade rapidly, and within a few hours to days, much of the RNA and DNA will be highly fragmented. However, if the samples are in very cold regions such that they are frozen and remain frozen, the nucleic acids can remain intact, sometimes for millions of years. Tissues that originate from desiccated tissues, even in warm areas, also can contain intact DNA for thousands of years. This is the case for natural mummifications where organisms (including humans) have died in dry areas, such as deserts, where they have been rapidly desiccated and remained so for long periods of time. Ice (e.g., in glaciers) is one of the best matrices for nucleic acid preservation. Nucleic acids are well preserved not only in ice or frozen deserts because of the cold temperatures but also because of the low concentrations of free water, which is needed for mobilization and activity of nucleases and other degradative molecules.

SELECTION OF TISSUES

As stated above, the selection of tissues is vital to obtaining usable nucleic acids. If the cells have been dead for some time (e.g., as in postmortem or forensic examinations), the nucleic acids may already be highly degraded. However, by using a larger piece of tissue, there may be enough fragments of longer DNAs or RNAs to be useful in further processes. If the tissues are in poor condition, then the sample must either be frozen rapidly for later extraction or the initial stages of extraction should be undertaken on site. Procedures to extract nucleic acids *in situ* have been published that require minimal equipment and reagents.[18] Once the nucleic acids are in an alcohol solution (70% or higher) or partially dried, they can be safely transported back to a lab, where the final steps can be completed. The partially extracted nucleic acids can also be sent to another location (including worldwide) without observable damage to them. While larger pieces of tissue ensure that a sufficient amount of nucleic acids will result, pieces as small as a few milligrams (and sometimes smaller) can often yield amounts of DNA or RNA that can be used for sequence analyses. Entire genomes can be determined from individual cells, although the success rates are higher when larger assemblages of cells are used.

DESCRIPTION OF METHODS

Success in the extraction of usable nucleic acids is dependent on several factors, including condition of tissue, size of sample, genome size, presence of inhibiting/interfering compounds/structures, and method used. Samples are usually ground frozen using dry ice or liquid nitrogen in a mortar and pestle, blender, or small grinder (e.g., coffee or spice mill), followed by the addition of hot (e.g., 65°C) extraction buffer.[1–11] Tissues can also be ground directly in the extraction buffer. However, grinding while frozen is preferred because the cells are broken while all of the nucleases are inactive. Grinding in room temperature or hot buffer can allow some activity of the nucleases prior to their inactivation by the extraction buffer. Usually, the buffer used for extraction contains a Tris (tris(hydroxymethyl)amino-methane) buffer solution (usually 10 mM) that is at pH 7.5–8.0, 1 mM EDTA (eth-ylenediaminetetraacetic acid), a detergent (usually SDS, sodium dodecyl sulfate, also called sodium lauryl sulfate; or CTAB, cetyltrimethylammonium bromide, also called hexadecyltrimethylammonium bromide), and a simple salt (e.g., NaCl, KCl, LiCl, or NH_4Cl). The Tris maintains the pH of the solution between 7.5 and 8.0, which is necessary because high pH denatures DNA and degrades RNA, while low pH degrades both RNA and DNA. The EDTA chelates magnesium ions (Mg^{2+}), which are cofactors for many of the nucleases. Thus, EDTA causes inac-tivation of nucleases. The detergents are used to disrupt the membranes to lyse the cells and the membrane-bound organelles, including nuclei. While the long hydrocarbon tail of SDS (the dodecyl sulfate) is anionic and will not associate with the DNA, the long hydrocarbon tail of the CTAB (the cetyltrimethylammo-nium) is cationic and will associate with the negatively charged phosphate groups of the nucleic acids, as well as negatively charged groups on proteins (including nucleases) and other compounds. These interactions are important to the func-tion of CTAB extraction buffers. The simple salts separate into their anions and cations, and the cations (most often Na^+) can bind to the DNA and RNA to form salts of those molecules, which can be readily used in molecular biological appli-cations and procedures.

Most extraction methods that use SDS-based extraction buffers are used for bacteria, animal cells, and generally for organism that can be lysed readily with a detergent. Often, bacteria need a pretreatment with an enzyme that will soften or break down the cell walls. Following lysis of the cells, alcohol (e.g., methanol, ethanol, butanol, and octanol have been reported) is added to the solutions. Sodium, potassium, lithium, and ammonium salts of DNA and RNA are insoluble in solutions of alcohol above 60–70%. The precipitated nucleic acid salts can be pelleted by centrifugation or looped out on a glass rod, washed with additional alcohol, partially dried, and then hydrated in a TE (Tris-EDTA; 10 mM Tris plus 1 mM EDTA; or 0.1X TE, which is 1 mM Tris plus 0.1 mM EDTA) solution to the desired concentration.

For organisms with tough cell walls (e.g., plants, fungi, and many protists), compounds that inhibit reagents used in molecular biological processes (e.g., polysaccharides, chitin, and melanin), or contain a mixture of cell types (e.g., environmental or stool samples), pretreatments or alternative methods (including use of purification columns) are necessary. To assure cell breakage of cells with tough cell walls, grinding while frozen is effective. Additional agents can be utilized, such as sand or glass beads. Filtering, centrifugation, affinity columns, and/or settling are effective for some types of samples. Some of the most effective methods for the removal of inhibiting compounds are those that employ CTAB. The cetyltrimethylammonium cation forms ionic bonds with the anionic phosphates of the nucleic acid polymers. Prior to precipitation of the nucleic acids, the proteins (including nucleases) are removed by mixing with chloroform or phenol (or both). Most proteins dissolve in the chloroform, while the nucleic acids remain in the aqueous layer. The CTA$^+$/DNA$^-$ salts in the aqueous phase are insoluble under low salt conditions (below about 0.4 M NaCl). By lowering the salt concentration, they can be precipitated away from other compounds that stay in solution under the same conditions (e.g., polysaccharides). The precipitated nucleic acids can then be rehydrated in a high salt (e.g., NaCl) buffer, which causes the sodium cations to replace the cetyltrimethylammonium cations on the nucleic acids. The nucleic acids can then be separated from the CTAB by adding alcohol, which causes precipitation of the sodium salts of the nucleic acids. They can then be pelleted, washed, partially dried, and rehydrated in TE or 0.1X TE for use in molecular biological procedures.

TRIzol (Guanidinium isothiocyanate, phenol, chloroform, at low pH) is a reagent based on a method that has been used successfully for decades to isolate high-quality RNA.[19] It can also be used to isolate DNA, but other methods (described above) usually result in higher yields, so TRIzol is primarily utilized for RNA extraction rather than DNA extraction. The guanidinium isothiocyanate rapidly inactivates enzymes, including nucleases, such as RNases. Phenol denatures proteins, so the two reagents essentially inactivate all of the proteins, while leaving the nucleic acids unmodified. Chloroform is then added, which causes a phase separation, resulting in an aqueous phase containing the nucleic acids and the chloroform phase containing most of the inactivated and denatured proteins. The RNA can then be further purified by precipitation with alcohol (usually ethanol), pelleting by centrifugation, partial drying of the pellet, and rehydration in a buffer (usually a Tris-EDTA solution). The RNA is immediately usable in a variety of molecular biological procedures, such as gel electrophoresis, cDNA synthesis and sequencing, hybridization assays, and others.

In addition to the test tube and microfuge tube protocols (described above), several methods have been developed that utilize materials that bind nucleic acids, which are packed into columns.[20–24] The most common column-based method for purifying nucleic acids employs silica, in combination with a salt, usually NaCl. Polymers of silica are produced that are bound or packed into a column (usually a small plastic tube) that resembles a pipette. The columns are activated by adding alcohol, which is allowed to pass completely through the column. The sample is combined with a buffer that includes NaCl and then loaded onto the column. Inside the column, the Na^+ cations form a bridge between the negatively charged nucleic acids and the negatively charged silica, binding the nucleic acids to the column. Additional buffer is allowed to flow through the column in order to remove debris and contaminating compounds. Then, an elution buffer is added that removes the Na^+ bridges, thus releasing the nucleic acids from the column. They can then be precipitated with alcohol (as described above) and immediately utilized or frozen.

YIELDS

While the methods described above have been successful for extractions from a broad diversity of tissues, nucleic acid yields are affected by a number of factors. As mentioned above, the condition of the sample is important to the success of extraction of usable nucleic acids, and this is especially true for extraction of RNA. If RNA is the desired class of nucleic acids, the sample must be freshly obtained from living tissue, or it can originate from a frozen sample (either from a freezer or from an environmental frozen source). DNA is much more stable than RNA, so a small amount of tissue degradation is not a major concern. DNA has been obtained from samples in glacial ice up to several millions of years old. It has also been obtained from mummified corpses, as well as from bone marrow and tooth pulp up to tens of thousands of years old. All methods have some strengths and weaknesses. Tube methods take more time, but the yields are generally higher than with columns. Some tissues contain higher amounts of salt, or vary in pH, and therefore adjustments must be made for many types of samples. Some samples contain large amounts of water, while mummified tissues are nearly completely desiccated. Adjustments must be made for each of those situations. Stool samples have a mixture of microbes, degraded food, water, enzymes, and cells from the person/animal. Kits specifically made for stool samples are available commercially to mitigate the problems associated with this type of mixture of components. Some tissues have high amounts of inhibiting molecules (e.g., polysaccharides), and methods must be chosen (e.g., CTAB methods) that separate those molecules from the nucleic acids. Commercially available extraction kits are designed to work for tissues and species that are commonly the focus of DNA and RNA studies and evaluations. However, they sometimes produce low yields from uncommonly used species and samples. Therefore, for specialized applications, it is advantageous to optimize customized extraction procedures rather than relying on commercially available kits.

CONCLUSIONS

Nucleic acids can be recovered from nearly any sample, but the yield and condition of the nucleic acids can vary greatly. Careful consideration of the degree of tissue degradation, as well as the quantity of nucleic acids needed, must be made prior to choosing a method of preservation and extraction in order to achieve the highest yield of usable nucleic acids for the planned use.

REFERENCES

1. Kim, W.K.; Mauthe, W.; Hausner, G.; Klassen, G.R. Isolation of high molecular weight DNA and double-stranded RNAs from fungi. Can. J. Bot. **1990**, *68*, 1898–1902.
2. Rogers, S.O. Phylogenetic and taxonomic information from herbarium and mummified DNA. In *Conservation of Plant Genes II: Utilization of Ancient and Modern DNA*; Adams R.P.; Miller J.; Golenberg E.; Adams J.E., Eds.; Missouri Botanical Gardens Press: St. Louis, MO, **1994**; 47–67.
3. Rogers, S.O.; Bendich, A.J. Extraction of DNA from milligram amounts of fresh, herbarium and mummified plant tissues. Plant Mol. Biol. **1985**, *5*, 69–76.
4. Rogers, S.O.; Bendich, A.J. Extraction of DNA from plant tissues. In *Plant Molecular Biology Manual*; Gelvin, S.B.; Shilperoort R.A., Eds.; Kluwer Academic Publishers, Boston, MA, **1988**; *A6*, 1–10.
5. Rogers, S.O.; Bendich, A.J. Extraction of total cellular DNA from plants, algae and fungi. In *Plant Molecular Biology Manual*, 2nd Ed.; Gelvin S.B.; Shilperoort R.A., Eds.; Kluwer Academic Press, Dordrecht, the Netherlands, **1994**; D1, 1–8.
6. Rogers, S.O.; Kaya, Z. DNA from ancient cedar wood from King Midas' Tomb, Turkey, and Al-Aksa Mosque, Israel. Silvae Genet. **2006**, *55*, 54–62.
7. Rogers, S.O.; Rehner, S.; Bledsoe, C.; Mueller, G.J.; Ammirati, J.F. Extraction of DNA from Basidiomycetes for ribosomal DNA hybridizations. Can. J. Bot. **1989**, *67*, 1235–1243.
8. Sidiq, F.; Hoostal, M.; Rogers, S.O. Rapid identification of fungi in culture-negative clinical blood and respiratory samples by DNA sequence analysis. BMC Res. Notes **2016**, *9*, 293; doi: 10.1186/s13104-016-2097-0
9. Shivji, M.S.; Rogers, S.O.; Stanhope, M.J. Rapid isolation of high molecular weight DNA from marine macroalgae. Mar. Ecol. Prog. Ser. **1992**, *84*, 197–203.
10. Venkateswaren, K.; Vaishampayan, P.; Cisneros, J.; Pierson, D.L.; Rogers, S.O.; Perry, J. International space station environmental microbiome – microbial inventories of ISS filter debris. Appl. Microbiol. Biotechnol. **2014**, *98*, 6453–6456.
11. Vingataramin, L.; Frost, E.H. A single protocol for extraction of gDNA from bacteria and yeast. BioTechniques **2018**, *58*; doi:10.2144/000114263
12. Hug, L.; Baker, B.J.; Anantharaman, K.; Brown, C.T.; Probst, A.J.; Caselle, C.J.; Butterfield, C.N.; Hernsdorf, A.W.; Amano, Y.; Ise, K.; Suzuki, Y.; Dudek, N.; Relman, D.A.; Finstad, K.M.; Amundson, R.; Thomas, B.C.; Banfield, J.F. A new view of the tree of life. Nat. Microbiol. **2016**, *1*, 16048; doi:10.1038/nmicrobiol.2016.48
13. Castello, J.D.; Rogers, S.O.; Bachand, G.; Fillhart, R.; Murray, J.S.; Wiedmann, K.; Bachand, M.; Almond, M.A. Detection and partial characterization of tenuiviruses from black spruce. Plant Dis. **2000**, *84*, 143–147.
14. Ma, L.; Rogers, S.O.; Catranis, C.; Starmer, W.T. Detection and characterization of ancient fungi entrapped in glacial ice. Mycologia **2000**, *92*, 286–295.
15. Rogers, S.O.; Shtarkman, Y.M.; Koçer, Z.A.; Edgar, R.; Veerapaneni, R.; D'Elia, T. Ecology of subglacial Lake Vostok (Antarctica), based on metagenomic/metatranscriptomic analyses of accretion ice. Biology **2013**, *2*, 629–650.

16. Shtarkman, Y.M.; Koçer, Z.A.; Edgar, R.; Veerapaneni, R.S.; D'Elia, T.; Morris, P.F.; Rogers, S.O. Subglacial Lake Vostok (Antarctica) accretion ice contains a diverse set of sequences from aquatic, marine and sediment-inhabiting Bacteria and Eukarya. PLoS ONE **2013**, *8*(7), e67221; doi:10.1371/journal.pone.0067221

17. Ghatak, S.; Muthukumaran, R.B.; Nachimuthu, S.K. A simple method of genomic DNA extraction from human samples for PCR-RFLP analysis. J. Biomol. Tech. **2013**, *24*(4), 224–231.

18. Abu Almakarem, A.S.; Heilmann, K.L.; Conger, H.L.; Shtarkman, Y.M.; Rogers, S.O. Extraction of DNA from plant and fungal tissues *in situ*. BMC Res. Notes **2012**, *5*, 266; doi:10.1186/1756-0500-5-266

19. Chomczynski, P.; Sacchi, N. Single-step method of RNA isolation by acid guanidinium thiocyanate-phenol-chloroform extraction. Anal. Biochem. **1987**, *162*(1), 156–159.

20. Cady, N.C.; Stelick, S.; Batt, C.A. Nucleic acid purification using microfabricated silicon structures. Biosens. Bioelectron. **2003**, *19*, 59–66.

21. Karp, A.; Isaac, P.G.; Ingram, D.S. Isolation of nucleic acids using silica-gel based membranes: Methods based on the use of QIAamp spin columns. In Molecular Tools for Screening Biodiversity; Karp, A.; Issac, P.G.; Ingram, D.S., Eds; Springer, Dordrecht, **1998**, 59–63; doi:10.1007/978-94-009-0019-6_14

22. Liu, L.; Guo, Z.; Huang, Z.; Zhuang, J.; Yang, W. Size-elective separation of DNA fragments by using lysine-functionalized silica particles. Sci. Rep. **2016**, *6*, 22029.

23. Melzak, K.A.; Sherwood, C.S.; Turner, R.F.B.; Haynes, C.A. Driving forces for DNA adsorption to silica in perchlorate solutions. J. Colloid Interface Sci. **1996**, *181*, 635–644.

24. Tian, H.; Hühmer, A.F.; Landers, J.P. Evaluation of silica resins for direct and efficient extraction of DNA from complex biological matrices in a miniaturized format. Anal. Biochem. **2000**, *283*(2), 175–191.

2 DNA Extraction from Mummified Tissues

Yayoi Sato
Chiba University, Chiba, Japan

CONTENTS

INTRODUCTION

Nucleic acids extracted from mummified tissues are considered valuable materials for the identification of forensic cases and for the study of ancient human beings in the field of archaeoanthropology and the emergence of pathogens in Paleontology.

After the powerful polymerase chain reaction (PCR) technology was developed, amplification of a minute amount of DNA extracted from various archived tissues could be performed. In the analysis of degraded or ancient DNA, the extraction process represents one of the critical stages. However, significant difficulty in extracting nucleic acids from mummified tissues has been reported due to chemical modification, degradation, small amounts, inhibition, and contamination over a long period. These problems can make amplification of the nucleic acids very difficult. From a variety of tissue samples including aorta, brain, skin, genitals, ilium, penis, lung, muscle, connective tissue, DNA extraction was carried out using either organic solvents (such as phenol/chloroform) or nonorganic solvents (salt) method. The amount of DNA retained in the mummy is assumed to be minute and degraded, so several modification methods are applied so as to achieve efficient extraction. Here, several

DOI: 10.1201/9781003247432-2

representative protocols for extracting DNA from mummified tissues are presented and the precautions in the analysis are described.

SAMPLES

A variety of tissue samples from mummies have been subjected to DNA extraction, including aorta,[1] brain,[2,3] skin,[4,5] genitals, ilium, penis,[6,7] lung,[8] muscle, connective tissue.[9]

SAMPLE PREPARATION

Samples should be obtained with extreme caution to avoid contamination from other DNA sources. Pre- and post-PCR operations should be performed in separate rooms, and a laminar flow cabinet is used during extraction and purification steps. In each run, there is a negative control (water) used to detect contamination in the reagents.

The surfaces of the biopsy sites are cleaned and tissue samples are obtained from deep sites. After washing with sterile distilled water, tissues are cut into small pieces and then homogenized with phosphate-buffered saline (PBS)[5,6] or pulverized with a mill after lyophilization in the liquid nitrogen.[1,6,8] The sample sizes were 1.6 g, 0.5–1 g, 5-mm³ size,[2,4,6] and 50 mg, 0.5 g, 1 g,[1,5,6] respectively.

DNA EXTRACTION

DNA extraction from mummified tissues is carried out using either organic solvents (such as phenol/chloroform) or nonorganic solvents (salt) to remove proteins. The amount of DNA retained in the mummy is assumed to be minute and degraded, so several modification methods are applied so as to achieve efficient extraction. The representative methods are described here. The sample size is determined depending on the degree of degradation and the amount of retained DNA in the mummified tissues.

METHOD 1 (NONORGANIC SOLVENTS METHOD: PUREGENE™ DNA ISOLATION KIT)

The PUREGENE™ DNA Isolation kit (Gentra Systems Minneapolis, MN) is used to extract DNA from mummified tissues.[1,10]

In this kit, purification of genomic DNA is carried out using ammonium acetate as a substitute for toxic organic solvents in the protein precipitation step. The pulverized tissue is mixed with cell lysis solution containing Tris–HCl, ethylenediaminetetraacetic acid (EDTA), and sodium dodecyl sulfate (SDS), and incubated at 65°C for 60 min. After proteinase K solution (20 mg/mL) was added to the lysate, incubation at 55°C is carried out until tissue particulates have dissolved. RNase A solution (4 mg/mL) is added to the lysate and incubated at 37°C for 60 min. Protein precipitation solution (ammonium acetate) is added to the solution and vortexed vigorously then centrifuged. The DNA is precipitated with 100% isopropanol (2-propanol). After centrifugation, the pellet is washed several times with 70% ethanol and then air-dried. Purification is carried out with a Microcon™ YM-100 microconcentrator (Millipore, Bedford, MA). Finally, the DNA is hydrated with sterile distilled water. The extract is quality-checked by agarose gel electrophoresis and determined by a UV spectrophotometer. The E_{260}/E_{280} ratio is recommended to be greater than 1.5, as the success rate of PCR is thought to depend on the purity. When the DNA yield is expected to be low (<1 μg), DNA carrier such as glycogen solution (20 mg/mL) is added.

METHOD 2 (ORGANIC SOLVENTS METHOD: PHENOL/CHLOROFORM EXTRACTION)

This method has been used conventionally and is still widely used for degraded and aged samples. Many modifications are presented for the kind of organic solvents or reagents of digestion and so on. Here the representative method is described as applied to the extraction of mummified tissues.

5 M GTC Digestion Method

Guanidium thiocyanate (GTC) is a chaotropic agent commonly used for tissue lysis of archived specimens.[6,7] After homogenization, the pellets are lysed in a 5 M GTC buffer containing 5 M GTC, 0.5% bovine serum albumin, 80 mM EDTA, 400 mM Tris–HCl (pH 7.5), and 0.5% sodium N-lauroyl sarcosine at 60°C for 1 hr and then at 37°C overnight. DNA is extracted twice with phenol–chloroform at a 1:1 ratio, followed by chloroform once, and then precipitated by the addition of a 1/10 volume of 3 M sodium acetate (pH 5.2) and 2.5 volumes of absolute ethanol. The pellets are washed with 70% ethanol and air-dried. They are then dissolved in Tris–EDTA (10 mM Tris–HCl [pH 8.0], 1 mM EDTA) buffer.

Proteinase K Digestion Method

Samples are put into digestion buffer (10 mM Tris–HCl, pH 8.0, 10 mM EDTA, 50 mM NaCl, 2% SDS) containing proteinase K solution (10 mg/mL) and incubated for several hours at 56°C.[1,8] Additional aliquots of proteinase K are added to achieve complete digestion. Next, the solution is extracted with phenol/chloroform/isoamyl alcohol (25:24:1) by a mild mixer. After centrifugation, the upper phase is transferred to a new tube. The extraction is repeated until the interface is clear. Finally, one more extraction is performed with chloroform to remove phenol, and the aqueous phase is put into another tube. After the addition of 5 M NaCl solution, DNA is precipitated with 100% ethanol. The recovered DNA is washed three times with 70% ethanol and air-dried. It is hydrated with sterile distilled water.

Microconcentrator-Based Method

Samples are put into the solution containing 10 mM Tris–HCl (pH 8.0), 2 mM EDTA, and 10 mM NaCl.[2,8,11] The tissue is broken up into small pieces by vortexing, and then collagenase is added. The tubes are incubated at 37°C with slow agitation for 3 hr. SDS is then added as well as dithiothreitol (DTT) and proteinase K. Incubation is continued for approximately 20 hr. The solution is extracted twice with the water-saturated phenol of neutral pH and once with chloroform/isoamyl alcohol (24:1). It was concentrated by a Centricon 30 microconcentrator.

Silica-Based Method

Extracts of DNA are made from each sample by a silica-based method that is highly efficient in the retrieval of ancient DNA.[5,12] DNA extraction is carried out with the supernatants in an automated nucleic acid extractor, starting with proteinase K digestion at 56°C for 1 hr. A standard phenol/chloroform extraction is carried out, followed by mixing the samples with a silica powder (glass milk, Dianova). This process produces a binding of DNA to the glass beads in the presence of isopropanol and sodium acetate (2.0 M, pH 4.5). In the last phase of the automated extraction procedure, the DNA/glass milk samples are collected on filtration membranes and washed with ethanol. Finally, the DNA is manually eluted from the silica beads with sterile water and run on a 3% agarose gel.

CONCLUSION

The results obtained from the method described above are listed in Table 2.1. A PEREGENE™ DNA Isolation kit is useful for extracting DNA from relatively fresh mummified tissues, compared to the archaeological samples, and it presents no hazard.[1,10] For tissues preserved for thousands of years, the phenol/chloroform method may be superior to the kit. Konomi et al.[6] have compared nine extraction methods

TABLE 2.1
DNA Extraction from Mummified Tissues

Sample	Time after Death	Extraction Method	PCR Success Rate	Ref.
Aorta	1.5 years old	NO	1/1 (HLA)	[1]
Skin of leg	2400 years old	PC(P)	1/23 (cloned and sequenced)	[4]
Brain	7000 years old	PC(PM)	1/1 (mtDNA)	[2]
Skin of hip, extremities	>500 years old	PC(PS)	8/17 (Y-specific 3.4 kb repeat sequence)	[5]
Genitals, ilium, penis	>800 years old	PC(G)	12/12 (GAPDH), 0/12 (hepatitis viruses)	[6]
Lung	3600 years old	PC(P)	25/48 (*Mycobacterium tuberculosis*)	[8]
Brain	8000 years old	PC(P)	1/1 (mtDNA)	[3]
Genitals	>800 years old	PC(G)	12/12 (GADPH), 2/12 (*M. tuberculosis* complex)	[7]

Abbreviations: (G)—GTC digestion; GAPDH—glyceraldehydes 3-phosphate dehydrogenase; NO—nonorganic method; (P)—proteinase K digestion; PC—phenol/chloroform method; (PM)—proteinase K digestion + microconcentrator; (PS)—proteinase K digestion + silica.

and concluded that GTC-based methods are more efficient than proteinase K/detergent-based methods for the recovery of both DNA and RNA from mummified tissues. However, proteinase K is a powerful protein lysis reagent and has been widely used. It is not clear which is better for digestion, as both are quite useful.

Although mummified tissues are well-preserved for thousands of years, the integrity of nucleic acids for molecular biological study still remains an issue.[13] Cooper and Wayne[14] commented that the extracted DNA from a 2400-year-old mummy was, in hindsight, likely a contaminant both because of its size and nature. Furthermore, it has been shown that more than 99% of DNA isolated from mummified tissues is depolymerized and chemically modified.[2,4,15]

It is important to prevent contamination, remove the inhibitors, achieve efficient recovery, and confirm authenticity when DNA is extracted from mummified tissues preserved for a long time. The study of ancient DNA poses numerous problems, some of which are presented below.

SMALL SIZE OF DNA

The DNA recovered from ancient tissues is severely degraded, so it is difficult to amplify successfully more than approximately 200-bp-long fragments.[2,12] In the typing of HLA (human leukocyte antigen) alleles from aorta tissue of an approximately 1.5-year-old mummy, we have been able to obtain specific PCR fragments sized approximately 100–200-bp long, but have sometimes failed in amplifying a 429-bp fragment used as a positive control.[1]

LOW AMOUNT OF DNA

DNA yields of approximately 70 ng/μL can be measured for bone samples, and yields of less than 2 ng/μL can be measured for soft-tissue samples.[5] We have recovered 23 ng of DNA per milligram of dried aorta tissue naturally preserved for approximately 1.5 years. PCR amplification fails when the amount of available DNA is too minute.[16] Both bone and soft-tissue samples are suitable for ancient DNA analysis, but that skeletal remains should be given preference when source materials are severely degraded.

AUTHENTICITY OF AMPLIFIED DNA

It is important to confirm whether the recovered DNA is authentic or not. For example, it should be noted whether it was derived from an authentic human, a modern human, an animal, a pathogen, or some other source. It remains possible that a sample can be contaminated during the experiment. It is therefore necessary for rigid quality criteria to be established to avoid more spurious and unsubstantiated reports.

Marota et al. have reported that DNA would be entirely degraded after slightly more than 800 years, and this result provides an indirect argument against the reliability of claims regarding the recovery of authentic DNA from Egyptian mummies and bone remains.[13]

Inhibitors

DNA extracts from ancient tissue remains often contain an unidentified component that inhibits polymerase activity. Inhibition can result from either reducing sugars[17] or an excess of coextracted microbacterial DNA,[18] the remains of porphyrines,[19] degraded nucleic acids,[20–22] and soil components such as humic and fulvic acids, tannins, or iron ions (Fe^{2+}),[23–25] or collagen type.[26] This inhibition may be overcome by the addition of bovine serum albumin and increased amounts of enzyme. Through the purification step by the microconcentrator device described above, the success rate of PCR is increased.

REFERENCES

1. Sato, Y.; Hayakawa, M.; Nakajima, T.; Motani, H.; Kiuchi, M. HLA typing of aortic tissues from unidentified bodies using hot start polymerase chain reaction-sequence specific primers. Legal Med. **2003**, *5*, S191–S193.
2. Pääbo, S.; Gifford, J.A.; Wilson, A.C. Mitochondrial DNA sequences from a 7000-year old brain. Nucleic Acids Res. **1988**, *16*(20), 9775–9787.
3. Doran, G.H.; Dickel, D.N.; Ballinger, W.E.; Agee, O.F.; Laipis, P.J.; Hauswirth, W.W. Anatomical, cellular and molecular analysis of 8000-yr-old human brain tissue from the Windover archaeological site. Nature **1986**, *323*, 803–806.
4. Pääbo, S. Molecular cloning of ancient Egyptian mummy DNA. Nature **1985**, *314*, 644–645.
5. Lassen, C.; Hummel, S.; Herrmann, B. Comparison of DNA extraction and amplification from ancient human bone and mummified soft tissue. Int. J. Legal Med. **1994**, *107*, 152–155.
6. Konomi, N.; Lebwohl, E.; Zhang, D. Comparison of DNA and RNA extraction methods for mummified tissues. Mol. Cell. Probes **2002**, *16*, 445–451.
7. Konomi, N.; Lebwohl, E.; Mowbray, K.; Tattersall, I.; Zhang, D. Detection of mycobacterial DNA in Andean mummies. J. Clin. Microbiol. **2002**, *40*(12), 4738–4740.
8. Zink, A.R.; Sola, C.; Reischl, U.; Grabner, W.; Rastogi, N.; Wolf, H.; Nerlich, A.G. Characterization of *Mycobacterium tuberculosis* DNAs from Egyptian mummies by spoligotyping. J. Clin. Microbiol. **2003**, *41*(1), 359–367.
9. Handt, O.; Richards, M.; Trommsdorff, M.; Kilger, C.; Simanainen, J.; Georgiev, O.; Bauer, K.; Stone, A.; Hedges, R.; Scaffner, W.; Utermann, G.; Sykes, B.; Pääbo, S. Molecular genetic analyses of the Tyrolean Ice Man. Science **1994**, *264*(5166), 1775–1778.
10. *PUREGENE™. DNA Isolation Kit Protocol*; Gentra Systems: Minneapolis, MN, **2000**.
11. Foo, I.; Salo, W.L.; Aufderheide, A.C. PCR libraries of ancient DNA using a generalized PCR method. BioTechniques **1992**, *12*(6), 811–815.
12. Kelman, L.M.; Kelman, Z. The use of ancient DNA in paleontological studies. J. Vertebr. Paleontol. **1999**, *19*, 8–20.
13. Marota, I.; Basile, C.; Ubaldi, M.; Rollo, F. DNA decay rate in papyri and human remains from Egyptian archaeological sites. Am. J. Phys. Anthropol. **2002**, *117*, 310–318.
14. Cooper, A.; Wayne, R. New uses for old DNA. Curr. Opin. Biotechnol. **1998**, *9*, 49–53.
15. Krings, M.; Capelli, C.; Tschentscher, F.; Geisert, H.; Meyer, S.; von Haeseler, A.; Grossschmidt, K.; Possnert, G.; Paunovic, M.; Pääbo, S. A view of Neanderthal genetic diversity. Nat. Genet. **2000**, *26*(2), 144–146.
16. Schmerer, W.M. Extraction of ancient DNA. Methods Mol. Biol. **2003**, *226*, 57–61.

17. Pääbo, S. Amplifying Ancient DNA. In *PCR Protocols*; Innis, M.A., Gelfand, D.H., Sninsky, J.J., White, T.J., Eds.; Academic Press: San Diego, CA, **1990**; 159–166.
18. Jeffreys, A.J.; Maxine, J.A.; Hagelberg, E.; Sonnberg, A. Identification of the skeletal remains of Josef Mengele by DNA analysis. Forensic Sci. Int. **1992**, *56*, 65–76.
19. Higuchi, R.G. Dr. Russ' problem corner. Ancient DNA Newsl. **1992**, *1*, 6–8.
20. Rogan, P.K.; Salvo, J.J. Study of nucleic acids isolated from ancient remains. Yearb. Phys. Anthropol. **1990**, *33*, 195–214.
21. Abrol, S.; Chaudhary, V.K. Excess of PCR primers inhibit DNA cleavage by some restriction endonucleases. BioTechniques **1993**, *15*, 630–632.
22. Pikaart, M.J.; Villeponteau, B. Suppression of PCR amplification by high levels of RNA. BioTechniques **1993**, *14*, 24–25.
23. Kumar, S.S.; Nasidze, I.; Walimbe, S.R.; Stoneking, M. Discouraging prospects for ancient DNA from India. Am. J. Phys. Anthropol. **2000**, *113*(1), 129–133.
24. Hagelberg, E.; Clegg, J. Isolation and characterization of DNA from archaeological bone. Proc. R. Soc. B. **1991**, *244*, 45–50.
25. Tuross, N. The biochemistry of ancient DNA in bone. Experientia **1994**, *50*, 530–535.
26. Scholz, M.; Giddings, I.; Pusch, C.M. A polymerase chain reaction inhibitor of ancient hard and soft tissue DNA extracts is determined as human collagen type. Anal. Biochem. **1998**, *259*, 283–286.

3 Commercial DNA Extraction Kits

Stuart C. Clarke
Stobhill Hospital, Glasgow, U.K.

CONTENTS

INTRODUCTION

Genomic DNA is used in a wide range of applications in molecular biology including disease diagnosis, food microbiology, and environmental microbiology. However, DNA analysis in many aspects of medicine and biology relies on the extraction of DNA from complex samples or environments. These may possess numerous contaminants depending on the sample type. In medicine, the extraction of human genomic DNA from tissue samples may be contaminated with commensal bacterial flora. In environmental microbiology, the analysis of water or soil samples may be contaminated by bacterial DNA, plant DNA, and minerals. Interestingly, the analysis of DNA has led to the revised classification of some soil bacteria. Even the complex environment of human blood hampers the detection of infection because of the presence of red blood cells, white blood cells, transient contaminant bacteria, and numerous components of the immune system. Therefore, the technique used for DNA extraction is important, as are the steps performed during the process such as cell lysis, DNA recovery, and DNA purification. In medical microbiology, there is a growing interest in molecular methods for the laboratory confirmation of infection and advances in molecular biology have promoted the routine use of techniques such as polymerase chain reaction (PCR) and DNA sequencing. However, difficulties exist with the extraction of genomic DNA from bodily fluids. Sensitive DNA extraction methods are required but the DNA extracted must be of sufficient quality to be used for the improved diagnosis and surveillance of microbial diseases. In all of the areas above, traditional phenol–chloroform extraction is not usually practical

DOI: 10.1201/9781003247432-3

TABLE 3.1

Examples of Use of DNA Extraction Methods

Type of DNA Required	Sample	Reason	Reference
Human genomic DNA	Paraffin-embedded tissue	Disease diagnosis	[9,10]
Bacterial or parasitical genomic DNA	Feces	Confirmation of bacterial or parasitic infection	[8,11]
Bacterial genomic DNA	Soil	Bacterial quantification	[1]
Viral DNA	Dried blood spots	Confirmation of viral infection	[12,13]
Bacterial DNA	Blood or CSF	Confirmation of bacterial infection	[4,14,15]

because of the time and safety constraints of the method. Here DNA extraction systems are compared in terms of ease of use, sensitivity, and their ability for automation. An emphasis is placed on their use in clinical microbiology, particularly in relation to the use of whole blood for the diagnosis of microbial infections.

SAMPLES

As described above, there are many areas where purified genomic DNA is required for analysis (Table 3.1).[1–8] In medicine, samples may include body fluids such as whole blood, serum, cerebrospinal fluid (CSF), and tissue among others. DNA purified from these sources must be pure, nuclease free, and of good quality (i.e., not sheared). Furthermore, the DNA must be of a quality to provide long-term storage as clinical samples, and indeed others, are often unrepeatable.

MAGNETIC BEAD DNA EXTRACTION METHODS

Some methods for DNA extraction utilize silica-coated magnetic beads onto which the DNA binds and allows extraction, washing, and elution of pure genomic DNA. Such methods include the Bilatest beads DNA extraction kit and Promega bead kit. They are particularly useful for extracting bacterial genomic DNA from whole blood. Typically, a lysis mixture is prepared by mixing magnetic particle suspension with proteinase K and lysis buffer. Premixed lysis solution is then added to the sample and incubated at 55°C to achieve lysis of the red and white blood cells. This is usually performed in a microtitre plate which is then placed on a magnetic separator to allow the metallic bead–DNA complex to sediment. The supernatant is discarded, the DNA washed, and the microtitre placed on the magnetic separator to allow the particle–DNA complex to sediment, and the supernatant discarded. The DNA is then reeluted by adding sterile distilled water and incubating at 65°C.

SILICA MEMBRANE DNA EXTRACTION METHODS

These kits can be used for the extraction of genomic DNA from whole blood or other bodily fluids and are again based on the principle that DNA binds to silica. They are available in single tube or 96-well plate formats. Lysis buffer is added to the sample in order to lyse red and white blood cells and incubated at 70°C. After incubation, ethanol is added, the sample vortexed, and placed on a vacuum manifold to draw the liquid phase through the filter. Washing is repeated and the liquid phase is again removed by vacuuming. Elution buffer is added directly to the silica membrane and incubated at room temperature briefly followed by vacuuming to elute the DNA.

FILTER-BASED DNA EXTRACTION METHODS

These kits are operationally similar to the silica membrane-based methods but, instead of the DNA binding to the silica membrane, the DNA is held on the filter by size exclusion. As for the other methods, lysis solution is added to the bodily fluid and left at room temperature for cell lysis to take place. After incubation, neutralization solution is added to the tube and mixed again. The lysate is transferred from the sample tube and dispensed into the corresponding well of a clearing plate placed on a vacuum manifold. The liquid phase is then drawn through the filter into the corresponding well of a collection plate and the filtrate is discarded. The DNA remaining in the clearing plate is washed by adding wash solution followed by vacuuming, and again the filtrate is discarded. A new collection plate is placed inside the vacuum manifold and the DNA is recovered by adding elution buffer.

AUTOMATION

Many methods in molecular biology require high-throughput sample processing, and therefore manual extraction of DNA extraction is not possible.[5] Although many commercial DNA extraction methods can be purchased in tube format, these still require a significant amount of hands-on time in order to add reagents, remove reagents, and perform centrifugation or vacuum steps.[4,5] Efficient methods of DNA extraction that produce pure, high-quality DNA are crucial to the success of PCR and sequencing reactions and the subsequent treatment of disease.[8,16] Current manual methods of DNA extraction are simple and reliable, and are suitable for the extraction of low numbers of samples.[17] As the demand for molecular tests increases, new automated methods of DNA extraction will have to be developed to handle larger numbers of clinical samples.[18] There are a number of DNA extraction kits available, based on a 96-well plate format, which allow integration into the workstation of a robotic liquid handler. This allows a much higher throughput of samples, is less work intensive, and produces PCR-ready bacterial DNA.[4,16,18] A number of commercial robotic systems are available for such purposes from companies such as MWG Biotech, Tecan, Qiagen, Beckman, and Roche.

COMPARISON OF DNA EXTRACTION METHODS

Purified bacterial DNA is required for many procedures in modern molecular biology, but clinical microbiology now routinely utilizes DNA for the laboratory confirmation of various infections.[4,8,12] The infections requiring a laboratory diagnosis are primarily of bacterial, viral, or parasite etiology. As such, infections of public health importance are now often diagnosed by PCR, and even typing information is provided on the organism causing infection. One such example ismeningococcal disease. For this infection, rapid, sensitive, and specific methods are required so that case contacts can be managed appropriately. Bacterial DNA can be amplified from clinical samples by PCR to characterize the infecting bacterium.[19–23] As well as dedicated DNA extraction robot systems, a number of commercially available DNA extraction kits can be automated[4,14] and can be modified appropriately to various platforms. The overall efficiency of such methods can be determined by throughput time, DNA yield, and labor intensity.[4] All three main methods of DNA extraction can be automated, but their overall efficiency differs according to manufacturer. Some commercial kits, such as the 96-well plate format of the Promega SV96 kit, allow easy adaptation for use on a liquid handling robot. A large number of samples can then be processed with little or no manual intervention. Such kits are extremely simple to use and can extract up to 96 samples in under 20 min. The kits based on the use of metallic beads are also very simple to automate. Some magnetic bead kits are very efficient and can rapidly extract high-quality DNA from a large number of samples automatically. Some kits produce more DNA per reaction than others although these may not necessarily be the best for automation and test specificity. Overall, it is difficult to compare methods and decide on the best because it depends on the sample type, downstream method required, and the assay type being performed. However, kits have been automated for the extraction of bacterial genomic DNA from whole blood samples and the sensitivity, specificity, and efficiency of the kits have been compared. In that study, the Bilatest Beads DNA 2 kit and the Promega Wizard SV96 kit were the most successful kits in all areas[4] and have subsequently been used successfully for the laboratory confirmation of septicemia and meningitis caused by *Neisseria meningitidis, Streptococcus pneumoniae*, and *Haemophilus influenzae*.[14] Ultimately, the choice of the kit depends on the needs of users, often whether they prefer a 96-well plate format or a metallic bead system.

CONCLUSION

DNA extraction methods are now becoming an essential component of molecular biology methods for the sensitive and specific amplification of genomic DNA. Numerous commercial kits are now available which utilize three main technologies for the purification of DNA, and these can be used in diverse applications within the specialties of medicine, environmental biology, and food microbiology. DNA extraction methods can also be automated, which can provide high-throughput analysis of samples.

ACKNOWLEDGMENTS

The author would like to thank Karen Smith, Strathclyde University, for useful comments on the manuscript. The author's laboratory is funded by the National Services Division of the Scottish Executive, Chief Scientist's Office of the Scottish Executive Health Department, and Meningitis Association (Scotland).

REFERENCES

1. Courtois, S.; Frostegard, A.; Goransson, P.; Depret, G.; Jeannin, P.; Simonet, P. Quantification of bacterial subgroups in soil: Comparison of DNA extracted directly from soil or from cells previously released by density gradient centrifugation. Environ. Microbiol. **2001**, *3*, 431–439.
2. Sadler, F.; Borrow, R.; Dawson, M.M.; Kaczmarski, E.B.; Cartwright, K.; Fox, A.J. Improved methods of detection of meningococcal DNA from oropharyngeal swabs from cases and contacts of meningococcal disease. Epidemiol. Infect. **2000**, *125*, 277–283.
3. Guiver, M.; Levi, K.; Oppenheim, B.A. Rapid identification of *Candida* species by TaqMan PCR. J. Clin. Pathol. **2001**, *54*, 362–366.
4. Smith, K.; Diggle, M.A.; Clarke, S.C. Comparison of commercial DNA extraction kits for extraction of bacterial genomic DNA from whole-blood samples. J. Clin. Microbiol. **2003**, *41*, 2440–2443.
5. Clarke, S.C. Nucleotide sequence-based typing of bacteria and the impact of automation. Bioessays **2002**, *24*, 858–862.
6. Guiver, M.; Borrow, R.; Marsh, J.; Gray, S.J.; Kaczmarski, E.B.; Howells, D.; Boseley, P.; Fox, A.J. Evaluation of the applied biosystems automated Taqman polymerase chain reaction system for the detection of meningococcal DNA. FEMS Immunol. Med. Microbiol. **2000**, *28*, 173–179.
7. Olive, D.M.; Bean, P. Principles and applications of methods for DNA-based typing of microbial organisms. J. Clin. Microbiol. **1999**, *37*, 1661–1669.
8. McOrist, A.L.; Jackson, M.; Bird, A.R. A comparison of five methods for extraction of bacterial DNA from human faecal samples. J. Microbiol. Methods **2002**, *50*, 131–139.
9. Bielawski, K.; Zaczek, A.; Lisowska, U.; Dybikowska, A.; Kowalska, A.; Falkiewicz, B. The suitability of DNA extracted from formalin-fixed, paraffin-embedded tissues for double differential polymerase chain reaction analysis. Int. J. Mol. Med. **2001**, *8*, 573–578.
10. Cao, W.; Hashibe, M.; Rao, J.Y.; Morgenstern, H.; Zhang, Z.F. Comparison of methods for DNA extraction from paraffin-embedded tissues and buccal cells. Cancer Detect. Prev. **2003**, *27*, 397–404.
11. Fretz, R.; Svoboda, P.; Ryan, U.M.; Thompson, R.C.; Tanners, M.; Baumgartner, A. Genotyping of *Cryptosporidium* spp. isolated from human stool samples in Switzerland. Epidemiol. Infect. **2003**, *131*, 663–667.
12. Knepp, J.H.; Geahr, M.A.; Forman, M.S.; Valsamakis, A. Comparison of automated and manual nucleic acid extraction methods for detection of enterovirus RNA. J. Clin. Microbiol. **2003**, *41*, 3532–3536.
13. Fischer, A.; Lejczak, C.; Lambert, C.; Servais, J.; Makombe, N.; Rusine, J.; Staub, T.; Hemmer, R.; Schneider, F.; Schmit, J.C.; Arendt, V. Simple DNA extraction method for dried blood spots and comparison of two PCR assays for diagnosis of vertical human immunodeficiency virus type 1 transmission in Rwanda. J. Clin. Microbiol. **2004**, *42*, 16–20.
14. Smith, K.; Diggle, M.A.; Clarke, S.C. Automation of a fluorescence-based multiplex PCR for the laboratory confirmation of common bacterial pathogens. J. Med. Microbiol. **2004**, *53*, 115–117.
15. Exner, M.M.; Lewinski, M.A. Isolation and detection of *Borrelia burgdorferi* DNA from cerebral spinal fluid, synovial fluid, blood, urine, and ticks using the Roche MagNA Pure system and real-time PCR. Diagn. Microbiol. Infect. Dis. **2003**, *46*, 235–240.
16. Read, S.J. Recovery efficiencies on nucleic acid extraction kits as measured by quantitative LightCycler PCR. Mol. Pathol. **2001**, *54*, 86–90.

17. Smit, M.L.; Giesendorf, B.A.; Heil, S.G.; Vet, J.A.; Trijbels, F.J.; Blom, H.J. Automated extraction and amplification of DNA from whole blood using a robotic workstation and an integrated thermocycler. Biotechnol. Appl. Biochem. **2000**, *32*(Pt. 2), 121–125.

18. Harris, D.; Engelstein, M.; Parry, R.; Smith, J.; Mabuchi, M.; Millipore, J.L. High-speed plasmid isolation using 96-well, size-exclusion filter plates. BioTechniques **2002**, *32*, 626–628.

19. Borrow, R.; Claus, H.; Guiver, M.; Smart, L.; Jones, D.M.; Kaczmarski, E.B.; Frosch, M.; Fox, A.J. Non-culture diagnosis and serogroup determination of meningococcal B and C infection by a sialyltransferase (siaD) PCR ELISA. Epidemiol. Infect. **1997**, *118*, 111–117.

20. Borrow, R.; Claus, H.; Chaudhry, U.; Guiver, M.; Kaczmarski, E.B.; Frosch, M.; Fox, A.J. siaD PCR ELISA for confirmation and identification of serogroup Y and W135 meningococcal infections. FEMS Microbiol. Lett. **1998**, *159*, 209–214.

21. Urwin, R.; Kaczmarski, E.B.; Guiver, M.; Fox, A.J.; Maiden, M.C. Amplification of the meningococcal porB gene for non-culture serotype characterization. Epidemiol. Infect. **1998**, *120*, 257–262.

22. Clarke, S.C.; Diggle, M.A.; Edwards, G.F. Automated nonculture-based sequence typing of meningococci from body fluids. Br. J. Biomed. Sci. **2001**, *58*, 230–234.

23. Clarke, S.C.; Diggle, M.A.; Edwards, G.F. Semiautomation of multilocus sequence typing for the characterisation of clinical isolates of *Neisseria meningitidis*. J. Clin. Microbiol. **2001**, *39*, 3066–3071.

4 Automated Nucleic Acid Extraction

Juergen Loeffler, Kathrin D. Schmidt,
Holger Hebart, Hermann Einsele
Eberhard-Karls-Universität Tuebingen,
Tuebingen, Germany

CONTENTS

INTRODUCTION

In the past, the extraction and purification of nucleic acids had been a complicated, time-, and labor-intensive process. Manual extraction of nucleic acids requires repeated centrifugation steps followed by removal of supernatants and, depending on the type of specimen, additional mechanical treatment, e.g., for the lysis of blood cells, organ biopsies, or bacterial and fungal cell walls. In addition, many extraction methods suffer from the presence of strong inhibitors that are present in different clinical specimens, such as sera and stool, and may lead to false-negative results. Furthermore, many manual protocols rely on the use of toxic chemicals, such as guanidinium thiocyanate or phenol–chloroform.

Robotic workstations for the extraction and purification of nucleic acids should fulfill a true walk-away automation. The technology should allow a high throughput of samples; the yield, purity, reproducibility, and scalability of the RNA and DNA as well as the speed, accuracy, and reliability of the assay should be maximal, whereas the risk of cross-contamination must be minimal.

Here we give an overview of commercially available workstations for an automated extraction of nucleic acids.

DOI: 10.1201/9781003247432-4

BioRobot WORKSTATIONS

The BioRobot EZ1 workstation (Qiagen, Hilden, Germany) purifies nucleic acids from many clinically relevant specimens; one to six samples can be processed within a time period of 15 min. This rapid purification is feasible using a magnetic particle technology: nucleic acids in sample lysates are isolated in one step through their binding to the silica surface of magnetic particles. The automated protocols for nucleic acid purification are provided on special preprogrammed cards.

The BioRobot M48/M96 workstations run the identical magnetic particle technology but the pipet tips function as reaction chambers. However, up to 48/96 samples can be processed per run in increments of six samples. The pipettor head contains six syringe pumps, which operate simultaneously to allow aspiration or dispensing of volumes between 25 and 1000 µL. The workstation uses UV light to decontaminate surfaces between runs.

Finally, the BioRobot 9604 workstation allows automated nucleic acid purification for up to 2×96 samples in parallel.

Xu et al.[1] describe an automated assay for the extraction of HIV RNA by using the BioRobot 9604. They conclude that this procedure yields comparable results to a manual RNA extraction method with improved precision. The assay was found to be linear down to 50 copies.

For the extraction of hepatitis C virus RNA, it was shown that it is possible to achieve a detection level of 12.8 IU/mL (95% confidence). Cross-contamination studies have confirmed that the use of the BioRobot 9604 does not pose a detectable contamination risk.[2]

For hepatitis B virus detection, Mitsunaga et al. describe that it was possible to quantify DNA in all samples extracted by the BioRobot 9604 which contained more than 500 genome equivalents/mL. Extraction of 96 samples could be completed within 2 hr.[3]

COBAS AmpliPrep

The COBAS AmpliPrep robotic workstation (Roche Diagnostics, Mannheim, Germany) is a fully automated system that isolates DNA and RNA targets. It is based on silico-coated glass particles which bind the extracted nucleic acids. Internal controls allow target recovery monitoring. Up to 72 samples (three racks of 24 specimens each) can be loaded on the system simultaneously. Reagents are delivered in sealed, bar-coded, and ready-to-use cassettes. The pipetting integrity check ensures the accurate pipetting of samples and reagents, and clot detection reduces the risk of incorrect results. Besides kits for the extraction of generic DNA, introductory kits for hepatitis C virus and human immunodeficiency virus type 1 detection can be acquired. This product is not available in the United States.

In a preliminary study, the COBAS AmpliPrep was evaluated for the preparation of human immunodeficiency virus RNA and was compared to a manual sample preparation protocol.[4] The authors assessed the reproducibility by analyzing 584 plasma samples from infected patients. They achieved an interassay coefficient of variation between 39.4% and 48.4% and an intra-assay coefficient of variation from 6.2% to 58.0%. The mean viral load of 152 samples, which were found to be in the linear range of both tests, was 3.7 log(10) HIV copies/mL by the COBAS AmpliPrep compared to 3.73 log(10) copies/mL by the manual test (no significant difference). The authors concluded that the COBAS AmpliPrep assay is reproducible and sensitive.

Miyachi et al.[5] evaluated the COBAS AmpliPrep protocol for hepatitis C virus RNA extraction. Assay linearity was observed to range between 500 and 850,000 IU/mL by using serial dilutions. The comparison of the COBAS AmpliPrep test results to those obtained by a manual extraction method showed a good correlation [$R(2) = 0.972$, $n = 86$]. When known polymerase chain reaction (PCR) inhibitors such as heparin, dextran sulfate, hemoglobin, and bilirubin were added to the samples prior to extraction, the automated assay successfully eliminated the inhibitory effects.

KINGFISHER ML

The KingFisher technology (Thermo Electron, Waltham, USA) uses the shifting of magnetic particles through the different phases of the purification process. Volumes from 50 to 1000 µL can be processed, and because of the 3×5 fixed magnetic rods and the 3×5 separate disposable tube strips, a maximum capacity of 15 samples per run can be achieved. Besides RNA and genomic DNA, the KingFisher instrument enables the isolation of proteins from a variety of starting volumes.

Saukkoriipi et al.[6] compared the KingFisher instrument to a manual DNA extraction method for the purification of *Streptococcus pneumoniae* DNA from 50 nasopharyngeal swab samples. All pneumococcal culture-positive samples (44%) were PCR-positive regardless of the extraction method. Additionally, from the culture-negative specimens, 71% of the manually extracted and 82% of the samples purified with the KingFisher instrument were PCR-positive.

M1000

The m1000 sample preparation platform (Abbott Laboratories, Abbott Park, USA) uses a magnetic particle technology for the extraction and purification of RNA and DNA. It provides a fully automated sample preparation and is capable of processing 48 samples within 2 hr for a sample volume of 1 mL. Analyte-specific reagents (ASR) may be used for the preparation of samples for hepatitis C virus analysis as well as for the extraction of human DNA for HLA typing. The robot contains liquid-level sensing and a clot detection system. Standard primary bar-coded tubes are used.

Silva et al.[7] evaluated the performance of the m1000 instrument for the extraction of genomic human DNA from frozen blood specimens followed by HLA-DRB sequencing-based typing. Their results showed that the DNA yield was consistently high with an average concentration of 20 ng/µL (from 0.2 mL blood). The average (260:280 nm) ratio was 1.7, and the purified DNA was free of inhibitors; all DNA samples could successfully be amplified and sequenced.

MagNA Pure LC INSTRUMENT

Magnetism is the underlying principle of the automated nucleic acid isolation performed by the MagNA Pure system (Roche Diagnostics, Mannheim, Germany). The instrument has a completely closed housing as well as an automatic clot and tip loss detection. No filtration, centrifugation, or vacuum pumps are necessary, which minimizes the risk of cross-contamination. Up to 32 nucleic acid isolations (the eight-nozzle pipette head allows a variable number of samples from 1 to 32 per run) can be performed within one run. Extraction of human, viral, bacterial, and fungal DNA, RNA, and mRNA is possible. The protocol is applicable to a broad variety of samples such as blood, cultured cells, biopsies, sputum, urine, stool, plant tissue, and food products. After the extraction procedure, isolated nucleic acids and PCR mixes can be pipetted automatically into PCR tubes, 96-well plates, LightCycler capillaries (Roche Diagnostics), or COBAS A-Rings (Roche Diagnostics). Sample information can be entered manually or via a barcode reader. Samples are loaded into nuclease-free sample cartridges with a capacity of 4×8 wells. The initial sample volume is 20–1000 μL (or up to 5×10^6 cells or 10 mg of homogenized tissue); the dispensable volume is 5–1000 μL with a 2%–3% variance.

For the isolation of DNA or total nucleic acids, the sample material is initially incubated with a lysis buffer and proteinase K, and then magnetic glass particles are added. Nucleic acids bind to the surface of the magnetic glass beads and are thus separated from the sample remnants. For the extraction of mRNA, DNA is removed by DNase digestion. Streptavidin-coated magnetic beads and biotin-labeled oligo (dT) nucleotides are added to the lysed sample. The mRNA binds specifically to the glass beads through the oligo (dT) nucleotides.

In addition, since 2003, a smaller version named MagNA Pure Compact System is available. This system allows the extraction of nucleic acids from eight samples in parallel using bar-coded prefilled reagent cartridges and assembled tip trays.

Comparison of the MagNA Pure and a manual extraction protocol (phenol/chloroform) showed equivalent detection sensitivities for the extraction of *Borrelia burgdorferi* DNA from a variety of specimens such as urine, blood, and cerebral spinal fluid. All 80 positive samples (as determined by an independent method) were also tested positive by MagNA Pure extraction.[8]

Wolk et al.[9] describe a protocol for the extraction of DNA from *Encephalitozoon* species. These species cause microsporidiosis, a disease of which the prevalence is likely to be underestimated because of the labor-intensive, insensitive, and non-specific conventional methods used for the diagnosis of the spores in feces. The detection limit of the assay was 100–10,000 × lower compared to microscopy with trichrome stain.

Our group[10] demonstrated that the MagNA Pure technique provides rapid automated DNA isolation from numerous pathogenic fungi revealing high sensitivity and purity. Although the fungal cell wall is highly resistant to mechanical, chemical, and enzymatical treatment, we could achieve a sensitivity of 1 colony forming unit/mL blood for *Candida albicans*.

In addition, there is a need for simple and sensitive assays for mRNA detection, e.g., to assess innate and adaptive immune responses[11] or to determine different subpopulations of cells.[12] The MagNA Pure protocol allowed extraction of mRNA from 32 samples in parallel within 1.5 hr, compared to at least 3 hr using manual methods. The amount (3–6 µg RNA from 10^6 peripheral blood mononuclear cells) and purity (260:280 nm ratio in spectral photometer 1.8–1.9) of RNA were comparable or higher than those achieved by manual extraction.[12]

Finally, Williams et al. compared the BioRobot 9604 workstation to the MagNA Pure instrument, extracting genomic DNA from 106 blood samples followed by genotyping for factor V Leiden.[13] The comparison showed that both methods were similar in DNA yields (8.1 ± 2.3 µg by MagNA Pure, 7.1 ± 2.3 µg by BioRobot), although the DNA purity was higher in samples extracted by MagNA Pure (ratio 1.9 ± 0.11 vs. 1.6 ± 0.24). The extraction failure rate was 0.7% by MagNA Pure and 2.0% by BioRobot. Furthermore, with both assays the analysis of DNA extracted from sodium citrate and heparin anticoagulated blood samples was feasible. Finally, five heterozygous samples were analyzed in two separate runs. Statistical analysis using an F-test indicated that the ΔT_ms for the MagNA Pure were more consistent ($p < 0.001$) which may be due to differences in sample purity.

NucliSens EXTRACTOR

This instrument (bioMérieux, Marcy l'Etoile, France) is able to extract DNA and RNA, and has been designed for a highly variable initial sample volume from 10 up to 2000 µL. The concentration factor can be up to 55-fold, e.g., a 2000-µL sample volume input can be eluted in 35 µL. The closed system minimizes carryover and environmental template contamination. Plasma, serum, and whole blood as well as urine, semen, feces, and sputum can be analyzed within 45 min per run, and up to 10 samples can be processed within one run.

The extraction protocol is based on a method described by Boom et al.[14] After lysis of the samples and addition of silica, nucleic acids bound to the silica are captured on a filter by applying air pressure on the closed sample chamber. The silica particles remain on top of the filter. Through a hollow air-pen, multiple wash cycles are performed. After drying of the silica at 56°C, the nucleic acids are eluted into an elution tube by means of air pressure.

Jongerius et al.[15] used this technology for the extraction of hepatitis C virus RNA. They found a cutoff value of 16 genome equivalents (geq)/mL with a 100% hit rate, demonstrating a high sensitivity of this assay for the extraction of RNA.

The comparison of the BioRobot 9604 to the NucliSens Extractor for the isolation of hepatitis C virus RNA showed a lower detection limit of 82 vs. 12 geq/mL.[16]

CONCLUSION

Current manual methods for the extraction of nucleic acids are suitable for low-throughput analysis.[17] However, these methods are unable to cope with the increasing demand for molecular diagnosis of infectious diseases and genetic analysis, necessitating the development of new, fully automated, and reliable methods. All automated workstations described reveal high yield, purity, reproducibility, scalability,

TABLE 4.1

Comparison of Eight Different Workstations for Automated Nucleic Acid Extraction Concerning Assay Duration, Maximal Number of Samples, Processing Volume, and Price

Workstation	Duration per Run (Maximal Sample Load)	Maximal Number of Samples per Run	Processing Volume (μL)	Approximate Price of Workstation
EZ1 (Qiagen)	15 min	6 samples	25–1000[a]	US$35,000
BioRobot 9604 (Qiagen)	2 hr	2 × 96 samples		US$100,000
COBAS AmpliPrep (Roche)	3 hr	3 × 24 samples	200–1000	US$50,000
KingFisher mL (Thermo Electron)	45 min	15 samples	50–1000	US$15,000
m1000 (Abbott)	2 hr	48 samples	200 or 1000	US$110,000
MagNApure LC (Roche)	1.5 hr	32 samples	20–1000	US$90,000
MagNAPure Compact (Roche)	30 min	8 samples	100–1000	US$35,000
NucliSens Extractor (bioMérieux)	45 min	10 samples	10–2000	n.a.

Abbreviation: n.a.—not available.

[a] Depending on the workstation.

accuracy, and reliability for the extraction of nucleic acids (Table 4.1). However, as different laboratories may claim unique performances of a workstation, the individual choice of an instrument should depend on the type and volume of samples to be analyzed and on the throughput of samples per day.

REFERENCES

1. Xu, M.; Chan, Y.; Fischer, S.H.; Remaley, A.T. Automated procedure for improving the RNA isolated step in viral load testing for human immunodeficiency virus. J. Clin. Microbiol. **2004**, *42* (1), 439–440.
2. Grant, P.R.; Sims, C.M.; Krieg-Schneider, F.; Love, E.M.; Eglin, R.; Tedder, R.S. Automated screening of blood donations for hepatitis C virus RNA using the Qiagen BioRobot 9604 and the Roche COBAS HCV Amplicor assay. Vox Sang. **2002**, *82* (4), 169–176.
3. Mitsunaga, S.; Fujimura, K.; Matsumoto, C.; Shiozawa, R.; Hirakawa, S.; Nakajima, K.; Tadokoro, K.; Juji, T. High-throughput HBV DNA and HCV RNA detection system using a nucleic acid purification robot and real-time detection PCR: Its application to analysis of posttransfusion hepatitis. Transfusion **2002**, *42* (1), 100–106.

4. Berger, A.; Scherzed, L.; Sturmer, M.; Preiser, W.; Doerr, H.W.; Rabenau, H.F. Evaluation of the Cobas AmpliPrep/Cobas Amlicor HIV-1 monitor ultrasensitive test: Comparison with the Cobas Amplicor HIV-1 monitor test (manual specimen preparation). J. Clin. Virol. **2002**, *25* (3), 103–107.

5. Miyachi, H.; Masukawa, A.; Asai, S.; Miura, T.; Tamatsukuri, S.; Hirose, T.; Ando, Y. Quantitative assay of hepatitis C virus RNA an automated extraction system for specific capture with probes and paramagnetic particle separation. J. Clin. Microbiol. **2003**, *41* (2), 572–575.

6. Saukkoriipi, A.; Kaijalainen, T.; Kuisma, L.; Ojala, A.; Leinonen, M. Isolation of pneumococcal DNA from nasopharyngeal samples for real-time, quantitative PCR: Comparison of three methods. Mol. Diagn. **2003**, *7* (1), 9–15.

7. Silva, J.T.; Wong, C.Y.; Dileanis, J.L.; Dunn, C.M.; Impraim, C.C. Automated DNA extraction from frozen blood using the Abbott m1000™ for HLA-DRB sequencing-based typing. Hum. Immunol. **2003**, *64* (10), 95.

8. Exner, M.M.; Lewinski, M.A. Isolation and detection of *Borrelia burgdorferi* DNA from cerebral spinal fluid, synovial fluid, blood, urine, and ticks using the Roche MagNA Pure system and real-time PCR. Diagn. Microbiol. Infect. Dis. **2003**, *46* (4), 235–240.

9. Wolk, D.M.; Schneider, S.K.; Wengenack, N.L.; Sloan, L.M.; Rosenblatt, J.E. Real-time PCR method for detection of *Encephalitozoon intestinalis* from stool specimens. J. Clin. Microbiol. **2002**, *40* (11), 3922–3928.

10. Loeffler, J.; Schmidt, K.; Hebart, H.; Schumacher, U.; Einsele, H. Automated extraction of genomic DNA from medically important yeast and filamentous fungi by using the MagNA Pure LC system. J. Clin. Microbiol. **2002**, *40*, 2240–2243.

11. Stordeur, P.; Zhou, L.; Byl, B.; Brohet, F.; Burny, W.; de Groote, D.; van der Poll, T.; Goldman, M. Immune monitoring in whole blood using real-time PCR. J. Immunol. Methods **2003**, *276* (1–2), 69–77.

12. Loeffler, J.; Swatoch, P.; Akhawi-Araghi, D.; Hebart, H.; Einsele, H. Automated RNA extraction by MagNA Pure followed by rapid quantification of cytokine and chemokine gene expression using fluorescence resonance energy transfer. Clin. Chem. **2003**, *49*, 955–958.

13. Williams, S.M.; Meadows, C.A.; Lyon, E. Automated DNA extraction for real-time PCR. Clin. Chem. **2002**, *48* (9), 1629–1630.

14. Boom, R.; Sol, C.J.; Salimans, M.M.; Jansen, C.L.; Wertheim-van Dillen, P.M.; van der Noordaa, J. Rapid and simple method for purification of nucleic acids. J. Clin. Microbiol. **1990**, *28* (3), 495–503.

15. Jongerius, J.M.; Bovenhorst, M.; van der Poel, C.L.; van Hilten, J.A.; Kroes, A.C.; van der Does, J.A.; van Leeuwen, E.F.; Schuurman, R. Evaluation of automated nucleic acid extraction devices for application in HCV NAT. Transfusion **2000**, *40* (7), 871–874.

16. Beld, M.; Habibuw, M.R.; Rebers, S.P.; Boom, R.; Reesink, H.W. Evaluation of automated RNA-extraction technology and a qualitative HCV assay for sensitivity and detection of HCV RNA in pool-screening systems. Transfusion **2000**, *40* (5), 575–579.

17. Chomczynski, P.; Sacchi, N. Single step method of RNA isolation by acid guanidinium thiocyanate–phenol–chloroform extraction. Anal. Biochem. **1987**, *162*, 156–159.

5 Forensic DNA Samples
Collection and Handling

Mark Benecke
International Forensic Research & Consulting,
Cologne, Germany

CONTENTS

INTRODUCTION

Under dry and cool conditions, but sometimes even at temperatures over 50°C, DNA is quite a stable molecule (e.g., on paper, in hair, as skin flakes, blood, saliva and sperm, from old fingerprints, and out of bones and teeth[1–12]) in induced anhydrobiosis utilizing storage matrices like Samplematrix™,[13,14] or DNAstable™ plates, but also home-(lab-)made trehalose and polyvinyl alcohol (PVA) plates.[8] DNA in epithelial cells can survive soapy water.[15,16] Under indoor conditions, 10 minutes rinsing of clothes under the tap, or one week in the bathtub, skin cell DNA can be recovered from clothing. Outdoors, temperature and movement of water (pond vs. river) are of

DOI: 10.1201/9781003247432-5
This chapter has been made available under a CC BY NC SA 4.0 license.

relevance. Complete STR (short tandem repeats) profiles were found after two weeks in a sample of skin cells in a pond during cold winter vs. only 4 hours in hot summer conditions.[17,18] Domestic machine washing and drying often leads to transfer of foreign DNA onto freshly laundered items. After swabbing, around 1 ng of foreign DNA was found ('indirect' or 'innocent' transfer). In most cases, the wearer is the major or sole contributor,[19–21] however, interpretation of such stains is a delicate matter. On semen-stained UK school uniforms (T-shirts, trousers, tights) that were stored in a wardrobe for eight months, 6–18 µg of DNA were found after washing them multiply at 30°C or 60°C. Unstained socks that had been washed together with the stained items contained one-tenth of the sperm donor's DNA.[22] Spermatozoa persist on cotton and terry towels for at least six wash cycles.[15] Vaginal secretions also leave amounts of DNA on clothing laundered at 30°C that are sufficient to produce complete genetic profiles.[23,24]

From a technical and criminalistic point of view, DNA can be collected and stored like most visible biological stains. Crucial considerations in the examination of evidence include photographic documentation, and careful storage of the samples under dry and cool conditions. Special aids such as sexual assault kits, swabs, drying devices, and filter paper treated with denaturants are available and should be used. However, DNA collection in forensic environments is not a merely technical but also a criminalistic task. Two questions are of special importance: (1) whether a stain is of relevance to the actual crime, e.g., if it could have been left at the scene some time ago by persons who are not related to the crime, and (2) if a stain should be used for extraction straight away, or stored as long as possible for morphological measurements and crime (scene) reconstruction, e.g., the form of blood stains on wallpaper, the exact location of sperm stains on clothing, or the exact location of skin cells found on furniture.

EVIDENCE EXAMINATION

Irrespective of possible chain of custody rules, examination of evidence starts with photographic and/or drawn descriptions of the items received by the forensic biologist. In every photograph, an absolute scale must be visible (millimeters/centimeters; no pennies, no pens). Resolution should be ≥3266 × 2449 pixels = 8 MPixel to allow blowing up of the pictures. Since this size is now easily achieved by most cell phones, these phones may be used after appropriate training. They also work for microscopic photographs (Figure 5.1). Use of a flash should be avoided because brighter parts of the objects often 'flash out' (become white). Biological stains that were detected either by their surface properties (detection by touch: e.g., sperm stains on dark clothing),[25] monochromatic light (e.g., saliva), or regular bright light (e.g., hair or small blood stains) are circled and numbered by use of a water-resistant pen or neon color.

COLLECTION OF BIOLOGICAL STAINS

SWABS

Practically all stains can be collected by rubbing them off with a cotton swab.[26,27] Stains on fabric should be cut out first (Figure 5.2). Swabs are soaked with one drop of fresh distilled sterile water. After transfer of the stain to the swabs, they must be

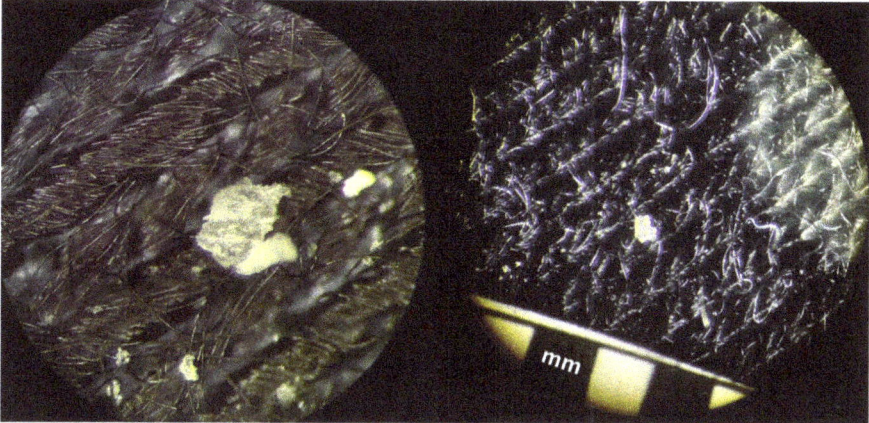

FIGURE 5.1 Freehand micrographs captured during initial examination of a mask. The micrographs were captured using an iPhone 11 Pro and dissecting microscope (Leica Mz 12.5 at magnifications 25× and 100×, respectively).

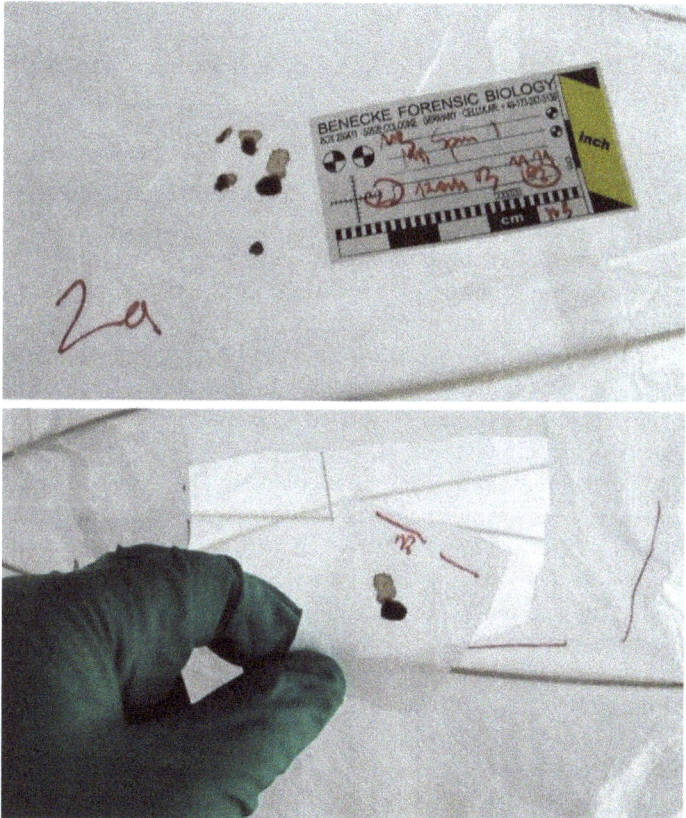

FIGURE 5.2 Removal of stains by cutting stained materials.

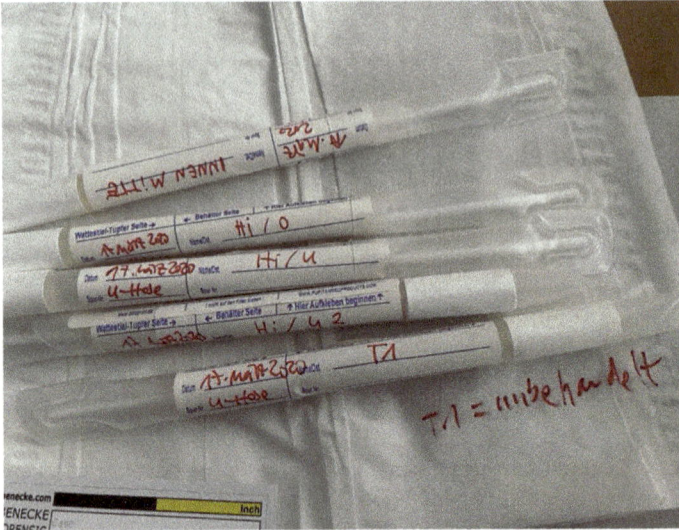

FIGURE 5.3   Swab collection tubes.

dried immediately.[28] DNA sampling tools that offer rapid drying can significantly improve the preservation of DNA collected on a swab, increasing the quantity of DNA available for subsequent analysis. In saliva samples, slow drying of swabs in storage tubes leads to a decrease from a yield of 95% recoverable DNA to only 12% recoverable DNA.

After collection, the swabs remain inside of their tubes and are put in protective containers (Figure 5.3). A convenient way to dry swabs is to put them into a closed cardboard box at room temperature (Figure 5.4). There, they can neither touch neighboring objects nor develop mold.[29–32] Swab tubes consisting of a paper wall stabilized with plastic are preferable since nothing has to be assembled, i.e., the risk for contamination is lowered. Cardboard boxes are a good place to store evidence because residual moisture, especially from clothing, can easily evaporate (Figure 5.5). Touched objects may be rubbed with cotton tips moistened with a 2% SDS (sodium dodecyl sulfate) solution.[33]

In professional forensic environments, contamination caused by airflow during the drying process has not been reported to be a problem. Some field laboratory manuals ask for drying in closed cupboards (Figure 5.6) or under sterile laminar airflow. If used, cupboards must never be tightly closed to avoid building up of humidity and mold. If minute amounts of DNA, especially mtDNA or Y-chromosomal DNA may be of relevance, freestanding cupboard drying must be avoided. Under field conditions in poor countries, it is still an option when only STRs are used.

Swabbing is performed by *intense, multiple* rubbing of the stained surface to collect a maximal amount of DNA. Inside of oral cavities, the cotton swab is rubbed against the mucous membrane; saliva alone may not contain enough cells. After

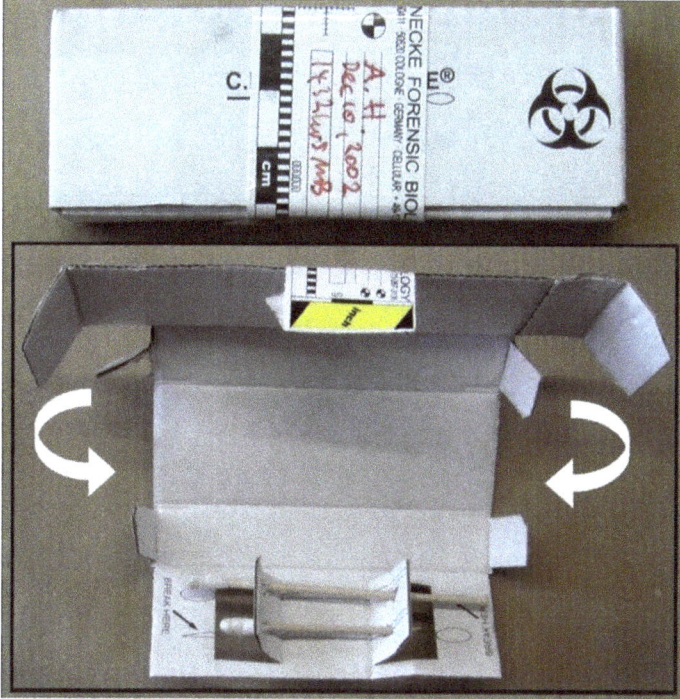

FIGURE 5.4 Cardboard box allows simple and safe drying and storage of swabs.

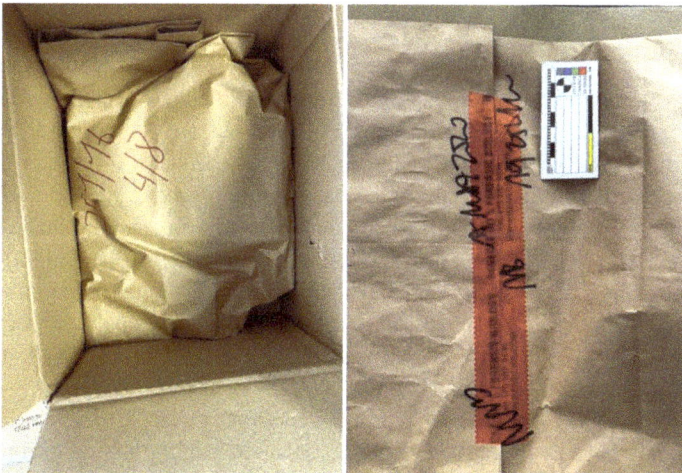

FIGURE 5.5 Samples are wrapped in paper ready for transport or storage.

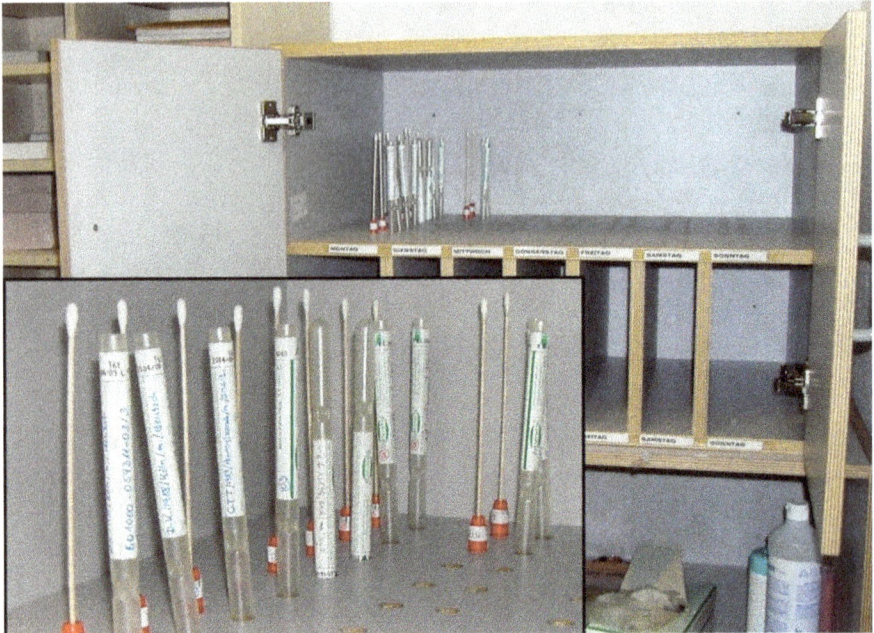

FIGURE 5.6 Under extreme field conditions, cupboards may be used for drying of swabs. Contamination is not a problem as long as the swabs do not touch each other or the wall. Note that the cupboard must not be tightly closed to avoid building up of humidity.

complex shooting situations, used bullets can be matched to the victims by swabbing off traces of tissue that remain on the bullet once it enters the body.[34]

EARLY SWABBING

Swabbing of clothing items, especially of skin, should be performed as soon as possible in forensic and police investigations. For example, DNA typing was possible in the following cases where swabs had been collected early at the scene of the crime. Before swabbing, intelligent criminalistic assumptions concerning the location of the invisible yet possible stains had been made. The swabs may be made of cotton or synthetic material.[35] They must be thoroughly checked to be DNA-free in the laboratory, because in some circumstances, they may become contaminated. In a very severe German serial homicide case, contamination misled the prosecution and police for years – the alleged culprit was an old woman working in the (German) cotton swab factory who had touched the material at times.[36]

In contrast to common belief, corneocytes contain DNA. Therefore, all surfaces that may have been touched by an offender (through grabbing of ropes, wearing of baseball caps, hitting a person, inside of gloves) may be swabbed (or lifted, see 'Touch DNA') successfully.[37–39] Epithelial cells of an unknown suspect were swabbed off the front side of a collar of a polo-neck pullover. The victim had been stabbed, but the stains had not been visible on the collar.[40] Alternatively, single cells can be observed microscopically, taken off either with a pair of forceps or a vacuum device, and then used for single-flake amplification.[12,41] DNA contained in epithelial cells that had been transferred by saliva of an offender was swabbed off the skin of an experimental victim that had showered. Amplification of the offender's STRs and Y haplotype was successful up to several hours after transfer of his saliva to the skin of the victim.[42]

Early swabbing is also necessary whenever cells from the top edge of bottles, beer cans, etc. are collected. Collection of the complete bottle or can frequently leads to spilling of its contents and dilution or washing off the cells. If early swabbing is not possible, the liquid must be drained out of the container by drilling a hole in its bottom.

DOUBLE SWABBING

Double swabbing is the use of a wet cotton swabs first followed by rubbing with a dry one. It often leads to better results in cases of touched objects (sweat) and bite marks (saliva).[43,44] After double swabbing, around 1/5 of the expected alleles can be amplified even from bullet cartridges that have been fired – a particularly challenging surface.[45]

FILTER PAPER

Liquid blood can be stored on filter paper that is then dried in the same way as cotton swabs (Figure 5.4). Filter paper that contains denaturants, buffer, and a free radical trap (e.g., FTA paper™)[2,46,47] will lyse the blood cells and immediately deactivate blood-borne pathogens such as herpes, cytomegalovirus, and HIV. Filter paper can also be used to store saliva and liquids from decomposed bodies, especially tissue (cells) from internal organs.[48] If the DNA is too degraded, regular STRs may be subsequently substituted by massive parallel sequencing (MPS). This allows detections of numerous single nucleotide polymorphisms, which are suitable for identification of body parts, for example.[49]

In automated laboratories, standard-sized filter paper is the preferred option. Pieces can easily be punched out of it by a machine and subsequently processed by a DNA extraction and polymerase chain reaction (PCR) robot. The advantage of FTA paper over regular filter paper is that it can be used for multiple PCR reactions. Template DNA will stick to FTA paper after washing off the PCR products and can then be reused. It is also suitable for touch DNA. On steering wheels, FTA paper collects a two-fold amount of DNA compared to double swabbing or tape lifting.[50]

ELECTROSTATIC SAMPLING

For the sampling of trace DNA from clothing, electrostatic dust print lifters (DPL) have the same success rate as sampling with wet cotton swabs. However, in single aggressor cases, almost no mixed aggressor-victim profiles suitable for database entry can be established, which is sometimes necessary to better understand the case criminalistically.[51]

URINE AND FECES

Because feces are found especially at scenes of (serial) burglaries, it should be collected irrespective of its repulsive nature. Fresh feces as well as liquid urine should be frozen below −20°C to avoid bacterial activity. DNA typing of urine is successful especially if it was excreted in the morning (when the highest number of epithelial cells are found compared to the rest of the day)[33,52] and from feces after PCR inhibitors are removed. To recover the cells, urine needs to be centrifuged (cells are located in the sediment), whereas stool samples can be extracted straightaway or from swabbing with mini spin columns. The estimated number of up to 6×10^5 pg human DNA/mg stool is never reached in practice because of bacterial and digestive action. Nevertheless, up to 170 pg DNA/mg stool were successfully extracted and amplified under case work conditions.[53,54]

SEXUAL ASSAULT KITS

After sexual assaults, biological stains are often collected in a hospital environment, at home, at a general practitioner's office, or at a police station. To avoid contamination of the samples and to allow full collection following a checklist, sexual assault kits are available. Their use is generally and strongly recommended to guarantee collection of all stains in the best possible way even under highly stressful conditions or in cases where lay personnel have to collect the evidence.[27] The kits consist of prepacked envelopes in a cardboard box, which can be stored and stacked at room temperature (e.g., Sexual Assault Care Kit, University of Bern, Figure 5.7). The envelopes contain swabs, combs for hair (head and pubic), filter paper, sterile distilled water ampoules, large paper bags, and standardized protocol sheets.[30]

CLASSIC FINGERPRINTS AND 'TOUCH DNA'

In fingerprints, the DNA loss ranges from half to three quarters of the DNA compared to the amount of cells transferred during the touch event.[55,56] However, DNA is resistant to many histological stains, including substances used to develop fingerprints (or other skin lines). DNA typing was successful from developed skin line prints after cyanoacrylate (super glue fume) or color reagents such as amido black, leucomalachite green, Hungarian Red, DFO, or luminol had been applied.[57–60]

Developed skin line prints should first be documented with a high-resolution camera. The original skin line prints can then be submitted to DNA storage and extraction like any other biological stain. The stronger the initial fingerprint or 'touch', the more likely a DNA profile may be obtained.[11,61–63]

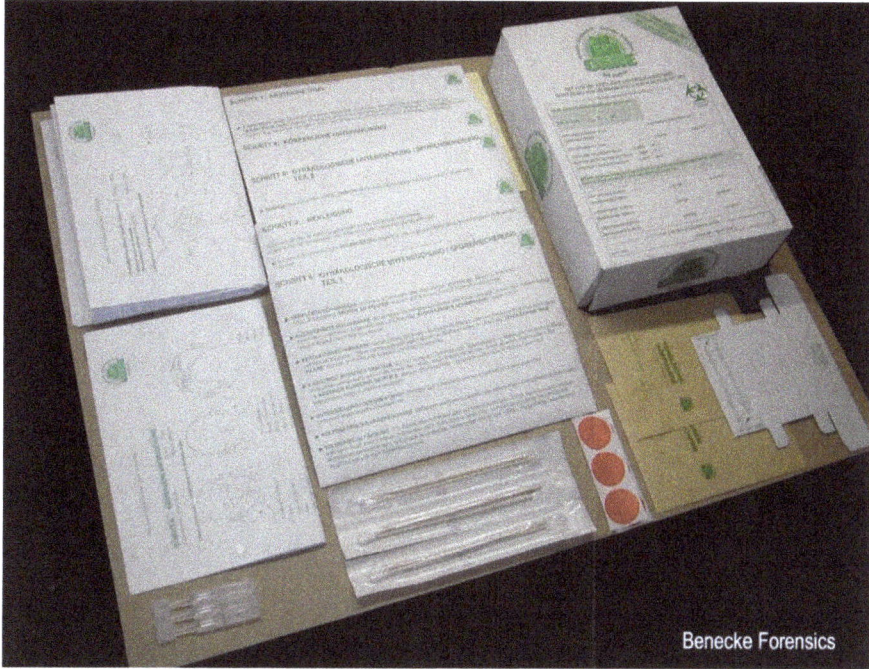

FIGURE 5.7 Sexual assault kit: Standardized descriptions and checklists are printed on the actual envelopes that contain the collection materials.

Since humans shed around 400,000 skin cells daily, it takes only two seconds of handling time to transfer enough DNA ('touch DNA') onto clothing to obtain a complete profile.[64–67] Full DNA profiles can be obtained from the cells of a person sleeping in a bed for a night or after ten minutes of wearing a sweatband.[68,69] During wearing for a day, the DNA quantity – mostly DNA from the wearer of a shirt – increases eightfold. On the back area of a shirt, DNA mixtures are more likely than on the front.[18]

When targeting for such trace DNA, the sample area should be narrowed down as much as possible to maximize target DNA recovery (Figure 5.8).[66,70,71] Care must be taken not to destroy the evidence during such narrowing down because any type of air movement or vibration (tram tracks, laboratory equipment) as well as static electricity that builds up between plastic forceps and some types of garment may cause skin flakes to jump out of sight.[72] If no skin particles are visible, small segments of the garment should be either rubbed, taped, or used as a whole to avoid DNA mixtures from neighboring areas. The same is true for fingernail clippings where each nail clipping might be cut down into thin segments, each for one PCR tube.

Swabbing of 'touch DNA' is often successful as well as direct extraction (STR and mtDNA).[73] However, taping is more precise and is performed by gel-film or single or double-sided sticky tape to lift off the material from the fabric.[74–77] An

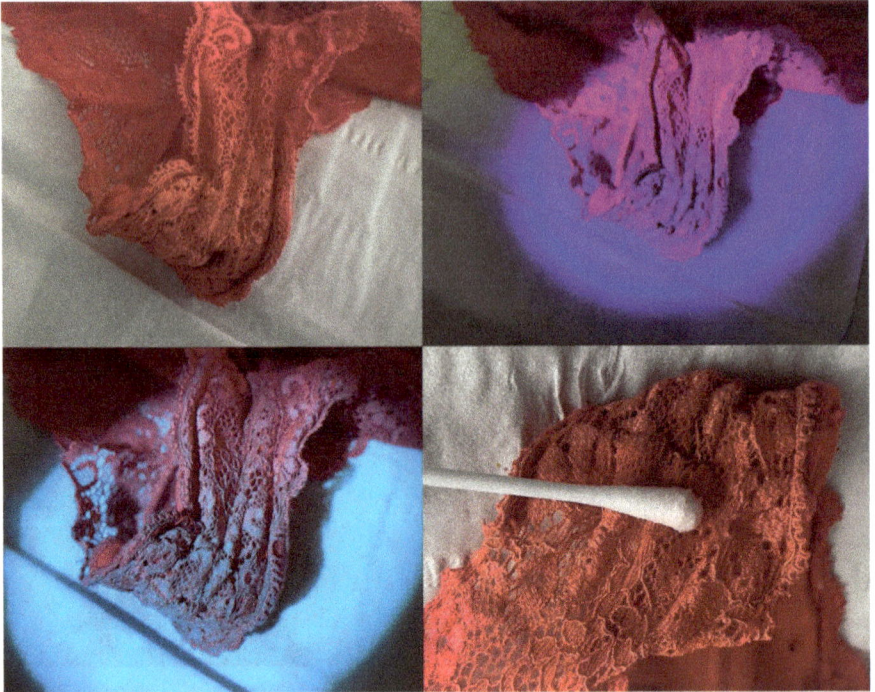

FIGURE 5.8   Narrowing down the location of sperm on washed or unwashed clothing by use of forensic light source. Note that 'innocent' transfer might have taken place; therefore document every single step of the examination and all locations photographically.

advantage of this method is that the tape may be brought onto a microscopic glass slide where it cannot only be inspected and stained but the cells can also be cut out without danger of jumping off or mixing with DNA from other sources. When it comes to skin on clothing, tape or 'mini-tape lifting' often leads to better results than swabbing.[78]

SINGLE SPERM AND MICRODISSECTION

Isolation of DNA from single sperm within vaginal cell mixtures is possible by preferential extraction methods (i.e., differential lysis: vaginal cells are digested more easily). If there are less than 250 pg of DNA available, e.g., on few sperm heads on a microscope slide, the use of laser capture microdissection for the isolation of spermatozoa is preferable (Figure 5.9). Such slides are often available in laboratories which perform a quick visual check of fresh swabs from rape and other sexual assault cases. Sperm heads are usually robust and even survive heating on a heat plate to fix them to the slide and staining, e.g., with Kernechtrot (nuclear fast red) and picroindigocarmine ('Christmas tree staining' due to the green and red colors) often used to visualize sperm heads and vaginal cells microscopically.[79]

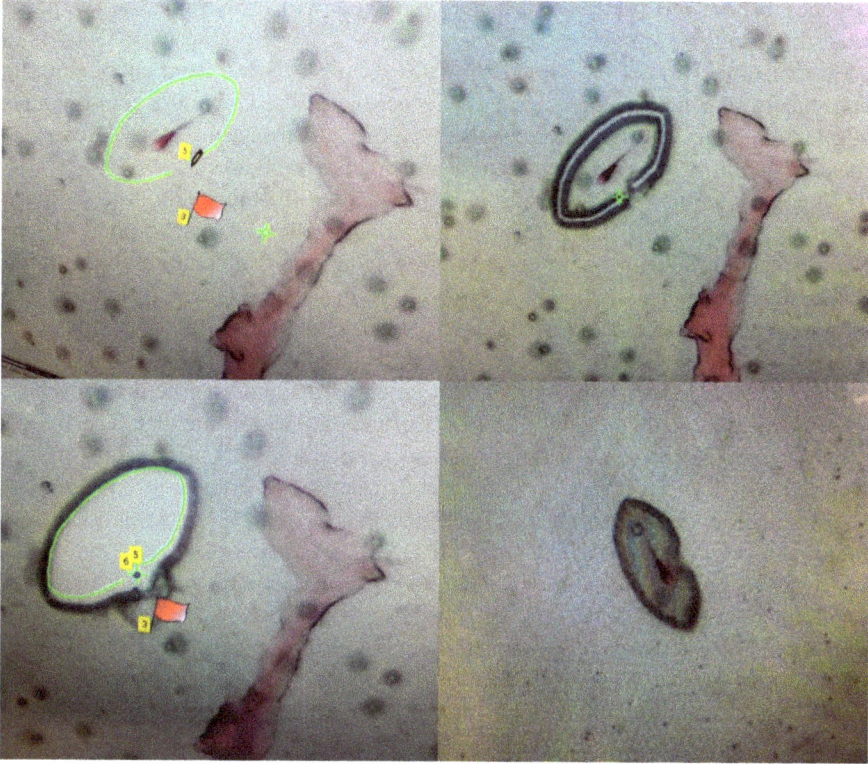

FIGURE 5.9   Laser microdissection of sperm head out of the matrix after embedding and 'Christmas tree' staining. The single cell cut out is shot directly into the DNA free reaction tube.

After laser dissection, low copy number PCR may help in obtaining a suitable DNA profile.[80]

At wavelengths of 320–400 nm (compared to infrared 812 nm), the dissection is precise and cutting enables single cell and subcellular microdissection. After photovolatilization of the cells, the layer containing the cells is ejected against gravity and either simply falls (by the force of gravity) or is directed by electrostatic forces into the reaction tube. Since the absorption maxima of DNA, RNA (and proteins) lie outside the operating wavelength, no harm to DNA and RNA occurs.[81–84]

STORAGE AND EXTRACTION

Dried biological samples should be stored in standardized paper bags (envelopes, brown paper bags) in a dry and cool environment (Figure 5.5). This will preserve the DNA over months to years. If dry samples need to be stored for more than 2 years, freezing below −20°C is recommended. To avoid paper layers sticking to each other in the freezer, the envelopes should be put into plastic bags. Never write on plastic surfaces that become frozen because any type of ink will easily come off. Use paper labels instead.

In temperate parts of the world, DNA was successfully extracted out of clothing and smears on slides that had been stored more than 10 years in dark environments at room temperature. In tropical countries, freezing is always necessary because of the high humidity, which allows bacteria and mold to build up.

Biological stains on glass slides, either embedded (histological tissue samples) or just regular smears (vaginal smears or blood), generally lead to good extraction results. The slides should be stored in standard cases for microscopic slides. Alternatively, they can be fixed with sticky tape inside of a paper envelope. Traces of dust generally do not affect the quality of dry stains but should obviously be avoided.

Insects collected at crime scenes or from corpses should not be dried because museum beetles will frequently destroy the samples within months. The insects should be preserved in 90% EtOH. At room temperature, DNA extraction of such material will then be possible up to several weeks after storage; at temperatures below −20°C, extraction will be successful for several years.[58,85,86] Never use formalin to preserve samples; it will degrade the DNA.

Cigarette butts, envelopes with stamps, fingernail clippings, and dried nasal secretions should be stored dry in paper bags, envelopes, or cardboard boxes. Fingernails can be thoroughly swabbed if clipping is not an option.[87–89]

Because telogenic hair and broken-off hair shafts have been successfully used for DNA extraction, hair should be carefully stored, e.g., by attaching one end of every hair with sticky tape to the inside of an envelope or between two layers of filter paper. If hair is collected by the police using sticky tape for fiber collection, all material (fibers, lint, and hair) should remain on the tape until extraction becomes necessary.[3,4]

If the samples are 'challenged', i.e., either not clean or stick to tape, quick extraction with lysis buffers like BTA™ may be used to get access to the DNA by destroying matrices of teeth and bone as well as of adhesive-containing substrates including chewing gum, cigarette butts and tape lifts.[90,91]

The success of later DNA typing depends on the number of cells transferred to and from the material used as evidence. Cigarettes, bloodstains, and headwear have high success rates for DNA extraction, even after prolonged dry storage. Cartridge cases, crowbars, and tie-wraps are less successful. If the DNA concentration decreases below 6 pg/μL, only 5% of the extracts provide meaningful DNA profiling data in a standard STR setting. Traces with a concentration above 100 pg/μL generally result in DNA profiles that can be used for DNA database storage.[92]

It should be noted that in forensic laboratories with many different types of mostly swabbed stains, simple procedures with few steps may be superior to commercial extraction kits as well as protocols with many manipulations. For example, a successful extraction in a high-throughput yet not automated laboratory in a metropolitan city is a 30-minute incubation of parts of the swab with 0.01% SDS and proteinase K at 56°C, followed by an incubation at 100°C for 10 minutes. Chelex-100™ may be added but may pose problems for automated liquid handling systems, cause loss of DNA, and is therefore not necessary in high-throughput method of stains containing low amounts of DNA. The extracts are then concentrated and SDS removed at the same time by one centrifugation step in Microcon™ 100 tubes; 1 ng poly(A) RNA helps to increase the amount of DNA recovered.[93]

EXTRACTED DNA STORED IN BUFFERS

Depending on the applied extraction method, DNA stored in TE [10 mM Tris–HCl (pH 7.5), 1 mM EDTA] or similar buffers may be stable for weeks (after Chelex extraction) or months (formerly, phenol/chloroform extraction or today, use of spin columns) in the refrigerator at +4°C to +12°C. Freezing of extracted DNA in TE buffer below −20°C will preserve the sample for years. Before freezing, it is strongly recommended to distribute the DNA in small aliquots (e.g., 10 µL each) to avoid repetitive thawing and freezing of single samples. Dry storage is preferred for original stains and swabs.

EMERGENCY BUFFERS

Under difficult field conditions, a possible standard storage buffer for extracted DNA is TE buffer. Before its use, the buffer is autoclaved or cleaned with a sterile filter. It can then be stored at room temperature. Under extreme conditions, if drying of the samples is impossible (because of dust, humidity, chaotic mass disaster environments), TE can be used to collect samples by aliquoting 1 mL TE into sterile, DNA free 1.5-mL plastic tubes. The collected biological stains can be put inside these emergency containers. DNA must still be extracted as soon as possible, and the samples should be stored as cool as the situation allows.

If more than a few hours are expected to pass before freezing or drying is possible, any solid biological sample in the field should be stored in centrifuge tubes containing aliquots of 95% EtOH. At room temperature, this will preserve the sample's DNA for weeks.[94] Recently, 100% isopropyl alcohol or 70% ethanol (v/v) alcohol were successfully used for storage of swabs that were in danger of developing mold.[95]

MAIN DESTRUCTIVE INFLUENCES ON DNA

Under the influence of UV light (including sunlight) and acids, DNA contained in biological stains as well as extracted DNA breaks into pieces (degrades). Depending on the intensity of fragmentation, PCR is often possible. Humidity does not directly affect DNA but will allow mold and bacteria to destroy the sample including the DNA within days. Frequent freezing and unfreezing of stains or extracted DNA will also lead to degradation. Household use of detergents and cleaners does not necessarily destroy DNA.[96]

Sperm heads on fabric may survive machine washing at 30°C–40°C if no bleach was used. However, for detection of such stains, narrowband, fixed-wavelength lighting is minimally successful at higher washing water temperatures, probably because most of the seminal fluid is dissolved during washing whereas sperm heads stick to or in between the fibers. In the beginning, acid phosphatase tests with an extended cutoff are still highly sensitive. In still ocean water, spermatozoa on cotton fabric are undetectable after 12 hours, in swimming pool water after one week, yet with no upper limit of detectability for tap or river water, even though a decreasing trend overtime occurs.[97] Semen-stained underwear DNA led to recovery of between 13 and 55 ng/μL DNA with successful STR typing in all such cases. When semen-stained underwear is washed after a month at 30°C, some semen stains can still be detected by narrowband forensic light sources or prostate specific antigen, and all stains can be successfully DNA typed.[98]

CONTAMINATION

Under conditions of normal case work, contamination is only observed after careless manipulation or purposeful spraying of high (nanogram) amounts of DNA near or directly into open tubes before PCR. Secondary transfer via door handles, etc. is only a problem under extremely careless, unprofessional conditions.[99–101]

Obviously, mixtures of DNA might be present in the samples themselves. Mixtures of epithelial cells with sperm can be separated by differential lysis (separation of sperm from epithelial cells).[102] Other mixtures may show distinctively different peak heights after electrophoretic separation of the PCR products. For example, an object at a crime scene may have been touched by Person A days before a biological stain (such as blood) of Person B was deposited on the same surface. In that case, a DNA mixture might be present later. It can often be detected by the different peak heights of the STR alleles. Mouth-to-mouth-kissing is a lesser practical problem since the other person's DNA inside of the oral cavity of the other person will not show in STR systems already 1 minute after the end of the kiss.[103]

Irrespective of the possible presence of mixtures, swabbing is always recommended if the items cannot be moved, are bulky, or if the stain is located on a person. Subsequent procedures like differential lysis should not be performed before DNA extraction becomes necessary. Generally, once evidence examination is completed, all biological samples should simply be stored cool and dry, and left intact as long as possible.

Care must however be taken to avoid contamination and misinterpretation of DNA that was transferred by persons not related to the (criminal) event, e.g., persons present at parties (skin), newspaper or journal readers, clothing stored on the same shelf, etc. (secondary transfer).[17,104–107] Tertiary as well as non-transfer is also possible. Accused persons sometimes argue that their garment was used by the 'real' offender who did not leave traces. In sweat bands, three subsequent wearers leave their respective, full profiles on the outside (67%) and on the inside (80%) whereas profiles of only the first wearer are hardly found (one of 200 cases); a single profile of only the second wearer may be apparent in 7% of samples. It is therefore highly unlikely to wear/use a piece of clothing for even a short period of time without leaving DNA behind.[69]

There are numerous guidelines concerning the handling of DNA evidence but they are sometimes limited by local regulations, education of personnel and agencies involved.[108–110] General standardization of evidence examination and procedures in the form of International Organization for Standardization (ISO) guidelines are nearly impossible due to the different composition of stains, laboratories, and evidence examination teams. A DIN/ISO (Deutsches Institut für Normung) attempt to standardize 'recognition, recording, recovering, transport, and storage of material' was retracted.[111] Even though laboratory gloves are not a main source for contamination, other surfaces are, including the body and clothing of the person collecting the stain. Continuous training to keep avoiding contamination in the light of today's single cell DNA approaches is necessary.

WITHDRAWAL OF SAMPLES OUT OF STORAGE

If parts of a stored biological sample need to be withdrawn for DNA extraction, forceps and scissors must be wiped with paper towels and 70% EtOH (or methylated spirits) every time they are used. In routine use, cross-contamination caused by wiped, smooth-surface forceps has not been observed. An exception is forceps with grooves. They must be autoclaved before every use because the groves quickly fill up with contaminants.

Still, especially during evidence examination and withdrawal, it is essential to take care of cross-contamination caused by contaminated distilled water, touching the swabs with used gloves, etc. Standard bacteriological procedures are an optimal guide.

SAMPLE RETAINMENT

It is recommended to always retain at least half of a stain in storage. One reason is that extracted DNA in liquid buffers is less durable than the original, dried stain. In addition, the defense should have a chance to reexamine the stain beginning with the original sample, not the extracted DNA. Only if DNA extraction and PCR seem to fail because of low amounts of DNA, stored samples be used up completely. This needs the consent of the prosecutor's (D.A.'s) office. Even in these cases, at least a minute amount of the original material should be stored so that future DNA technologies may be applied later on.

CONCLUSION

Collection of biological stains should be documented by photographs and drawings. Dry and cool storage will allow biological samples to be stored over years.

Extraction of DNA should be performed only if necessary for a current investigation. The original stains should never be extracted completely. Contamination in the laboratory does not occur if the sampling is performed by trained personnel. Because many surfaces and even stains like fingerprints (skin lines), corneocytes on ropes, telogenic hair, the surface of skin after showering, etc., may contain material that is suitable for DNA typing, intelligent criminalistic decisions have to be made before collecting the evidence.

Intense swabbing and the use of sexual assault kits are simple yet very important procedures that guarantee maximum yield of DNA and collection of biological material even if it is not visible at the moment of collection. Even difficult stains such as feces can be extracted and should be stored frozen whenever possible. Under extreme field conditions, 90% EtOH may be used as a collection and storage liquid.

REFERENCES

1. Alessandrini, F.; Cecati, M.; Pesaresi, M.; Turchi, C.; Carle, F.; Tagliabracci, A. Fingerprints as evidence for a genetic profile: morphological study on fingerprints and analysis of exogenous and individual factors affecting DNA typing. J. Forensic Sci. **2003**, *48*(3), 586–592.

2. GE Healthcare Life Sciences. *Application Note 28-9822-22 AA, Reliable Extraction of DNA from Whatman™ FTA™ Cards*, **2010**, https://us.vwr.com/assetsvc/asset/en_US/id/16147319/contents

3. Hellmann, A.; Rohleder, U.; Schmitter, H.; Wittig, M. STR typing of human telogenic hairs—a new approach. Int. J. Legal Med. **2001**, *114*, 269–273.

4. Higuchi, R.; von Beroldingen, C.H.; Sensabaugh, G.F.; Erlich, H.A. DNA typing from single hairs. Nature **1988**, *332*, 543–546.

5. Hochmeister, M.N.; Budowle, B.; Jung, J.; Borer, U.V.; Comey, C.T.; Dirnhofer, R. PCR-based typing of DNA extracted from cigarette butts. Int. J. Legal Med. **1991**, *104*, 229–233.

6. Hochmeister, M.; Rudin, O.; Ambach, E. PCR analysis from cigarette butts, postage stamps, envelope sealing flaps, and other saliva-stained material. In *Forensic DNA Profiling Protocols. Methods in Molecular Biology*; Lincoln P; Thomson J., Eds.; Humana Press: Totowa, NJ, **1998**, 98, 27–32. doi: 10.1385/0-89603-443-7:27

7. Irwin, J.; Edson, S.; Loreill, O.; Just, R.; Barritt, S.; Lee, D.; Holland, T.; Parsons, T.; Leney, M. DNA identification of 'Earthquake McGoon' 50 years post-mortem. J. Forensic Sci. **2007**, *52*, 1115–1118.

8. Ivanova, N.; Kuzmina, M. Protocols for dry DNA storage and shipment at room temperature. Mol. Ecol. Resour. **2013**, *13*, 890–898. doi: 10.1111/1755-0998.12134

9. McLaughlin, P.; Hopkins, C.; Springer, E.; Prinz, M. Non-destructive DNA recovery from handwritten documents using a dry vacuum technique. J. Forensic Sci. **2021**, *66*, 1443–1451. doi: 10.1111/1556-4029.14696

10. Rahikainen, A.; Jukka, U.; de Leeuw, W.; Budowle, B.; Sajantila, A. DNA quality and quantity from up to 16 years old postmortem blood stored on FTA cards. Forensic Sci. Int. **2016**, *261*, 148–153.

11. van Renterghem, P.; Leonard, D.; de Greef, C. Use of latent fingerprints as a source of DNA for genetic identification. In *Progress in Forensic Genetics*; Sensabaugh, G.F.; Lincoln, P.J., Eds.; Elsevier Science: Amsterdam, **2000**; 8, 501–503.

12. Schneider, H.; Sommerer, T.; Rand, S.; Wiegand, P. Hot flakes in cold cases. Int. J. Legal Med. **2011**, *125*, 543–548. doi: 10.1007/s00414-011-0548-7

13. Lee, S.; Clabaugh, K.; Silva, B.; Odigie, K.; Coble, M.; Loreille, O.; Scheible, M.; Fourney, R.; Stevens, J.; Carmody, G.; Parsons, T.; Pozder, A.; Eisenberg, A.; Budowle, B.; Ahmad, T.; Miller, R.; Crouse, C. Assessing a novel room temperature DNA storage medium for forensic biological samples. Forensic Sci. Int. Genet. **2012**, *6*, 31–40.

14. Roberts, K.; Johnson, D.J. *Investigations on the Use of Samplematrix to Stabilize Crime Scene Biological Samples for Optimized Analysis and Room Temperature Storage.* U.S. Department of Justice, Document 237838; Award 2007-DN-BX-K172, **2012**. https://www.ncjrs.gov/pdffiles1/nij/grants/237838.pdf

15. Nolan, A.; Speers, S.; Murakami, J.; Chapman, B. A pilot study: The effects of repeat washing and fabric type on the detection of seminal fluid and spermatozoa. Forensic Sci. Int. **2018**, *289*, 51–56. doi: 10.1016/j.forsciint.2018.05.021

16. Nussbaumer, C.; Halama, T.; Ostermann, W. DNA-typing of sperm collected from skin and clothes after washing in proceedings of the third European Academy of Forensic Science Meeting. Forensic Sci. Int. **2003**, *136*(Suppl. 1), 47.

17. Helmus, J.; Bajanowski, T.; Poetsch, M. DNA transfer-a never ending story. A study on scenarios involving a second person as carrier. Int. J. Legal Med. **2016**, *130*, 121–125. doi: 10.1007/s00414-015-1284-1

18. Helmus, J.; Zorell, S.; Bajanowski, T.; Poetsch, M. Persistence of DNA on clothes after exposure to water for different time periods—a study on bathtub, pond, and river. Int. J. Legal Med. **2018**, *132*, 99–106.

19. Magee, A.; Breathnach, M.; Doak, S.; Thornton, F.; Noone, C.; McKenna, L. Wearer and non-wearer DNA on the collars and cuffs of upper garments of worn clothing. Forensic Sci. Int. Genet. **2018**, *34*, 152–161.

20. Ruan, T.; Barash, M.; Gunn, P.; Bruce, D. Investigation of DNA transfer onto clothing during regular daily activities. Int. J. Legal Med. **2017**, *132*, 1035–1042. doi: 10.1007/s00414-017-1736-x

21. Szkuta, B.; Ansell, R.; Boiso, L.; Connolly, E.; Kloosterman, A.; Kokshoorn, B.; McKenna, L.; Steensma, K.; Steensma, K.; van Oorschot, R. DNA transfer to worn upper garments during different activities and contacts: An inter-laboratory study. Forensic Sci. Int. Genet. **2020**, *46*, 102268. doi: 10.1016/j.fsigen.2020.102268

22. Brayley-Morris, H.; Sorrell, A.; Revoir, A.; Meakin, G.; Syndercombe Court, D.; Morgan, R. Persistence of DNA from laundered semen stains: Implications for child sex trafficking cases. Forensic Sci. Int. Genet. **2015**, *19*, 165–171. doi: 10.1016/j.fsigen.2015.07.016

23. Kafarowski, E.; Lyon, A.; Sloan, M. The retention and transfer of spermatozoa in clothing by machine washing, Can. Soc. Forensic Sci. J. **1996**, *29*, 7–11. doi: 10.1080/00085030.1996.10757042

24. Noël, S.; Lagacé, K.; Rogic, A.; Granger, D.; Bourgoin, S.; Jolicoeur, C.; Séguin, D. DNA transfer during laundering may yield complete genetic profiles. Forensic Sci. Int. Genet. **2016**, *23*, 240–247. doi: 10.1016/j.fsigen.2016.05.004

25. Fiedler, A.; Rehdorf, J.; Hilbers, F.; Johrdan, L.; Rehdorf, J.; Benecke, M. Detection of semen (human and boar) and saliva on fabrics by a very high powered UV-/VIS-light. Open Forensic Sci. J. **2008**, *1*, 12–15. doi: 10.13140/2.1.2897.1848

26. Adamowicz, M.; Stasulli, D.; Sobestanovich, E.; Bille, T. Evaluation of methods to improve the extraction and recovery of DNA from cotton swabs for forensic analysis. PLoS One **2014**, *9*, e116351. doi: 10.1371/journal.pone.0116351

27. Benecke, M. A Routine Rape Case that Became Tricky (and Educational). Proc. Am. Acad. Forensic Sci. 2021, 73rd Annual Scientific Meeting, American Academy of Forensic Sciences, Colorado Springs, CO, **2021**, B47, 192.

28. Garvin, A.M.; Holzinger, R.; Berner, F.; Krebs, W.; Hostettler, B.; Lardi, E.; Hertli, C.; Quartermaine, R.; Stamm, C. The ForensiX evidence collection tube and its impact on DNA preservation and recovery. BioMed Res. Int. **2013**, *2013*, 105797. doi: 10.1155/2013/105797

29. Hochmeister, M.; Eisenberg, A.; Budowle, B.; Binda, S.; Serra, A.; Whelan, M.; Gaugno, D.; Fitz, J.; Dirnhofer, R. The Development of a New Sexual Assault Kit for the Optimization of Collection in Handling and Storage of Physical and Biological Evidence, Proceedings of the First European Symposium on Human Identification, Toulouse, France, Promega: Madison, WI, **1996**.

30. Lovell, R.E.; Singer, M.; Flannery, D.; McGuire, M. The case for "investigate all": assessing the cost-effectiveness of investigating no CODIS hit cases in a sexual assault kit initiative. J. Forensic Sci., **2021**, *66*(4), 1316–1328. doi: 10.1111/1556-4029.14686

31. Hochmeister, M.; Rudin, O.; Meier, R.; Peccioli, M.; Borer, U.; Eisenberg, A.; Nagy, R.; Dirnhofer, R. A foldable cardboard box for drying and storage of by cotton swab collected biological samples (German). Arch Kriminol. **1997**, *200*, 113–120.

32. van Oorschot, R.; Ballantyne, K.; Mitchell, R. Forensic trace DNA: a review. Investig. Genet. **2010**, *1*, 14. doi: 10.1186/2041-2223-1-14

33. Thomasma, S.; Foran, D. The influence of swabbing solutions on DNA recovery from touch samples. J. Forensic Sci. **2013**, *58*, 465–469.

34. Benecke, M.; Schmitt, C. Five cases of forensic short tandem repeat DNA typing. Electrophoresis **1997**, *18*, 690–694.

35. Mulligan, C.; Kaufman, S.; Quarino, L. The utility of polyester and cotton as swabbing substrates for the removal of cellular material from surfaces. J. Forensic Sci. **2011**, *56*(2), 485–490.

36. Stockrahm, S. *"Geschlampt und dumm angestellt". Das DNA-Phantom von Heilbronn.* Die Zeit, March 27, **2009**. https://www.zeit.de/online/2009/14/benecke-interview-wattestaebchen/komplettansicht

37. Wiegand, P.; Heide, S.; Stiller, D.; Kleiber, M. DNA typing from epithelial cells recovered from the pullover and neck of the victim. Rechtsmedizin **1998**, *8*, 226–228.

38. Hammer, U.; Bulnheim, U.; Karstädt, G.; Meissner, N.; Wegener, R. DNA typing of epidermal cells transferred after physical violence. Rechtsmedizin **1997**, *7*, 180–183.

39. Banaschak, S.; Baasner, A.; Driever, F.; Madea, B. Interpretation von DNA-Spuren auf einem Körper. Rechtsmedizin **2001**, *11*, 148.

40. Wiegand, P.; Bajanowski, T.; Brinkmann, B. DNA typing of debris from fingernails. Int. J. Legal Med. **1993**, *106*, 81–83.

41. Jiang, X. One method of collecting fallen off epithelial cell. Forensic Sci. Int. Genet. Suppl. Ser. **2009**, *2*, 193.

42. Williams, S.; Panacek, E.; Green., W.; Kanthaswamy, S.; Hopkins; Calloway, C. Recovery of salivary DNA from the skin after showering. Forensic Sci. Med. Pathol. **2015**, *11*, 29–34. doi: 10.1007/s12024-014-9635-7

43. Sweet, D.; Lorente, M.; Lorente, J.; Valenzuela, A.; Villanueva, E. An improved method to recover saliva from human skin: the double swab technique. J. Forensic Sci. **1997**, *42*, 320–322.

44. Pang, B.; Cheung, B. Double swab technique for collecting touched evidence. Legal Med. (Tokyo) **2007**, *9*(4), 181–184.

45. Horsman-Hall, K.; Orihuela, Y.; Karczynski, S.; Davis, A.; Ban, J.; Greenspoon, S. Development of STR profiles from firearms and fired cartridge cases. Forensic Sci. Int. Genet. **2009**, *3*, 242–250. doi: 10.1016/j.fsigen.2009.02.007

46. Burgoyne, L.A. Solid medium and method for DNA storage. US Patent 5,496,562, **1996**.

47. Mundorff, A.; Amory, S.; Huel, R.; Bilić, A.; Scott, A.; Parsons, T. An economical and efficient method for postmortem DNA sampling in mass fatalities. Forensic Sci. Int. Genet. **2018**, *36*, 167–175.

48. Green, H.; Tillmar, A.; Pettersson, G.; Montelius, K. The use of FTA cards to acquire DNA profiles from postmortem cases. Int. J. Legal Med. **2019**, *133*(6), 1651–1657. doi: 10.1007/s00414-019-02015-2

49. Tillmar, A.; Grandella, I.; Monteliusa, K. DNA identification of compromised samples with massive parallel sequencing. Forensic Sci. Res. **2019**, *4*, 331–336. doi: 10.1080/20961790.2018.1509186

50. Kirgiz, I.; Calloway, C. Increased recovery of touch DNA evidence using FTA paper compared to conventional collection methods. J. Forensic Leg. Med. **2017**, *47*, 9–15. doi: 10.1016/j.jflm.2017.01.007

51. Zieger, M.; Defaux, P.M.; Utz, S. Electrostatic sampling of trace DNA from clothing. Int. J. Legal Med. **2016**, *130*, 661–667. doi: 10.1007/s00414-015-1312-1

52. Brinkmann, B.; Rand, S.; Bajanowski, T. Forensic identification of urine samples. Int. J. Legal Med. **1992**, *105*, 59–61.

53. Martin, L. STR Typing of Nuclear DNA from Human Fecal Matter Using the Qiagen Qiaamp Stool Mini Kit, Twelfth International Promega Symposium on Human Identification, Biloxi, MS, Promega: Madison, WI, **2001**.

54. Prinz, M.; Grellner, W.; Schmitt, C. DNA typing of urine samples following several years of storage. Int. J. Legal Med. **1993**, *196*, 75–79.

55. Tang, J.; Ostrander, J.; Wickenheiser, R.; Hall, A. Touch DNA in forensic science: The use of laboratory-created eccrine fingerprints to quantify DNA loss. Forensic Sci. Int.: Synergy **2020**, *2*, 1–16.

56. Burrill, J.; Daniel, B.; Frascione, N. A review of trace "Touch DNA" deposits: variability factors and an exploration of cellular composition. Forensic Sci. Int. Genet. **2019**, *39*, 8–18. doi: 10.1016/j.fsigen.2018.11.019

57. Van Oorschot, R.A.H.; Jones, M.J. DNA fingerprints from fingerprints. Nature **1997**, *387*, 767.

58. Stein, C.S.; Kyeck, S.H.; Hennsge, C. DNA typing of fingerprint reagent treated biological stains. J. Forensic Sci. **1996**, *41*, 1012–1017.

59. Haines, A.; Tobe, S.; Kobus, H.; Linacre, A. Successful direct STR amplification of hair follicles after nuclear staining. Forensic Sci. Int. Genet. Suppl. Ser. **2015**, *5*, e65–e66.

60. Thamnurak, C.; Bunakkharasawat, W.; Riengrojpitak, S.; Panvisavas, N. DNA typing from fluorescent powder dusted latent fingerprints. Forensic Sci. Int. **2011**, *3*(1), e524–e525.

61. Bhoelai, B.; de Jong, B.J.; de Puit, M.; Sijen, T. Effect of common fingerprint detection techniques on subsequent STR profiling. Forensic Sci. Int. **2011**, *3*, e429–e430.

62. Quinones, I.; Daniel, B. Cell free DNA as a component of forensic evidence recovered from touched surfaces. Forensic Sci. Int. Genet. **2012**, *6*(1), 26–30.

63. Schulz, M.; Reichert, W. Archived or directly swabbed latent fingerprints as a DNA source for STR typing. Forensic Sci. Int. **2002**, *127*, 128–130.

64. Breathnach, M.; Williams, L.; McKenna, L.; Moore, E. Probability of detection of DNA deposited by habitual wearer and/or the second individual who touched the garment. Forensic Sci. Int. Genet. **2016**, *20*, 53–60.

65. Fonneløp, A.; Ramse, M.; Egeland, T.; Gill., P. The implications of shedder status and background DNA on direct and secondary transfer in an attack scenario. Forensic Sci. Int. Genet. **2017**, *29*, 48–60.

66. Sessa, F.; Salerno, M.; Bertozzi, G.; Messina, G.; Ricci, P.; Ledda, C.; Rapisarda, V.; Cantatore, S.; Turillazzi, E.; Pomara, C. Touch DNA: impact of handling time on touch deposit and evaluation of different recovery techniques: An experimental study. Sci. Rep. **2019**, *9*, 9542. doi: 10.1038/s41598-019-46051-9

67. Wickenheiser, R. Trace DNA: a review, discussion of theory, and application of the transfer of trace quantities of DNA through skin contact. J. Forensic Sci. **2002**, *47*, 442–450.

68. Petricevic, S.; Bright, J.; Cockerton, S. DNA profiling of trace DNA recovered from bedding. Forensic Sci. Int., **2006**, *159*(1), 21–26. doi: 10.1016/j.forsciint.2005.06.004

69. Poetsch, M.; Pfeifer, M.; Konrad, H.; Bajanowski, T.; Helmus, J. Impact of several wearers on the persistence of DNA on clothes—a study with experimental scenarios. Int. J. Legal Med. **2018**, *132*, 117–123. doi: 10.1007/s00414-017-1742-z

70. Hanson, E.; Ballantyne, J. 'Getting blood from a stone': Ultra sensitive forensic DNA profiling of microscopic bio-particles recovered from 'touch DNA' evidence. Methods Mol. Biol. **2013**, 1039, 3–17. doi: 10.1007/978-1-62703-535-4_1

71. Hellerud, B.; Johannessen, H.; Haltbakk, H.; Hoff-Olsen, P. Zip lock poly bags in drug cases – a valuable source for obtaining identifiable DNA results? Forensic Sci. Int. Genet. Suppl. Ser. **2008**, 1, 433–434.

72. Jiang, Z.; Zhang, X.; Deka, R.; Jin, L. Genome amplification of single sperm using multiple displacement amplification. Nucleic Acids Res. **2005**, 33, e91.

73. Tokutomi, T.; Takada, Y.; Kanetake, J.; Mukaida, M. Identification using DNA from skin contact: case reports. Legal Med. (Tokyo) **2009**, 1(11 Suppl.), S576–S577. doi: 10.1016/j.legalmed.2009.02.004

74. Barash, M.; Reshef, A.; Brauner, P. The use of adhesive tape for recovery of DNA from crime scene items. J. Forensic Sci. **2010**, 55, 1058–1064.

75. Farash, K.; Hanson, E.; Ballantyne, J. Enhanced genetic analysis of single human bioparticles recovered by simplified micromanipulation from forensic 'touch DNA' evidence. J. Vis. Exp. **2015**, 97, 52612. doi: 10.3791/52612

76. Hansson, O.; Finnebraaten, M.; Heitmann, I.K.; Ramse, M.; Bouzga, M. Trace DNA collection – performance of minitape and three different swabs. Forensic Sci. Int. Genet. Suppl. Ser. **2009**, 2, 189–190.

77. Stoop, B.; Defaux, P.; Utz, S.; Zieger, M. Touch DNA sampling with SceneSafe Fast™ minitapes. Legal Med. **2017**, 29, 68–71.

78. Hess, S.; Haas, C. Recovery of trace DNA on clothing: a comparison of mini-tape lifting and three other forensic evidence collection techniques. J. Forensic Sci. **2017**, 62(1), 187–191.

79. NIJ (National Institute of Justice). Protocol 2.05: Semen Stain Identification: Kernechtrot Picoindigocarmine Stain (KPIC) (Identification). In President's DNA Initiative: DNA Analyst Training Laboratory Training Manual, **2019**. https://static.training.nij.gov/lab-manual/Linked%20Documents/Protocols/pdi_lab_pro_2.05.pdf

80. Elliott, K.; Hill, D.S.; Lambert, C.; Burroughes, T.R.; Gill, P. Use of laser microdissection greatly improves the recovery of DNA from sperm on microscope slides. Forensic Sci. Int. **2003**, 137, 28–36.

81. Domazet, B.; Maclennan, G.T.; Lopez-Beltran, A.; Montironi, R.; Cheng, L. Laser capture microdissection in the genomic and proteomic era: targeting the genetic basis of cancer. Int. J. Clin. Exp. Pathol. **2008**, 1(6), 475–488.

82. Ladanyi, A.; Sipos, F.; Szoke, D.; Galamb, O.; Molnar, B.; Tulassay, Z. Laser microdissection in translational and clinical research. Cytom. A. **2006**, 69(9), 947–960. doi: 10.1002/cyto.a.20322

83. Niyaz, Y.; Stich, M.; Sägmüller, B.; Burgemeister, R.; Friedemann, G.; Sauer, U.; Gangnus, R.; Schütze, K. Noncontact laser microdissection and pressure catapulting: sample preparation for genomic, transcriptomic, and proteomic analysis. Methods Mol. Med. **2005**, 114, 1–24.

84. Vandewoestyne, M.; Deforce, D. Laser capture microdissection in forensic research: a review. Int. J. Legal Med. **2010**, 124(6), 513–521. doi: 10.1007/s00414-010-0499-4

85. Butler, J.M. Forensic DNA Typing: Biology & Technology behind STR Markers. Academic Press: San Diego, CA, **2001**.

86. Benecke, M.; Wells, J. Molecular techniques for forensically important insects. In Entomological Evidence: Utility of Arthropods in Legal Investigations; Byrd, J.H.; Castner, J.L., Eds.; CRC Press: Boca Raton, FL, **2000**; 341–352.

87. Brinkmann, B.; Carracedo, A., Eds.; Progress in Forensic Genetics; Excerpta Medica International Congress Series 1239, Elsevier: Amsterdam, **2003**, 9.

88. De Stefano, F.; Bruni, G.; Casarino, L.; Costa, M.G.; Mannucci, A. Multiplexed DNA markers from cigarette butts in a forensic casework. Adv. Forensic Haemogenet. **1996**, 6, 252–254.

89. Wiegand, P.; Bajanowski, T.; Brinkmann, B. DNA typing of debris from fingernails. Int. J. Legal Med. **1993**, *196*, 81–83.

90. Barbaro, A.; Cormaci, P.; Falcone, G. Validation of BTA™ lysis buffer for DNA extraction from challenged forensic samples. Forensic Sci. Int. Genet. Suppl. Ser. **2011**, *3*, E61–E62. doi: 10.1016/j.fsigss.2011.08.030

91. Stray, J.; Holt, A.; Brevnov, M.; Calandro, L.; Furtado, M.; Shewale, J. DNA from biological materials and calcified tissues. Forensic Sci. Int. Genet. Suppl. Ser. **2009**, *2*, 159–160. doi: 10.1016/j.fsigss.2009.08.086

92. Mapes, A.; Kloosterman, A.; van Marion, V.; de Poot, C. Knowledge on DNA success rates to optimize the DNA analysis process: from crime scene to laboratory. J. Forensic Sci. **2016**, *61*, 1055–1061.

93. Schiffner, L.; Bajda, E.; Prinz, M.; Sebestyen, J.; Shaler, R.; Caragine, T. Optimization of a simple, automatable extraction method to recover sufficient DNA from low copy number DNA samples for generation of short tandem repeat profiles. Croat Med. J. **2005**, *46*, 578–586.

94. Borowsky, R. Field preservation of fish tissues for DNA fingerprint analysis. Fingerprints News **1991**, *3*(2), 8–9.

95. Phuengmongkolchaikij, S.; Panvisavas, N.; Bandhaya, A. Alcohols as solution for delaying microbial degradation of biological evidence on cotton swabs. Forensic Sci. Int. Genet. Suppl. Ser. **2017**, *6*, e539–e541.

96. Van Oorschot, R.A.H.; Szepietowska, I.; Scott, D.L.; Weston, R.K.; Jones, M.K. Retrieval of genetic profiles from touched objects. In *First International Conference on Forensic Human Identification in the Millennium*; Forensic Science Service: London, **1999**.

97. Beckwith, S.; Murakami, J.; Chapman, B. The persistence of semen. Australian J. Forensic Sci. **2018**, *52*, 155–164. doi: 10.1080/00450618.2018.1484164

98. Karadayi, S.; Moshfeghi, E.; Arasoglu, T.; Karadayi, B. Evaluating the persistence of laundered semen stains on fabric using a forensic light source system, prostate-specific antigen Semiquant test and DNA recovery-profiling. Med. Sci. Law **2020**, *60*, 122–130. doi: 10.1177/0025802419896935

99. Scherczinger, C.A.; Ladd, C.; Bourke, M.T.; Adamowicz, M.S.; Johannes, P.M.; Scherczinger, R.; Beesley, T.; Lee, H.C. Systematic analysis of PCR contamination. J. Forensic Sci. **1999**, *44*, 1042–1045.

100. Ladd, C.; Adamowicz, M.S.; Bourke, M.T.; Scherczinger, C.A.; Lee, H.C. A systematic analysis of secondary DNA transfer. J. Forensic Sci. **1999**, *44*, 1270–1272.

101. Wiegand, P.; Heimbold, C.; Klein, R.; Immel, U.; Stiller, D.; Klintschar, M. Transfer of biological stains from different surfaces. Int. J. Legal Med. **2011**, *125*, 727–731.

102. Gill, P.; Jeffreys, A.J.; Werrett, D. Forensic application of DNA fingerprints. Nature **1985**, *318*, 577–579.

103. Banaschak, S.; Möller, K.; Pfeiffer, H. Potential DNA mixtures introduced through kissing. Int. J. Legal Med. **1998**, *111*, 284–285. doi: 10.1007/s004140050172

104. Benecke, M. Sexual assault & murder: When DNA does not help even though it is present. In *Handbook of DNA Profiling*; Dash, H.; Shrivatava, P.; Lorente, J., Eds.; Springer Nature: Singapore, **2021**.

105. Buckingham, A.K.; Harvey, M.L.; van Oorschot, R.A.H. The origin of unknown source DNA from touched objects. Forensic Sci. Int. Genet. **2016**, *25*, 26–33.

106. Butler, J. *Advanced Topics in Forensic DNA Typing: Methodologies*. Academic Press: Waltham, MA, **2012**.

107. Murphy, C.; Kenna, J.; Flannagan, L.; Gorman, M.; Boland, C.; Ryan, J. A study on the background levels of male DNA on underpants worn by females. J. Forensic Sci. **2019**, *65*, 399–405. doi: 10.1111/1556-4029.14198

108. Federal Bureau of Investigation. *Quality Assurance Standards for Forensic DNA Testing Laboratories*, **2020**. https://www.fbi.gov/file-repository/quality-assurance-standards-for-forensic-dna-testing-laboratories.pdf/view

109. *Scientific Working Group on DNA Analysis Methods (SWGDAM) Revised Validation Guidelines*, **2003**. https://strbase.nist.gov/validation/SWGDAM_Validation.doc

110. Butler, J. *National Institute of Standards and Technology (NIST) Applied Genetics Group: SWGDAM Guidelines*. CIB Forensic Science Center Training Seminar: Taipei, Taiwan, June 6–7, **2012**. https://strbase.nist.gov/training/Taiwan2012-SWGDAM.pdf

111. Deutsches Institut für Normung. *Forensic Analysis – Part 2: Recognition, Recording, Recovering, Transport and Storage of Material*. ISO/DIS 21043-2, **2017**.

6 DNA Preservation

C. William Kilpatrick
University of Vermont, Burlington, Vermont, U.S.A.

CONTENTS

INTRODUCTION

Early claims of DNA recovery from million-year-old fossils[1–4] are now widely regarded as modern contamination.[5–8] However, the successful extraction and amplification of DNA from fossils at least hundreds of thousands of years old[9–17] has led to the development of the field of ancient DNA (aDNA).[18,19] Sequence data obtained from fossil bones, teeth, coprolites,[19–21] preserved tissues at human burial sites,[22–26] and from museum specimens[27–37] have provided new insights to the relationships between extinct organisms and their contemporary relatives, historical population biology, and paleoecology. The advancements in the techniques of aDNA in combination with next-generation sequencing has led to a field of paleogenomics.[38] Degradation of DNA occurs through enzymatic digestion[39] and spontaneous decomposition primarily from depurination/β-elimination[40] and free radical oxidation pathways.[41] At moderate temperatures in solutions that will inhibit enzymatic digestion, DNA will spontaneously degrade to short fragments over a time period of several thousand years.[42] Cryogenic methods of preservation provide an environment that slows both this spontaneous decomposition as well as enzymatic digestion whereas non-cryogenic preservation methods block enzymatic activity by various approaches.

MECHANISMS OF DNA DEGRADATION AND LOSS

Degradation or loss of DNA during storage primarily results from actions of endogenous nucleases,[39] depurination/β-elimination,[40] free radical oxidation,[41] and binding of DNA to storage tubes.[43] Double-stranded DNA molecules are highly

DOI: 10.1201/9781003247432-6

charged and very hydrophilic whereas polypropylene that is generally used for making microfuge tubes and PCR plates is a very hydrophobic polymer. Although these characteristics should reduce the interaction between DNA and the plate or tube walls, it has been reported that DNA can bind to polypropylene.[44] At high ionic strengths, a large proportion of a DNA sample may be bound to the walls of polyethylene and most polypropylene tubes.[43] Under the conditions where DNA is more frequently stored, moderate or low ionic strength, most polypropylene tubes bind less the 5% of the DNA; however, some tubes may bind as much as 25% of the DNA and considerable variation was reported among batches from the same manufacturer.[43] A later study conducted by NIST (National Institute of Standards and Technology) verified the binding of DNA to polypropylene tubes.[45] The binding of DNA with tube walls can induce changes in conformation which can result in complete denaturation with strand separation.[46–48] While this strand separation is most frequently noticeable with short DNA fragments at high ionic strength, under these conditions longer fragment may form multi-stranded complexes.[49] Polyallomer tubes show no evidence of either binding or denaturing of DNA[43] and a number of manufacturers have developed "nonstick" tubes and plates. In addition, bioactive contaminants have been found to leach from plasticware[50] with some solutions such as DMSO (dimethyl sulfoxide) which is used as a method of non-cryogenic tissue preservation.

In an aqueous solution, DNA is known to degrade by the two-step process of depurination and β-elimination.[40] The depurination reaction is mainly acid-catalyzed under physiological conditions.[51] At sites of depurination, the DNA chain undergoes cleavage by a β-elimination process within a few days.[52] Rates of chemical degradation of DNA have largely been limited to classic kinetic experiments where rates of DNA decay have been found to be associated with changes in temperature and pH.[51,53] The pH of the storage solution is expected to have a large effect on DNA stability. Next-generation sequencing of complete genomes from ancient individuals and extinct species has generated genome-wide statistical distribution of nucleotide misincorporations originating from postmortem DNA damage.[54] Analyses of these genomes suggest that depurination is a key driving force of DNA degradation and guanosine residues show elevated rates of depurination.[54] The rate of depurination was estimated to be 5.5×10^{-6} per year from a series of New Zealand moa fossils with an effective burial temperature of 13.1°C[55] which is about 400 times slower than the published kinetic rates at pH 5.[51]

DNA bases and deoxyribose sugars are also susceptible to degradation by free radical oxidation[41] that may occur either during or after extraction[56] and with exposure to extreme heat.[57] Use of specific chelators, such as EDTA (ethylenediaminetetraacetic acid), has been found to be extremely effective at reducing free radical oxidation for at least 1 year of DNA stored at room temperature.[58]

Degradation of DNA in a sample may also result from the action of endogenous enzymes, endo- and exonucleases that cleave DNA strands[39,42], and in DNA extracts from nucleases left behind from the extraction procedure.[59] Temperature, pH, and salt concentration are known to influence enzymatic activity.[60] Nucleases are divalent cation-dependent enzymes, and their activity is much reduced at high concentrations of EDTA or diaminocyclohexanetetraacetate (CDTA).

STORAGE OF DNA EXTRACT

Traditionally, purified DNA has been stored in either nuclease-free water or TE (10 mM Tris, pH 7.5–8.0, 1 mM EDTA), the extraction buffer provided in many commercial kits. In either of these solutions, DNA may be stored for several months at either 4°C or –20°C with little degradation. Extracted DNA stored in TE at temperatures between –70°C and 37°C for 6 months demonstrated little degradation, whereas DNA stored at 45°C–65°C shows signs of degradation within 24 hour and was completely degraded by 8 days.[61] Overall TE outperforms nuclease free water regardless of the storage temperature.[62] Repeated freezing and thawing of DNA does not appear to result in degradation,[63] however the addition of 50% glycerol has been reported as being effective in reducing degradation from the formation of ice crystals during storage in some buffers.[64,65] For longer cryogenic storage, DNA should be stored in TE solutions at –70°C (ultrafreezers) to –196°C (liquid nitrogen).[66,67]

Alternatively, DNA may be precipitated and stored in ethanol often layered with chloroform at room temperature or 4°C. Trehalose is used for both cryopreservation and lyophilization of DNA.[68–70] Dried DNA samples are easily shipped,[71] but problems with rehydration after prolonged storage have been reported.[61] A number of commercial products for the storage of nucleic acids at room temperature have been evaluated for their preservation of DNA in human forensics[72–75] but have not been evaluated among a wider range of taxa. The extent to which dried DNA is protected from degradation is dependent both on the storage material (trehalose, polyvinyl alcohol, or Biomatrica's DNAstable) and the storage temperature.[71] Buffers containing EDTA have been used to store DNA at room temperature with little degradation for periods of time of less than a year[58], and the addition of ethanol[58] has been found to further enhance DNA stability.

STORAGE OF TISSUES FOR DNA EXTRACTIONS OR AMPLIFICATIONS

Although a few general guidelines for tissue preservation are available,[66,76,77] no single method of tissue preservation for DNA extraction has proven to be optimal for all taxa or under all conditions (field or laboratory). Recommended methods vary across taxa from mammals[66,78] and other vertebrates to insects[79,80] and other invertebrates.[81] Several studies[66,78,81,82] have addressed short-term storage of tissues in the field without refrigeration.

Some tissue and other samples have received special attention with regards to storage for the recovery of DNA. Blood samples in vacuum tubes containing EDTA stored for 4 weeks at temperature from 4°C to 37°C yielded high molecular weight DNA, although degradation was evident in samples stored at room or higher temperatures.[83] Little degradation was reported for samples stored at room temperature for less than a year.[84] Interestingly, blood samples stored at 37°C demonstrated less DNA degradation than those stored at 23°C.[83] Avian blood in Queen's lysis buffer without refrigeration has been a reliable storage method,[85] however difficulties have been reported for some avian species.[86] Genomic DNA in dried blood spots on filter paper[87,88] or commercially developed cards (Guthrie, FTA®, FTA-Elute®)[89,90] have been widely used in biomedical research to preserve genomic DNA from humans and associated pathogens at room temperature under dry conditions with a desiccant[87–92] and wildlife samples.[93–95] Blood spots on FTA cards have preserved DNA for 4 years at 22°C–24°C[96] and 11 years at ambient temperatures,[97] though some degradation has been reported.[93] Filter paper appears to be as efficient as FTA cards for DNA preservation[98] and DNA was detected after at least 9 months of storage at 37°C. Other fluids and tissues have been stored at room temperature on filter paper or FTA cards[99–101] and DNA in those materials have been reported to be stable for at least 7.5 years.[102]

Several studies[103–105] have been published on the storage of fecal material for the recovery of DNA and the comparison of storage methods.[106,107] While no single preservative method appears to be optimal,[106] ethanol and DET (20% DMSO, 0.25 M EDTA, 100 mM Tris pH 7.5, and NaCl to saturation)[85] have been suggested as preferable[106] or found to be superior to other preservatives compared.[107] Scats from animals with diets consisting of plant-derived foods are recommended to be stored in DET.[104,107] Scats for carnivores are recommended to be stored in ethanol, whereas DET and ethanol were equally efficient for the storage of scats from insectivores.[107] Although it has been suggested that lysis buffers should perform as well as ethanol,[108] both DET and ethanol were found to be far superior to lysis buffers for preserving fecal samples,[107] and fecal samples should not be stored in lysis buffers.

Hair is a widely used source of DNA,[106,109,110] and has been stored in silica desiccant or at −20°C.[111] Neither storage method has proven successful for long-term storage of DNA in hairs.[111]

CRYOPRESERVATION

Cryopreservation is an effective method of preservation of tissues for DNA extraction and is often reported as the preferred method;[66,67,77,85,112–120] however, freeze–thaw cycles have been found to produce degradation.[84,85] Although some degradation does occurs at temperatures of −20°C and −80°C compared to −180°C,[121] most tissues in research collections[122,123], https://www.idigbio.org/content/dna-banks-and-genetic-resources-repositories-united-states] are stored in ultrafreezers (−80°C).[124] For long-term storage of tissue samples and organisms for DNA preservation, liquid nitrogen appears to be the method of choice,[125] though temperature alone is usually not the only decisive factor.[126] The use of chemical agents to protect macromolecules from damage caused by ice formation and other crystallization events[127] are frequently used in preservation of cell lines and certain tissue samples. The most frequently used cryoprotectants[128] include various types of glycerol (ethylene glycol, propylene glycol, glycerol), DMSO, sugars (primarily trehalose), and other chemical agents.

Cryopreservation in the field can be accomplished either by dry ice (−78°C) or liquid nitrogen (−196°C) but both are difficult to use in many field situations requiring careful handling, special equipment, and strict regulations for transport by air.[85,113,116,129] Numerous non-cryogenic methods have been developed[77,130] and appear to be effective for the short-term preservation of tissues for DNA extraction.[80,81,85,114,115,118,131–137] These non-cryogenic methods are frequently used while in the field and either combined with[138] or replaced by cryogenic methods in the laboratory.[80,120,139]

NON-CRYOGENIC PRESERVATION

DEHYDRATION

Dehydration methods include silica gel beads that can be readily used in the field and allow for easy shipping and transport of tissue samples[113] and freeze-drying that has been recommended for shipping tissues to long distances.[137] Extracted DNA has been amplified from samples dried from up to a year from a variety of sources,[140–142] but PCR was not successful from pig tissue stored in silica gel after 6 months.[66] Only slight degradation was reported for liver freeze-dried samples after 4 years[143] and lyophilisation has been reported to improve the quality of DNA extracted from feces.[144,145] Wing punches from bats, dried with silica gel has been found to result in higher recovery of DNA than fluid preservation methods.[146] In well dried samples, humidity and oxygen are the two major factors resulting in DNA breakdown.[147,148] Other physical and chemical drying methods are available[77] but they are generally not used for the preservation of DNA.

FLUIDS FOR TISSUE PRESERVATION

Ethanol has been suggested[114,115,118,119,133,149–153] for the storage of tissues for DNA extraction and a concentration of 95%–99% has been found to be optimal.[152] Tissues stored in ethanol have been reported to yield highly degraded DNA fragments[44,154]; however, much of this degradation appears to occur during the extraction procedure rather than during storage.[134] Addition of EDTA to 95% ethanol[116] or diluting ethanol to 70% with 1xTE[77] appear to reduce DNA degradation. For short-term storage of up to 1 year, 70% ethanol appears to function as well as 100% ethanol; however, 100%–95% ethanol is recommended for storage exceeding 2 years.[155] A 3:1 to 5:1 ethanol to tissue ratio or higher is recommended.[77,134] Replacing ethanol after 1–2 hour to allow diffusion[116] and again 2–3 days later improves preservation.[77,134] Spiders and other arachnids to be used for DNA studies are typically stored cold (−20°C to −80°C) in 95%–100% ethanol[156–158] as storage in 70% ethanol at room temperature resulted in substantial DNA degradation.[159]

High proof drinking spirits (e.g., Everclear® or 151 white rum) are more effective for short-term storage of the genetic material in branchiopods than denatured 95% ethanol.[160] Isopropanol has been found to be as efficient as ethanol for storage up to 2 months but less efficient for longer storage[66] and a poor preservative for branchiopods' DNA.[160] Other organic solvents have been examined for DNA preservability of insects and their intracellular symbiotic bacteria.[161] At room temperature for 6 months of storage; acetone, diethyl ether, and ethyl acetate were as efficient as ethanol or isopropanol at DNA preservation, whereas chloroform and methanol were poor preservatives.[161] In addition, acetone was found to be more robust against water contamination than ethanol[161] and may prove to be a better storage solution than ethanol. However, acetone was reported to be a poor preservative of the DNA in branchiopods.[160]

Several fluids used to preserve specimens, including 4% formaldehyde, modified Carnoy's solution (acetic acid + TE buffer), and Carnoy's solution have been found not to be suitable for the preservation of DNA.[114,115,162] Over time formaldehyde in formalin is oxidized to formic acid that results in the degradation of DNA.[163] The smaller the specimen preserved in formalin the greater the degradation.[163] However, attempts to extract DNA from formalin preserved museum specimens have been successful.[164–167] Archived formalin-fixed paraffin-embedded blocks have proven to be a valuable resource for the extraction of DNA and no significant degradation was detected in blocks stored for over 12 years.[168]

Attempts have been made to find solutions that could be used in pit and other types of traps for the capture of invertebrate that would preserved the DNA of the trapped organisms for genetic analysis.[162] Ethylene glycol is one of the most common and preferred killing solutions for these traps[77,169] and has been reported to be an excellent preservative of insect DNA.[118] However, spider DNA is reported to be degraded substantially in specimens stored in ethylene glycol at room temperature.[159] Propylene glycol has been reported as a potential effective alternative to both ethylene glycol in traps[170] and to ethanol for DNA preservation.[77,80,169,171–173] It has been reported to be significantly better than ethanol at preserving high quality DNA in spiders when stored at −20°C to −80°C.[174]

The use of a dimethyl sulfoxide (DMSO) solution for the preservation of DNA in various tissues has been demonstrated to be effective for storage from 6 months to over 2 years.[66,80,81,85,119,134,153,155,175,176] DMSO rapidly penetrates the tissue and displaces the water preventing enzymatic activity. A strong amplification was reported from DNA extracted from bees stored in pure DMSO for over a year.[119] A solution of 20%–25% DMSO in a saturated salt (NaCl) solution was found to be a better preservative than other salt solutions or lysis buffers,[81] however, both ethanol and propylene glycol were reported to be better preservatives.[80] A solution of 20% DMSO, 0.25 M EDTA, and NaCl to saturation, pH 7.5[66,85,155] was found to be a more effective method of non-cryogenic storage for the prevention of DNA degradation compared to ethanol or a lysis buffer.[134] Storage of feces in a DMSO/EDTA/Tris/salt solution (DETs) was found to be more effective for preserving nuclear DNA than ethanol, freezing, and drying methods.[104] The transfer of DMSO salt solution preserved tissue into ethanol after shipping to lab has been reported to increase the DNA yield.[176] However, morphological distortion of invertebrate specimens stored in DMSO has been reported,[80] thus different preservatives may be needed for samples for morphological and those for DNA preservation for molecular analyses.

CTAB (cetyl trimethyl ammonium bromide) is a detergent that has been used for DNA extraction.[177–179] In a saturated NaCl solution, CTAB has been used to preserve DNA in leaves[112] and when compared with a few other preservatives it was not found to perform as well as a DMSO salt solution.[81] A 4M guanidium-thiocyanate buffer solution has been shown to preserve nucleic acids at room temperature for up to 41 days,[180] however its utility for longer term preservation is not known.

LYSIS BUFFERS AND OTHER SOLUTIONS

Several different "lysis buffers" have been developed for the storage of tissues at room temperature.[77,78,85,153,181–184] The various formulations for "lysis buffers" generally consist of a Tris buffer at a pH between 7.5 and 8.0, EDTA (chelator), and sodium dodecyl sulfate or *n*-lauroylsarcosine (detergents to lysis cells). These buffers have demonstrated to be effective at room temperature in the preservation of DNA in blood samples for 6 months[44] and tissue samples for 2 years or more.[134,184] DMSO with 0.5% SDS has been reported as a standard buffer for shipping of tissues[77] but does not appear to have been tested for longer term storage of DNA or compared with other preservation methods. Queen's lysis buffer[85] was not as effective as a DMSO salt solution[81] and quality problems have been reported with some bird taxa.[86] After storage of tissues in Longmire's buffer, much of the DNA was found in the solution rather than in the remaining tissue.[134] Longmire's buffer was reported to be as effective at DNA preservation at room temperature as ethanol or DMSO,[134] whereas NAP buffer was found to be slightly better than either cryopreservation or 95% ethanol.[78]

Several proprietary preservatives for nucleic acids are now available for storage at room temperature for up to 6 months including DNAgard Tissue®, DNAgard Blood®, and DNAstable® (Sigma Aldrich), DNA Genotek, and RNA*later* (Ambion). These fluid preservatives are convenient for field condition and transport to a low-temperature freezer[120] but most have not been compared with other non-cryogenic preservative methods. Both DNAgard and DNA Genotek were reported to preserved DNA in tissue stored at room temperature for up to 1 month.[153]

The utility of RNA*later* (Ambion), a storage reagent designed to protects cellular RNA in samples by rapidly penetrating fresh tissues and deactivating nucleases,[185] has been compared with a number of other methods.[78,80,174] This reagent is applicable for DNA preservation[77,186] and has been found to preserve DNA at room temperature without significant degradation for at least 2 months and at 5°C for 8 months.[187] However, RNA*later* has been reported to interfere with certain extraction procedures.[66] At room temperature RNA*later* was reported as a relatively poor solution for DNA preservation for insects,[80] but at temperatures of −20°C to −80°C a good preservative for spiders.[174] The NAP buffer preserved RNA as well as RNA*later*[78] and ethanol or propylene glycol were found to be better preservatives of insect DNA.[80] Increased DNA yield was reported from samples stored in the field in RNA*later* and transferred into ethanol in the lab.[176]

CONCLUSIONS

Molecular methods utilizing DNA analyses have been incorporated into many fields of biology from phylogenetics to ecology and behavior. Organisms from bacteria and viruses to extant and extinct mammals are the foci of these studies. With these developing and expanding fields utilizing molecular data, the ability to preserve DNA extracts or samples for future extractions become evermore important. Biobanks and tissue collections provide duplicate samples for replication or expansion of studies, forensic samples for criminal justice, voucher material for systematic revisions, historical samples for demographic studies, and samples associated with various phenotypes that may allow future determination of their underling genetic basis to name a few. The roles these collections may provide in resolving the sources of future pandemics are only now being investigated and appreciated.

Storage methods for DNA extracts and tissues or specimens for the extraction of DNA have been developed that prevent or greatly reduce the primary pathways of DNA degradation. The actions of endogenous nucleases and the rates of depurination/B-elimination or free radical oxidation pathways can be greatly reduced by freezing, dehydration, or storage in solutions that either displace the water (dehydration of tissue) or deactivate enzymatic activity by a variety of mechanisms. Extracted DNA is relatively stable under a variety of cryogenic and non-cryogenic storage methods.

Tissue or specimens stored for DNA extraction are more susceptible to degradation, although some of this degradation likely occurs during the extraction procedure. Cryopreservation and/or ethanol are the preferred methods by many laboratories and repositories. Cryogenic storage at −196°C is considered the "gold standard" for long-term tissue storage, most tissue collections are stored in ultrafreezers at −80°C due to funding limitations. Since one universal method for the storage of tissue and specimens for DNA extraction does not exist, it is important to continue to weigh and evaluate the benefits of alternative methods available for long-term storage of DNA samples and tissues from which DNA may be extracted.

REFERENCES

1. Goldenberg, E.M.; Giamasi, D.E.; Clegg, M.T.; Smiley, C.J.; Durbin, M.; Henderson, D.; Zurawski, G. Chloroplast DNA-sequences from a Miocene magnolia species. *Nature* **1990**, *44*, 656–658.
2. Desalle, R.; Gatesy, J.; Wheeler, W.; Grimaldi, D. DNS-sequences from a fossil termite in Oligomiocene amber and their phylogenetic implications. *Science* **1992**, *252*, 1933–1936.
3. Cano, R.J.; Poinar, H.N.; Pieniazek, N.J.; Acra, A.; Poiner, G.O. Amplification and sequencing of DNA from a 120-135-million-year-old weevil. *Nature* **1993**, *363*, 536–538.
4. Woodward, S.R.; Weyand, N.J.; Bunnell, M. DNA-sequences from Cretaceous period bone fragments. *Science* **1994**, *266*, 1229–1232.
5. Pääbo, S.; Wilson, A.C. Miocene DNA sequences: a dream come true? *Curr. Biol.* **1991**, *1*, 45–46.
6. Austin, J.J.; Ross, A.J.; Smith, A.B.; Fortey, R.A.; Thomas, R.H. Problems with reproducibility – does geologically ancient DNA survive in amber-preserved insects? *Proc. Royal Soc. B*, **1997**, *264*, 467–474.
7. Willerslev, E.; Hansen, A.J.; Poinar, H.N. Isolation of nucleic acids and cultures from fossil ice and permafrost. *Trends. Ecol. Evol.* **2004**, *19*, 141–147.
8. Binladen, J.; Gilbert, M.T.; Willerslev, E. 800,000 year old mammoth DNA, modern elephant DNA or PCR artefact? *Biol. Lett.* **2007**, *3*, 55–56.
9. Janczewski, D.N.; Yuhki, N.; Gilbert, D.A.; Jefferson, G.T.; O'Brien, S.J. Molecular phylogenetic inferences from saber-toothed cat fossils of Rancho La Brea. *Proc. Natl. Acad. Sci. U. S. A.* **1992**, *89*, 9769–9773.
10. Höss, M.; Pääbo, S. DNA extraction from Pleistocene bones by a silica-based purification method. *Nucleic Acids Res.* **1993**, *21* (16), 1914–3913.

11. Hagelberg, E.; Thomas, M.G.; Cook, C.E. Jr.; Sher, A.V.; Baryshnikov, G.F.; Lister, A.M. DNA from ancient mammoth bones. *Nature* **1994**, *370*, 333–334.

12. Hanni, C.; Laudet, V.; Stehelin, D.; Taberlet, P. Tracking the origins of the cave bear (*Ursus spelaeus*) by mitochondrial DNA sequencing. *Proc. Natl. Acad. Sci. U. S. A.* **1994**, *91*, 12336.

13. Hoss, M.; Pääbo, S.; Vereschagin, N.K. Mammoth DNA sequences. *Nature* **1994**, *370*, 333.

14. Höss, M.; Dilling, A.; Currant, A.; Pääbo, S. Molecular phylogeny of the extinct ground sloth *Mylodon darwinii*. *Proc. Natl. Acad. Sci. U. S. A.* **1996**, *93*, 181.

15. Taylor, P.G. Reproducibility of ancient DNA sequences from extinct Pleistocene fauna. *Mol. Biol. Evol.* **1996**, *13* (1), 283–285.

16. Yang, H.; Golenberg, E.M.; Shoshani, J. Phylogenetic resolution within the elephantidae using fossil DNA sequence from American mastodon (*Mammut americanum*) as an outgroup. *Proc. Natl. Acad. Sci. U. S. A.* **1996**, *93*, 1190.

17. Greenwood, A.D.; Capelli, C.; Possnert, G.; Pääbo, S. Nuclear DNA sequences from Late Pleistocene megafauna. *Mol. Biol. Evol.* **1999**, *16* (11), 1466–1473.

18. Hagelberg, E.; Hofreiter, M.; Keyser, C. Ancient DNA: the first three decades. *Phil. Trans. R. Soc. B* **2015**, *370*, 20130371.

19. Hagelberg, E.; Sykes, B.; Hedges, R. Ancient bone DNA amplified. *Nature* **1989**, *342*, 485.

20. Hofreiter, M.; Serre, D.; Poinar, H.N.; Kuch, M.; Pääbo, S. Ancient DNA. *Nat. Rev. Genet.* **2001**, *2*, 353–359.

21. Slatkin, M.; Racimo, F. Ancient DNA and human history. *Proc. Natl. Acad. Sci. U. S. A.* **2016**, *113(23)*, 6380–6387.

22. Doran, G.H.; Dickel, D.N.; Ballinger, W.E. Jr.; Agee, O.F.; Laipis, P.J.; Hauswirth, W.W. Anatomical, cellular and molecular analysis of 8,000-yr-old human brain tissue from the Windover archaeological site. *Nature* **1986**, *323*, 803–806.

23. Pääbo, S.; Gifford, J.A.; Wilson, A.C. Mitochondrial DNA sequences from a 7000-year old brain. *Nucleic Acids Res.* **1988**, *16* (20), 9775–9787.

24. Lawlor, D.A.; Dickel, C.D.; Hauswirth, W.W.; Parham, P. Ancient HLA genes from 7,500-year-old archeological remains. *Nature* **1991**, *349*, 785–788.

25. Kurosaki, K.; Matsushita, T.; Ueda, S. Individual DNA identification from ancient human remains. *Am. J. Hum. Genet.* **1993**, *53*, 638–643.

26. Hardy, C.; Casane, D.; Vigne, J.D.; Callou, C.; Dennebouy, N.; Mounolou, J.-C.; Monnerot, M. Ancient DNA from bronze age bones of European rabbit (*Oryctolagus cuniculus*). *Experientia* **1994**, *50*, 564–570.

27. Higuchi, R.; Wrischnik, L.A.; Oaks, E.; George, M.; Tong, B.; Wilson, A.C. Mitochondrial DNA of the extinct quagga: Relatedness and extent of postmortem changes. *J. Mol. Evol.* **1987**, *25*, 283–287.

28. Thomas, W.K.; Pääbo, S.; Villablance, F.X.; Wilson, A.C. Spatial and temporal continuity of kangaroo rat populations shown by sequencing mitochondrial DNA from museum specimens. *J. Mol. Evol.* **1990**, *23*, 101–112.

29. Krajewski, C.; Driskell, A.C.; Baverstock, P.R.; Braun, M.J. Phylogenetic relationships of the thylacine (Mammalia: Thylacinidae) among dasyuroid marsupials: Evidence from cytochrome *b* DNA sequences. *Proc. R. Soc. B* **1992**, *250*, 19–27.

30. Taylor, A.C.; Sherwin, W.B.; Wayne, R.K. Genetic variation of microsatellite loci in a bottlenecked species: The northern hairy-nosed wombat (*Lasiorhinus krefftii*). *Mol. Ecol.* **1994**, *3* (4), 277–290.

31. Yang, H.; Golenberg, E.M.; Shoshani, J. Proboscidean DNA from museum and fossil specimens: an assessment of ancient DNA extraction and amplification techniques. *Biochem. Genet.* **1997**, *35* (5/6), 165–179.

32. Weber, D.S.; Stewart, B.S.; Garza, J.C.; Lehman, N. An empirical genetic assessment of the severity of the northern elephant seal population bottleneck. *Curr. Biol.* **2000**, *10* (20), 1287–1290.

33. Pergams, O.R.W.; Barnes, W.M.; Nyberg, D. Rapid change of mouse mitochondrial DNA. *Nature* **2003**, *423*, 397.

34. Wisely, S.M.; Buskirk, S.W.; Fleming, M.A.; McDonald, D.B.; Ostrander. Genetic diversity and fitness in black-footed ferrets before and during bottleneck. *J. Hered.* **2002**, *93* (4), 232–237.

35. Larson, S.; Jameson, R.; Etnier, M.; Fleming, M.; Bentzen, P. Loss of genetic diversity in sea otters (*Enhydra lutris*) associated with the fur trade of the 18th and 19th centuries. *Mol. Ecol.* **2002**, *11* (10), 1899–1903.

36. Hale, M.L.; Lurz, P.W.; Shirley, M.D.; Rushton, S.; Fuller, R.M.; Wolff, K. Impacts of landscape management on the genetic structure of red squirrel populations. *Science* **2001**, *293*, 2246–2248.

37. Ramakrishnan, U.; Hadly, E.A. Using phylochronology to reveal cryptic population histories: revie and synthesis of 29 ancient DNA studies. *Mol. Ecol.* **2009**, *18* (7), 1310–1330.

38. Brunson, K.; Reich, D. The promise of paleogenomics beyond our own species. *Trends Genet.* **2019**, *35* (5), 319–329.

39. Thomas, C.A.,Jr. The enzymatic degradation of desoxyribose nucleic acid. *J. Am. Chem. Soc.* **1956**, *78* (9), 1861–1868.

40. Lindahl, T. Instability and decay of the primary structure of DNA. *Nature* **1993**, *362*, 709–715.

41. Demple, B.; Harrison, L. Repair of oxidative damage to DNA. *Annul. Rev. Biochem.* **1994**, *63*, 915–948.

42. Linn, S. Deoxyribonucleases: Survey and perspectives. In *The Enzymes*; Boyer, P.D., Ed.; Academic Press: New York, NY, **1981**; *14*, 121–135.

43. Gaillard, C.; Strauss, F. Eliminating DNA loss and denaturation during storage in plastic microtubes. Application Notes. *Am. Biotechnol. Lab.* **2000**, *18* (12), 24.

44. Gaillard, C.; Strauss, F. Avoiding adsorption of DNA to polypropylene tubes and denaturation of short DNA fragments. *Technical Tips Online* **1998**, *3*, 63–65.

45. Kline, M.C.; Duewer, D.L.; Redman, J.W.; Butler, J.M. Results from the NIST 2004 DNA quantitation study. *J. Forensic Sci.* **2005**, *50* (3), 571–578.

46. Belotserkovskii, B.P.; Johnson, B.H. Polypropylene tube may include denaturation and multimerisation of DNA. *Science* **1996**, *271*, 222–223.

47. Gaillard, C.; Strauss, F. Polypropylene tube surfaces may induce denaturation and multimerization of DNA – response. *Science* **1996**, *271*, 223.

48. Belotserkovskii, B.P.; Johnson, B.H. Denaturation and association of DNA sequences by certain polypropylene surfaces. *Anal. Biochem.* **1997**, *251*, 251–262.

49. Gaillard, C.; Strauss, F. Association of poly(CA).poly(TG) DNA fragments into four-strand complexes bound by HMG1 and 2. *Science* **1994**, *264*, 433–436.

50. McDonald, G.R.; Hudson, A.L.; Dunn, S.M.J.; You, H.; Baker, G.B.; Whittal, R.M.; Martin, J.W.; Jha, A.; Edmondson, D.E.; Holt, A. Bioactive contaminants leach from disposable laboratory plasticware. *Science* **2008**, *322* (5903), 917.

51. Lindahl, T.; Nyberg, B. Rate of depurination of native deoxyribonucleic acid. *Biochemistry* **1972**, *11* (19), 3610–3618.

52. Lindahl, T.; Andersson, A. Rate of chain breakage at apurinic sites in double-stranded deoxyribonucleic acid. *Biochemistry* **1972**, *11* (19), 3618–3623.

53. Lindahl, T.; Karlstrom, O. Heat-induced depyrimidination of deoxyribonucleic acid in neutral solution. *Biochemistry* **1973**, *12*, 5151–5154.

54. Overballe-Petersen, S.; Orlando, L.; Willerslev, E. Next-generation sequencing offers new insights into DNA degradation. *Trends Biotech* **2012**, *30* (7), 364–368.

55. Allentoft, M.E.; Collins, M.; Harker, D.; Haile, J.; Oskam, C.L.; Hale, M.L.; Campos, P.F.; Samaniego, J.A.; Gilbert, M.T.P.; Willerslev, E.; Zhang, G.; Scofield, R.P.; Holdaway, R.N.; Bunce, M. The half-life of DNA in bone: measuring decay kinetics in 158 dated fossils. *Proc. Royal Soc. B* **2012**, *279*, 4724–4733.
56. Kvam, E.; Tyrrell, R.M. Artificial background and induced levels of oxidative base damage in DNA from human cells. *Carcinogensis* **1997**, *18* (11), 2281–2283.
57. Bruskov, V.I.; Malakhova, L.V.; Masalimov, Z.K.; Chernikov, A.V. Heat-induced formation of reactive oxygen species and 8-oxoguanine, a biomarker of damage to DNA. *Nucleic Acid Res.* **2002**, *30* (6), 1354–1363.
58. Evans, R.K.; Xu, Z.; Bohannon, K.E.; Wang, B.; Bruner, M.W.; Volkin, D.B. Evaluation of degradation pathways for plasmid DNA in pharmaceutical formulations via accelerated stability studies. *J. Pharm. Sci.* **2000**, *89* (1), 76–87.
59. Lee, S.; Crouse, C.; Kline, M. Optimizing storage and handling of DNA extracts. *Forensic Sci. Rev.* **2010**, *22* (2), 131.
60. Dixon, M.; Webb, E.C. *Enzymes*; Academic Press: New York, NY, **1979**.
61. Madisen, L.; Hoar, D.I.; Holroyd, C.D.; Crisp, M.; Hodes, M.E. DNA banking: The effects of storage of blood and isolated DNA on the integrity of DNA. *Am. J. Med. Genet.* **1987**, *27* (2), 379–390.
62. Beach, L.R. Evaluation of storage conditions of DNA used for forensic STR analysis. MS thesis, Purdue Univ. **2014**, 149 pp.
63. Shikama, K. Effects of freezing and thawing on the stability of double helix of DNA. *Nature* **1965**, *207*, 529–530.
64. Schaudien, D.; Baumgarter, W.; Herten, C. High preservation of DNA standards diluted in 50% glycerol. *Diagn. Mol. Pathol.* **2007**, *16*, 153–157.
65. Roder, B.; Fruhwirth, K.; Rossmanith, P. Impacts of long-term storage stability of standard DNA for nucleic acid-based methods. *J. Clin. Micro. Biol.* **2010**, *48* (11), 4260–4262.
66. Michaud, C.L.; Foran, D.R. Simplified field preservation of tissues for subsequent DNA analysis. *J. Forensic Sci.* **2011**, *56*, 846–852.
67. Rissanen, A.J.; Kurhela, E.; Aho, T.; Oittinen, T.; Tiirola, M. Storage of environmental samples for guaranteeing nucleic acid yield for molecular microbiological studies. *Appl. Mic. Biotech.* **2010**, *88*, 977–984.
68. Smith, S.; Morin, P.A. Optimal storage conditions for highly dilute DNA samples: a role for trehalose as a preserving agent. *J. Forensic Sci.* **2005**, *50*, 1101–1108.
69. Zhu, B.; Furuki, T.; Okuda, T.; Sakurai, M. Natural DNA mixed with trehalose persists in β-form double-stranding even in the dry state. *J. Phys Chem. B* **2007**, *111*, 5542–5544.
70. Zhang, M.; Oldenhof, H.; Sydykov, B.; Bigalk, J.; Sieme, H.; Wolkers, W.F. Freeze-drying of mammalian cells using trehalose: preservation of DNA integrity. *Sci. Rep.* **2017**, *7*, 6198.
71. Ivanova, N.V.; Kuzmina, M.L. Protocols for dry DNA storage and shipment at room temperature, *Mol. Ecol. Resources* **2013**, *13*, 890–898.
72. Hernandez, G.; Mondala, T.; Head, S. Assessing a novel room-temperature RNA storage medium for compatibility in microarray gene expression analysis. *BioTechniques* **2009**, *47*, 667–670.
73. Wan, E.; Akana, M.; Pons, J.; Chen, J.; Musone, M.; Kwok, P.Y.; Liao, W. Green technology for room temperature nucleic acid storage. *Curr. Issues Mol. Biol.* **2010**, *13*, 135–142.
74. Frippiat, C.; Zorbo, S.; Leonard, D.; Marcotte, A.; Chaput, M.; Aelbrecht, C.; Noel, F. Evaluation of novel forensic DNA storage methodologies. *Forensic Sci. Intern.: Genet.* **2011**, *5*, 386–392.

75. Lee, S.B.; Clabaugh, K.C.; Silva, B.; Odigie, O.; Coble, M.D.; Loreille, O.; Scheible, M.; Fourney, R.M.; Stevens, J.; Carmody, G.R.; Parson, T.J.; Pozder, A.; Eisenberg, A.J.; Budowle, B.; Ahmad, T.; Miller, R.W.; Crouse, C.A. Assessing a novel room temperature DNA storage medium for forensic biological samples. *Forensics Sci. Inter.: Genet.* **2012**, *6*, 31–40.

76. Prendini, L.; Hanner, R.; DeSalle, R. Obtaining, storing and archiving specimens and tissues samples for use in molecular studies. In *Techniques in Molecular Systematics and Evolution*; DeSalle, R., Giribet, G., Wheeler, W., Eds.; Basle, Birkhäuser; **2002**; 176–248.

77. Nagy, Z.T. A hands-on overview of tissue preservation methods for molecular genetic analysis. *Org. Divers. Evol.* **2010**, *10*, 91–105.

78. Camacho-Sanchez, M.; Burraco, P.; Gomez-Mestre, I.; Leonard, J.A. Preservation of RNA and DNA from mammal sampled under field conditions. *Mol. Ecol. Resources* **2013**, *13*, 663–673.

79. Post, R.J.; Flook, P.K.; Millest, A. Methods for the preservation of insects for DNA studies. *Biochem Syst. Ecol.* **1993**, *21*, 85–92.

80. Moreau, C.S.; Wray, B.D.; Czekanski-Moir, J.E.; Rubin, B.E.R. DNA preservation: a test of commonly used preservatives for insects. *Invert. Syst.* **2013**, *27* (1), 81–86.

81. Dawson, M.N.; Raskoff, K.A.; Jacobs, D.K. Field preservation of marine invertebrate tissue for DNA analysis. *Mol. Mar. Biol. Biotechnol.* **1998**, *7* (2), 145–152.

82. Laulier, M.; Pradier, E.; Bigot, Y.; Periquet, G. An easy method for preserving nucleic acids in field samples for later molecular and genetic studies without refrigeration. *J. Evol. Biol.* **1995**, *8*, 657–663.

83. Towne, B.; Devor, E.J. Effect of storage time and temperature on DNA extracted from whole blood. *Hum. Biol.* **1990**, *62* (2), 301–306.

84. Lahiri, D.K.; Schnabel, B. DNA isolation by a rapid method from human blood samples: Effects of $MgCl_2$, EDTA, storage time, and temperature on DNA yield and quality. *Biochem. Genet.* **1993**, *31* (7/8), 321–328.

85. Seutin, G.; White, B.N.; Boag, P.T. Preservation of avian blood and tissue samples for DNA analysis. *Can. J. Sci.* **1991**, *69*, 82–90.

86. Conrad, K.F.; Robertson, R.J.; Boag, P.T. Difficulties storing and preserving Tyrants flycatchers blood samples used for genetic analysis. *Condor* **2000**, *102*, 191–193.

87. McCabe, E.R.B.; Huang, S.-Z.; Seltzer, W.K.; Law, M.L. DNA microextractions from dried blood spots on filter paper blotters: potential applications to newborn screening. *Hum. Genet.* **1987**, *75*, 213–216.

88. McCabe, E.R. Utility of PCR for DNA analysis from dried blood spots on filter paper blotters. *PCR Methods Appl.* **1991**, *1*, 99–106.

89. McEwen, J.E.; Reilly, P.R. Stored Guthrie cards as DNA banks. *Am. J. Hum. Genet.* **1994**, *55* (1), 196–200.

90. Mullen, M.P.; Howard, D.J.; Powell, R.; Hanrahan, J.P. A note on the use of FTA technology for storage of blood samples for DNA analysis and removal of PCR inhibitors. *Irish J. Agric. Food Res.* **2009**, *48*, 109–113.

91. Hardin, J.; Finnell, R.H.; Wong, D.; Hogan, M.E.; Horovitz, J.; Shu, J.; Shaw, G.M. Whole genome microarray from neonatal blood paper cards. *BMC Genet.* **2009**, *10*, 38.

92. Ataei, S.; Nateshpour, M.; Hajjaran, H.; Edrissian, G.H.; Fotoushani, A.R. High specificity of seminested multiplex PCR using dried blood spots on DNA Banking cards for comparison with frozen liquid blood for detection of *Plasmodium falciparum* and *Plasmodium vivax. J. Clin. Anal.* **2011**, *25* (3) 185–190.

93. Galbraith, D.A.; Boag, P.T.; White, B.N. Drying reptile blood for DNA extraction. *Fingerpr. News* **1989**, *1*, 4–5.

94. Gutierrez-Corchero, F.; Arruga, M.V.; Sanz, L.; Garcia, C.; Hernandez, A.M.; Campos, F. Using FTA® cards to store avian blood samples for genetic studies. Their application in sex determination. *Mol. Ecol. Notes* **2002**, *2*, 75–77.

95. Smith, L.M.; Burgoyne, L.A. Collecting, archiving and processing DNA from wildlife samples using FTA® databasing paper. *BMC Ecol.* **2004**, *4*, 4.

96. Li, C.C.; Beck, I.A.; Seidel, K.D.; Frenkel, L.M.; Persistence of human immunodeficiency virus type I subtype B DNA in dried blood samples on FTA filter paper. *J. Clin. Microbiol.* **2004**, *42*, 3847–3849.

97. Chaisomchit, S.; Wichajarn, R.; Janejai, N.; Chareonsiriwatana, W. Stability of genomic DNA in dried blood spots stored on filter paper. *Southeast Asian J. Med. Public Health* **2005**, *36* (1), 270–273.

98. Michaud, V.; Gil, P.; Kwiatek, O.; Prome, S.; Dixon, L.; Romero, L.; Le Potier, M.-F.; Arias, M.; Couacy-Hymann, E.; Roger, F.; Libeau, G.; Albina, E. Long-term storage at tropical temperature of dried-blood filter papers for detection and genotyping of RNA and DNA viruses by direct PCR. *J. Virol. Methods* **2007**, *146*, 257–265.

99. Dove, C.J.; Dahlan, N.F.; Heacker, M.A.; Whatton, J.F. Using Whatmen FTA® cards to collect DNA from bird-strike identifications. *Human-Wildlife Interact.* **2010**, *5* (2), 218–223.

100. da Cunha Santos, G. FTA cards for preservation of nucleic acids for molecular analysis A review of the use of cytological and tissue samples. *Arch. Pathol. Lab. Med.* **2018**, *142*, 308–312.

101. Curry, P.S.; Ribble, C.; Sears, W.C.; Orsel, K.; Hutchins, W.; Godson, D.; Lindsay, R.; Dibernardo, A.; Campbell, M.; Kutz, S.J. Blood collected on filter paper for wildlife serology: evaluating storage and temperature challenges of field collections. *J. Wildlife Dis.* **2014**, *50* (2), 297–307.

102. Hansen, P.; Blakesley, R. Simple archiving of bacterial and plasmid DNAs for future use. *Focus* **1998**, *20* (3), 72–74.

103. Wasser, S.K.; Houston, C.S.; Koehler, G.M.; Cadd, G.G.; Fain, S.R. Techniques for application of fecal DNA methods to field studies of Ursids. *Mol. Ecol.* **1997**, *6* (11), 1091–1097.

104. Frantzen, M.A.J.; Silk, J.B.; Feguson, W.H.; Wayne, R.K.; Kohn, M.H. Empirical evaluation of preservative methods for fecal DNA. *Mol. Ecol.* **1998**, *7* (10), 1423–1428.

105. Murphy, M.A.; Waits, L.P.; Kendall, K.C.; Wasser, S.K.; Higbee, J.A.; Bogden, R. An evaluation of long-term preservation methods for brown bear (*Ursus arctos*) fecal DNA samples. *Cons. Genet.* **2002**, *3*, 435–440.

106. Waits, L.P.; Paetkau, D. Noninvasive genetic sampling tools for wildlife biologists: a review of applications and recommendation for accurate data collections. *J. Wildl. Manage.* **2005**, *69*, 1419–1433.

107. Panasci, M.; Ballard, W.B.; Breck, S.; Rodriguez, D.; Denssmore, L.D., III; Wester, D.B.; Baker, R.J. Evaluation of fecal DNA preservation techniques and effects of sample age and diet on genotyping success. *J. Wildl. Manage.* **2011**, *75* (7), 1616–1624.

108. Santini, A.; Lucchini, V.; Fabbri, E.; Randi, E. Ageing and environmental factors affect PCR success in wolf (*Canis lupus*) excremental DNA samples. *Mol. Ecol. Notes* **2007**, *7*, 955–961.

109. Goossen, B.; Waits, L.P.; Taberlet, P. Plucked hair samples as a source of DNA: reliability of dinucleotide microsatellite genotyping. *Mol. Ecol.* **1998**, *7*, 1237–1241.

110. Mowat, G.; Strobeck, C. Estimating population size of grizzly bears using hair capture, DNA profiling, and mark-recapture analysis. *J. Wildlife Manage.* **2000**, *64*, 183–193.

111. Roon, D.A.; Waits, L.P.; Kendall, K.C. A quantitative evaluation of two methods for preserving hair. *Mol. Ecol. Notes* **2003**, *3* (1), 163–166.

112. Rogstad, S.H. Saturated NaCl-CTAB solution as a means of field preservation of leaves for DNA analysis. *Taxon* **1992**, *41*, 701–708.

113. Chase, M.W.; Hills, H.H. Silica gel: an ideal material for field preservation of leaf samples for DNA studies. *Taxon* **1991**, *40* (2), 215–220.

114. Post, R.J.; Flook, P.K.; Millest, A.L. Methods for the preservation of insects for DNA studies. *Biochem. Syst. Ecol.* **1993**, *21*, 85–92.

115. Reiss, R.A.; Schwert, D.P.; Ashworth, A.C. Field preservation for molecular genetic analyses. *Physiol. Chem. Ecol.* **1995**, *24*, 716–719.

116. Dessauer, H.C.; Cole, C.J.; Hafner, M.S. Collection and storage of tissues. In *Molecular Systematics*, 2nd Ed.; Hillis, D.M., Moritz, C., Mable, B.K., Eds.; Sinauer Associates: Sunderland, MA, **1996**; 29–47.

117. Yates, T.L. Tissues, Cell Suspensions, and Chromosomes. In *Measuring and Monitoring Biological Diversity: Standard Methods for Mammals*; Wilson, D.E., Cole, F.R., Nichols, J.D., Rudran, R., Foster, M.S., Eds.; Smithsonian Institution Press: Washington, DC, **1996**; 275–278.

118. Dillon, N.; Austin, A.D.; Bartowsky, E. Comparison of preservation techniques for DNA extractions from hymenopterous insects. *Insect Mol. Biol.* **1996**, *5*, 21–24.

119. Frampton, M.; Conrad, S.; Prager, T.; Richards, M.H. Evaluation for specimen preservatives for DNA analyses of bees. *J. Hymenoptera Res.* **2008**, *17*, 195–200.

120. Wong, P.S.; Wiley, E.O.; Johnson, W.E.; Ryder, O.A.; O'Brien, S.J.; Haussler, D.; Koepfli, K.-P.; Houct, M.L.; Perelman, P.; Mastromonaco, G.; Bently, A.C.; Venkatesh, B.; Zhang, Y.-P.; Murphy, R.W. Tissue sampling methods and standards for vertebrate genomics. *GigaScience* **2012**, *1*, 8.

121. Florian, M.L. The effects of freezing and freeze-drying on natural history specimens. *Collec. Forum* **1990**, *6*, 45–52.

122. Dessauer, H.C.; Hafner, M.S. *Collections of Frozen Tissues.* In *Assoc. Syst. Collections*; Dessauer, H.C., Hafner, M.S., Eds; University of Kansas: Lawrence, **1984**; 74.

123. Catzeflis, F.M. Animal tissue collections for molecular genetics and systematics. *Trends Ecol. Evol.* **1991**, *6*, 168.

124. Hanner, R.; Corthals, A.; Dessauer, H. Salvage of genetically valuable tissues following a freezer failure. *Mol. Phylogenet. Evol.* **2005**, *34*, 452–455.

125. Jackson, J.A.; Laikre, L.; Baker, C.S.; Kendall, C. The Genetic Monitoring Working Group. Guideline for collecting and maintaining archives for genetic monitoring. *Con. Genet. Res.* **2011**, *4* (2), 527–536. DOI: 10.1007/s12686-011-9545-x

126. Corthals, A.; DeSalle, R. An application of tissue and DNA banking for genomics and conservation: The Ambrose Monell Cryo-Collection (AMCC). *Syst. Biol.* **2005**, *54* (5) 819–823.

127. Withers, L.A. Low temperature storage of plant tissue cultures. *Adv. Biomed. Engin./ Biotech.* **1990**, *18*, 101–150.

128. Karlsson, J.O.M.; Toner, M. Long-term storage of tissues by cryopreservation; critical issues. *Biomaterials* **1996**, *17*, 243–256.

129. Liston, A.; Riesberg, L.H. A method for collecting dried plant specimens for DNA and isozyme analyses and results of field test in Xinjiang, China. *Ann. Missouri Bot. Gard.* **1990**, *77*, 859–863.

130. Prendini, L.; Hanner, R.; DeSalle, R. Obtaining, storing and Archiving Specimens and Tissue Samples for use in molecular studies. In *Techniques in Molecular Systematics and Evolution*; DeSalle, R., Giribet, G., Wheeler, W.; Eds.; Birkhauser, Basle. **2002**; 176–248.

131. Pyle, M.M.; Adams, R.P. In-situ preservation of DNA in plant specimens. *Taxon* **1989**, *38*, 576–581.

132. Flournoy, L.E.; Adams, R.P.; Prandy, R.N. Interim and archival preservation of plant specimens in alcohols for DNA studies. *BioTechniques* **1996**, *7*, 1423–1428.

133. Quicke, D.L.J.; Lopez-Vaamonde, C.; Belshaw, R. Preservation of hymenopteran specimens for subsequent molecular and morphological studies. *Zool. Scripta.* **1999**, *28*, 261–267.

134. Kilpatrick, C.W. Noncryogenic preservation of mammalian tissues for DNA extraction: an assessment of storage methods. *Biochem. Genet.* **2002**, *40* (1/2), 53–62.

135. Morgan, C.A.; Herman, N.; White, P.A.; Vesey, G. Preservation of micro-organisms by drying; a review. *J. Micrbiol. Methods* **2006**, *66*, 183–193.

136. Mitchell, K.R.; Takacs-Vesbach, C.D. A comparison of methods for total community DNA preservation and extraction from various thermal environments. *J. Industrial Microbiol. Biotech.* **2008**, *35*, 1139–1147.

137. Straube, D.; Juen, A. Storage and shipping of tissue samples for DNA analyses: a case study on earthworms. *Eur. J. Soil Biol.* **2013**, *57*, 13–18.

138. Leal-Klevezas, D.S.; Martinez-Vazquez, I.O.; Cuevas-Hernandez, B.; Martinez-Soriano, J.P. Antifreeze solution improves DNA recovery by preserving the integrity of pathogen-infected blood and other tissues. *Clin. Vaccine Imunol.* **2000**, *7*, 945–946.

139. Rubink, W.L.; Murray, K.D.; Baum, K.A.; Pinto, M. Long term preservation of DNA from honey bees (*Apis mellifera*) collected in aerial pitfall traps. *Texas J. Sci.* **2003**, *55*, 159–168.

140. Pearson, G.; Lago-Leston, A.; Valente, M.; Serrao, E. Simple and rapid RNA extraction from freeze dried tissue of brown algae and seagrasses. *Eur. J. Phycol.* **2006**, *41*, 97–104.

141. Leboeuf, C.; Ratajczak, P.; Zhao, W.L.; Plassa, L.F.; Court, M.; Pisonero, H.; Murata, H.; Cayuela, J.M.; Ameisen, J.C.; Garin, J.; Janin, A. Long-term preservation at room temperature of freeze-dried human tumor samples dedicated to nucleic acid analysis. *Cell Preservation Tech.* **2008**, *6*, 191–197.

142. Simister, R.L.; Schmitt, S.; Taylor, M.W. Evaluating methods for the preservation and extraction of DNA and RNA for analysis of microbial communities in marine sponges. *J. Exp. Marine Biol. Ecol.* **2011**, *397*, 38–43.

143. Matsua, S.; Sugiyama, T.; Okuyama, T.; Yoshikawa, K.; Hpnda, K.; Takahashi, R.; Maeda, A. Preservationof pathological tissue specimens by freeze-drying for immunohistochemical staining and various molecular biochemical analyses. *Path. Int.* **1999**, *49*, 383–390.

144. Ruiz, R.; Robio, L.A. Lyophilisation improves the extraction of PCR-quality community DNA from pig fecal samples. *J. Sci. Food Agric.* **2009**, *89*, 723–727.

145. Murphy, M.A.; Waits, L.P.; Kendall, K.C. Quantitative evaluation of fecal drying methods for brown bear DNA analysis. *Wildl. Soc. Bull.* **2000**, *28* (4), 951–957.

146. Corthals, A.; Martin, A.; Warsi, O.M.; Woller-Skar, M.; Lancaster, W.; Russel, A.; Davalos, L.M. From the field to the lab: best practices for field preservation of bat specimens for molecular analyses. *PLoS ONE* **2015**, e0118994. DOI: 10.1371/journal.pone.0118994

147. Briggs, A.W.; Stenzel, U.; Johnson, P.L.F.; Green, R.E.; Kelso, J.; Prufer, K.; Meyer, M.; Krause, J.; Ronan, M.T.; Lachmann, M.; Paabo, S. Patterns of damage in genomic DNA sequences from a Neandertal. *PNAS, USA* **2007**, *104*, 14616–14621.

148. Colotte, M.; Coudy, D.; Tuffet, S.; Bonnet, J., Adverse effects of air exposure on the DNA stability of DNA stored at room temperature. *Biopreser. Biobank.* **2001**, *9*, 47–50.

149. Sibley, C.G.; Ahlquist, J.E. Instructions for specimen preservation for DNA extraction: A valuable source of data for systematics. *Assoc. Syst. Collect. Newsl.* **1981**, *9*, 44–45.

150. Nietfeldt, J.W.; Ballinger, R.E. A new method for storing animal tissue prior to mtDNA extraction. *Biotechniques* **1989**, *7*, 31–32.

151. Adams, R.P.; Zhong, M.; Fei, Y. Preservation of DNA in plant specimens: inactivation and re-activation of DNases in field specimens. *Mol. Ecol.* **1999**, *8*, 681–683.

152. King, J.R.; Porter, S.D. Recommendations on the use of alcohols for preservation of ant specimens (Hymenoptera, Formicidae). *Insect Sci.* **2004**, *51*, 197–202.

153. Allen-Hall, A.; Mcnevin, D. Non-cryogenic forensic tissue preservation in the field: A review. *Aust. J. Forensic Sci.* **2013**, *45* (4), 450–460.

154. Houde, P.; Braun, M.J. Museum collections as a source of DNA for studies of avian phylogenetics. *Auk* **1988**, *105*, 773–776.

155. Colton, L.; Clark, J.B. Comparison of DNA isolation methods and storage conditions for successful amplification of *Drosophila* genes using PCR. *Drosophila Inf. Serv.* **2001**, *84*, 180–182.

156. Hedin, M.C. Molecular phylogenetics at the population/species interface in cave spiders of the Southern Appalachians (Araneae: Nesticidae). *Mol. Biol. Evol.* **1997**, *14*, 309–324.

157. Wheeler, W.C.; Hayashi, C.Y. The phylogeny of the extant chelicerate orders. *Cladistics* **1998**, *14*, 173–192.

158. Vink, C.J.; Mitchell, A.D.; Paterson, A.M. A preliminary molecular analysis of phylogenetic relationships of Australasian wolf spiders (Araneae: Lycosidae). *J. Arach.* **2002**, *30*, 227–237.

159. A'Hara, S.; Harling, R.; McKinlay, R.G.; Topping, C.J. RAPD profiling of spiders (Araneae) DNA. *J. Arach.* **1998**, *26*, 397–400.

160. Wall, A.R.; Campo, D.; Wetzer, R. Genetic utility of natural history museum specimens: endangered fairy shrimp (Branchiopoda: Anostraca). *ZooKeys* **2014**, *457*, 1–4.

161. Fukatsu, T. Acetone preservation: A practical technique for molecular analysis. *Mol. Ecol.* **1999**, *8* (11), 1935–1945.

162. Gurdebeke, S.; Maelfait, J.-P. Pitfall trapping in population genetics studies; finding the right "solution". *J. Arach.* **2002**, *30*, 255–261.

163. Tang, E.P. Path to effective recovering of DNA from formalin-fixed biological samples in natural history collections: Workshop Summary. National Academies Press, Washington DC. **2006** http://www.nap.edu/catalog.php?record_id=11712

164. Stuart, B.L.; Dugan, K.A.; Allard, M.W.; Keamey, M. Extraction of nuclear DNA from skeletonized and fluid preserved museum specimens. *Syst. Biodiv.* **2006**, *4*, 133–136.

165. Michalik, S. Overcoming poor quality DNA. *Drug Disc. Devel.* **2008**, *11*, 24–30.

166. Campos, P.; Gilbert, T.P. DNA Extraction from formalin-fixed material. *Methods Mol. Biol.* **2012**; *840*, 81–85.

167. Hykin, S.M.; Bi, K.; McGuire, J.A. Fixing formalin: a method to recover genomic scale DNA sequences from formalin fixed museum specimens using throughput sequencing. *PLOS ONE* **2015**, *10*, e0141579.

168. Kokkat, T.J.; Patel, M.S.; McGarvey, D.; LiVolsi, V.A.; Baloch, Z.W. Archived formalin-fixed paraffin-embedded (FFPE) blocks: a valuable underexploited resource for extraction of DNA, RNA, and protein. *Biopreser. Biobank.* **2013**, *11*(2), 101–106.

169. Nakamura, S.; Tamura, S.; Taki, H.; Shoda-Kagaya, E. Propylene glycol: a promising preservative for insects, comparable to ethanol, from trapping to DNA analysis. *Entomol. Exp. Applicata* **2019**, *168* (2), 158–165. DOI: 10.1111/eea.12876

170. Hall, D.W. The environmental hazard of ethylene-glycol in insect pit-fall traps. *Coleopterists Bull.* **1991**, *45*, 193–194.

171. Ferro, M.L.; Park, J.S. Effects of Propylene glycol concentration on mid-term DNA preservation of Coleoptera. *Coleopterists Bull.* **2013**, *67*, 581–586.

172. Patrick, H.J.H.; Chomic, A.; Armstrong, K.F. Cooled propylene glycol as a pragmatic choice for preservation of DNA from remote field-collected Diptera for next-generation sequence analysis. *J. Econ. Entomol.* **2016**, *109*(3), 1469–1473.

173. Robinson, C.V.; Porter, T.M.; Wright, M.T.; Hajibabaei, M. Propylene glycol-based antifreeze is an effective preservative for DNA metabarcoding of benthic arthropods. *Freshwater Sci.* **2020**, *40* (1). DOI: 2020.02.28.970475

174. Vink, C.J.; Thomas, S.M.; Paquin, P.; Hayashi, C.Y.; Hedin, M. The effects of preservatives and temperatures on arachnid DNA. *Invert. Syst.* **2005**, *19*, 99–104.

175. Proebstel, D.S.; Evans, R.P.; Shiozawa, D.K.; Williams, R.N. Preservation of nonfrozen tissue samples from a salmonine fish *Brachymystax lenok* (Pallas) for DNA analysis. *J. Ichthyol.* **1993**, *32*, 7–17.

176. Williams, S.T. Safe and legal shipments of tissue samples: does it affect DNA quality. *J. Molluscan Stud.* **2007**, *73*, 416–418.

177. Doyle, J.J.; Doyle, J.L. A rapid DNA isolation procedure for small quantities of fresh leaf tissue. *Phytochem. Bull.* **1987**, *19*, 11–15.

178. Winnepenninckx, B.; Backeljau, T.; Dewachter, R. Extraction of high molecular weight DNA from molluscs. *Trends Genet.* **1993**, *9*, 407.

179. Shahjahan, R.M.; Hughes, K.J.; Leopold, R.A.; DeVault, J.D. Lower incubation temperatures increases yield of insect genomic DNA isolated by the CTAB method. *BioTechniques* **1995**, *19*, 332–334.

180. Laulier, M.; Pradier, E.; Bigot, Y.; Périquet, G. An easy method for preserving nucleic acids in field samples for later molecular and genetic studies without refrigerating. *J. Evol. Biol.* **1995**, *8*, 657–663.

181. Cockburn, A.F.; Seawright, J.A. Techniques for mitochondrial and ribosomal DNA analysis of anopheline mosquitoes. *J. Am. Mosq. Control Assoc.* **1988**, *3* (3), 261–265.

182. Longmire, J.L.; Maltbie, M.; Baker, R.J. Use of "lysis buffer" in DNA isolation and its implications for museum collections. *Occas. Papers Mus. Texas Tech. Univ.* **1997**, *163*, 1–3.

183. Muralidharan, K.; Wemmer, C., Transporting and storing field collected specimens for DNA without refrigeration for subsequent DNA extraction and analysis. *BioTechniques* **1994**, *17*, 420–422.

184. Asahida, T.; Kobayashi, T.; Saitoh, K.; Nakayama, I. Tissue preservation and total DNA extraction from fish stored at ambient temperature using buffers containing high concentration of urea. *Fisheries Sci.* **1996**, *62*, 727–730.

185. Ambion. Preserve RNA and tissue cell samples with RNAlater®. *Ambion TechNotes Newsletter* **1999**, *5*, 7–8.

186. Johnson, M.L.; Kim, S.H.; Emche, S.D. Storage effects on genomic DNA in rolled and mature coca leaves. *BioTechniques* **2003**, *35*, 310–316.

187. Gorokhova, E. Effects of preservation and storage of microcrustaceans in RNA*later* on RNA and DNA degradation. *Limnol. Oceanogr. Methods* **2005**, *3*, 143–148.

7 RNA Storage

Martin Hofmann
Institute of Virology and Immunoprophylaxis,
Mittelhaeusern, Switzerland

CONTENTS

INTRODUCTION

RNA can be isolated from many different types of cells from eukaryotic and prokaryotic origin, including animals, plants, and bacteria. Depending on the source and the type of sample (organ tissue, cultured cells), appropriate procedures must be applied for RNA extraction to (1) maintain the integrity of the RNA, (2) inhibit RNA degradation because of RNases, and (3) obtain clean RNA preparations by removing protein and DNA contaminations.

Because RNA extraction can be quite a lengthy procedure and, furthermore, in many cases, numerous samples have to be analyzed in parallel, it is necessary to store the isolated RNA. Again, storage conditions should be selected to preserve the RNA in such a way that storage does not modify the RNA in any way. However, the intended use of the stored RNA also determines the way how the RNA is to be stored.

In this article, an overview of currently used techniques to obtain RNA from various sources and a description of the factors to consider for RNA storage are presented.

HANDLING OF RNA: THE PROBLEM OF RIBONUCLEASE CONTAMINATION

The isolation of intact, clean RNA is crucial for downstream experiments such as reverse transcription of RNA for cDNA cloning, RNA fingerprinting, and secondary structure analysis, as well as for functional studies involving RNA metabolism. RNA stabilization is an absolute prerequisite for reliable gene-expression analysis.

Stabilization of RNA in biological materials immediately after harvesting the sample is necessary because changes in the gene-expression pattern occur because of specific and nonspecific RNA degradation as well as to transcriptional induction.

Even a single nick can render the RNA molecule nonfunctional. This has to be considered when gene expression is studied by performing in vitro transcription and translation experiments, or transfection of infectious, self-replicating viral RNA.[1]

Whereas double-stranded RNA is relatively resistant against most RNases,[2] particular care must be used when handling single-stranded RNA.

The most important factor determining the quality of the RNA to be studied is the ubiquitous presence of RNases. This is a group of RNA-degrading enzymes, each acting in a sequence-specific way on double- or single-stranded RNA. Furthermore, some RNases such as RNase A are extremely resistant to inactivation by heat. For this reason, a suitable method for RNA extraction must be chosen which allows instant inactivation of RNases. This is achieved by dissolving proteins in denaturing agents such as guanidinium thiocyanate, phenol, or beta-mercaptoethanol.

To avoid or minimize ribonuclease contamination problems, all materials used for working with RNA should be RNase-free. Water should be treated with diethylpyrocarbonate (DEPC) that inactivates nucleases or should be obtained by ultrafiltration through a high-efficiency organics removal system. Glassware must be baked at least for 4 hr at 180°C because autoclaving does not inactivate ribonucleases completely. Plasticware straight out of the package can generally be considered free from nuclease contamination and can be used without any pretreatment. However, reused plasticware such as electrophoresis tanks should be filled with 3% H_2O_2 solution for 10 min at room temperature, followed by thorough rinsing with RNase-free water. Because hands are a major source of RNase contamination, gloves must be worn and changed frequently. It is also a good idea to keep separate stocks of chemicals that are exclusively used for RNA work and are removed from the bottle only with RNase-free tools (e.g., baked spatula).

Several types of specific RNase inhibitors can be used. (1) Protein inhibitors of RNases, either isolated from human placenta or produced synthetically as a recombinant protein, can be added to the extracted RNA. Furthermore, because these RNase inhibitors generally do not interfere with reverse transcriptases and polymerases, they are frequently used to prevent RNA degradation during subsequent reactions such as reverse transcription. (2) Vanadyl–ribonucleoside complexes (VRC) bind to many RNases thereby inhibiting their activity. VRC can be added to the intact cells or tissue before lysis and remain in the RNA-containing fraction during all stages of RNA extraction and purification. However, VRC can only be used for certain in vitro enzymatic reactions (e.g., reverse transcription), whereas they interfere with others such as cell-free in vitro translation of mRNA.

Numerous commercial products for RNA extraction are available today. Most of them are either based on a combination of guanidinium isothiocyanate and phenol based on a single-step RNA isolation method originally described by Chomczynski and Sacchi[3] followed by RNA precipitation, or by binding of RNA in guanidinium lysed samples to a silica-gel membrane from which the RNA is subsequently eluted after washing. These methods allow disruption of the cells and simultaneously inactivate ribonucleases.

PRINCIPLES OF RNA STORAGE

Basically, RNA can be stored in purified form after its extraction or synthesis, or unprocessed samples containing the RNA of interest are stored. However, even in the latter case, proper care must be taken to avoid (re)contamination of RNA with RNases once it is extracted from the stored sample.

Little information is available on the long-term stability of isolated and purified RNA during storage. Most studies have been carried out focusing on various storage conditions of unprocessed samples before RNA extraction, or the material was stored at the stage of lysed cells or tissue but before RNA was isolated.

STORAGE OF EXTRACTED RNA

Depending on the extraction method used, purified RNA is either being eluted directly from a filter or column after ion exchange chromatography, or it is obtained as a pellet after precipitation with ethanol or isopropanol. Irrespective of the storage conditions, it is of great importance that all materials (glassware and plasticware, chemicals, water, etc.) are free from RNase contamination.

The most common way to store RNA is by dissolving it in RNase-free water and freezing at −70°C. This bears the advantage that the RNA can be used virtually for any subsequent experiments without having to worry about interference of the storage medium with the reaction conditions in the experiments. However, RNA stored in water is prone to degradation because of unfavorable pH, as well as traces of RNases. For this reason, RNA should be dissolved in or eluted with a low salt buffer such as 1 mM sodium citrate or TE (10 mM Tris, 1 mM EDTA) at pH < 7, thereby preventing base hydrolysis by chelating free cations present in the RNA preparation.

If the higher-order structural integrity of the RNA is crucial, it should be stored in a buffer with an increased salt concentration to maintain hybridization of double-stranded RNA as well as self-complementary sequence elements (stems, pseudoknots) in single-stranded RNA. This is particularly important if such RNA is to be analyzed for secondary structure elements by targeted chemical/enzymatic degradation.[4,5]

Stabilized formamide (FORMAzol®, Molecular Research Center, Inc., Cincinnati, OH) can be used to dissolve precipitated RNA. It allows to store the RNA for at least 2 years at −20°C, whereas RNA dissolved in aqueous solutions should always be kept at −70°C. However, formamide must be replaced by water if the RNA is to be used in any in vitro enzymatic reaction.

To avoid RNA degradation as a result of multiple freeze thawing, it is recommended to store the RNA in small aliquots rather than in one single volume.

As an alternative to storing in solution, RNA can also be kept frozen or even at room temperature as a precipitate that is often obtained anyway at the end of the extraction procedure. Aliquots are removed from the precipitate after vortexing, and the RNA is pelleted by centrifugation and redissolved under RNase-free conditions immediately before use.

STORAGE OF SAMPLES BEFORE RNA EXTRACTION UNDER RNA STABILIZING CONDITIONS

Samples containing the RNA of interest can also be stored prior to RNA extraction. Tissue or cells are allowed to be perfused by chemicals that inhibit RNase activity by denaturing all proteins. A rapid inactivation of enzymatic activity in the sample is required, in particular, in gene-expression studies where any induction or repression of mRNA transcription because of changed environmental conditions must be prevented immediately after specimen collection to reflect the expression profile of the intact tissue.

It has repeatedly been shown that Trizol®, a monophasic solution of phenol and guanidinium isothiocyanate (Invitrogen Life Technologies, Carlsbad, CA), can be used to store samples for extended periods of time even at ambient temperatures, without a significant decrease of the quantity or quality of RNA.[6,7] Therefore, storage of samples in Trizol should be considered whenever samples cannot be kept frozen immediately after collection.

Recently, another commercially available product, RNAlater® (Ambion, Austin, TX), has been introduced on the market. RNAlater® is a patented aqueous, nontoxic tissue and cell storage reagent that stabilizes and protects cellular RNA in intact, unfrozen tissue and cell samples by quickly permeating the tissue.[8] It makes possible the storage of collected tissue specimens and cells at 4°C or even at ambient temperature, eliminating the need for (1) immediate snap-freezing of tissue in liquid nitrogen after collection and (2) instant RNA extraction. Most of the commonly used techniques can be used to extract RNA from RNAlater®-treated samples. RNAlater® has been shown to be suitable for RNA stabilization in animal and plant tissue and cells as well as in bacteria. Further information is available from Ambion's website at http://www.ambion.com.

STORAGE OF UNPROCESSED SAMPLES

Tissue, cell, and serum samples can also be stored in a native, unprocessed form. However, to block RNA degradation as soon and as efficiently as possible, specimens should be frozen immediately after collection, and thawing and refreezing before RNA extraction must be avoided. It has been shown repeatedly that handling and storing conditions of diagnostic samples have a significant impact on the RNA levels detected by RT-PCR.[9–11] Tissue samples collected for gene-expression studies should be snap-frozen in liquid nitrogen. All samples should be stored at −70°C at all times.

CONCLUSION

There is no universally applicable "ideal" method for storing RNA. The procedure to be chosen is dependent on the source, the kind, and the intended use of the RNA. The most important factor that determines the quality of the stored RNA is the ability to protect the RNA from degradation by RNases. Therefore, if purified RNA is stored, RNase-free tubes and solutions are crucial, and the RNA should be stored frozen at temperatures as low as possible. Thawing and refreezing should be avoided.

If storage or shipment in a frozen state is not possible, samples from which RNA is to be extracted should be kept in a protein-denaturing agent to prevent RNase activity. For gene-expression studies, any ongoing mRNA transcription activity should be suppressed as soon as possible after sample collection by snap-freezing the samples or by storage in an excess volume of RNA-stabilizing agent.

Detailed protocols for RNA extraction and storage can be found in molecular biology laboratory manuals.[12,13]

REFERENCES

1. Lai, M.C. The making of infectious viral RNA: No size limit in sight. Proc. Natl. Acad. Sci. U. S. A. **2000**, *97* (10), 5025–5027.
2. Sorrentino, S.; Naddeo, M.; Russ, A.; D'Alessio, G. Degradation of double-stranded RNA by human pancreatic ribonuclease: Crucial role of noncatalytic basic amino acid residues. Biochemistry **2003**, *42* (34), 10182–10290.
3. Chomczynski, P.; Sacchi, N. Single-step method of RNA isolation by acid guanidinium thiocyanate-phenol-chloroform extraction. Anal. Biochem. **1987**, *162* (1), 156–159.
4. Knapp, G. Enzymatic approaches to probing RNA secondary and tertiary structure. Methods Enzymol. **1989**, *180*, 192–212.
5. Yu, H.Y.; Grassmann, C.W.; Behrens, S.E. Sequence and structural elements at the 3′ terminus of bovine viral diarrhea virus genomic RNA: Functional role during RNA replication. J. Virol. **1999**, *73* (5), 3638–3648.
6. Barbaric, D.; Dalla-Pozza, L.; Byrne, J.A. A reliable method for total RNA extraction from frozen human bone marrow samples taken at diagnosis of acute leukaemia. J. Clin. Pathol. **2002**, *55* (11), 865–867.
7. Hofmann, M.A.; Thuer, B.; Liu, L.; Gerber, M.; Stettler, P.; Moser, C.; Bossy, S. Rescue of infectious classical swine fever and foot-and-mouth disease virus by RNA transfection and virus detection by RT-PCR after extended storage of samples in Trizol®. J. Virol. Methods **2000**, *87* (1–2), 29–39.
8. Prediger, E. Get the RNA out intact. Ambion TechNotes **1999**, *6* (4), 1–2.
9. Ginocchio, C.C.; Wang, X.P.; Kaplan, M.H.; Mulligan, G.; Witt, D.; Romano, J.W.; Cronin, M.; Carroll, R. Effects of specimen collection, processing, and storage conditions on stability of human immunodeficiency virus type 1 RNA levels in plasma. J. Clin. Microbiol. **1997**, *35* (11), 2886–2893.
10. Gruber, A.D.; Moennig, V.; Hewickertrautwein, M.; Trautwein, G. Effect of formalin fixation and long-term storage on the detectability of bovine viral-diarrhoea-virus (BVDV) RNA in archival brain tissue using polymerase chain reaction. J. Vet. Med., Ser. B **1994**, *41* (10), 654–661.
11. Halfon, P.; Khiri, H.; Gerolami, V.; Bourliere, M.; Feryn, J.M.; Reynier, P.; Gauthier, A.; Cartouzou, G. Impact of various handling and storage conditions on quantitative detection of hepatitis C virus RNA. J. Hepatol. **1996**, *25* (3), 307–311.
12. Ausubel, F.M.; Brent, R.; Kingston, R.E.; Moore, D.D.; Seidman, J.G.; Smith, J.A.; Struhl, K. Current Protocols in Molecular Biology. John Wiley & Sons: Hoboken, NJ, **2002**; Vol. 1.
13. Sambrook, J.; Fritsch, E.F.; Maniatis, T. Molecular Cloning: A Laboratory Manual. Cold Spring Harbor Laboratory Press: New York, NY, **1989**; Vol. 1.

8 Disposable Electrochemical DNA Biosensors

Kagan Kerman[1,2] and Mehmet Ozsoz[1]
[1]Ege University, Izmir, Turkey
[2]Japan Advanced Institute of Science
and Technology, Ishikawa, Japan

CONTENTS

INTRODUCTION

Electrochemical DNA hybridization biosensors have been a major research field for over a decade, because their rapid, simple, and low-cost detection capabilities provide several advantages to clinical, forensic, and environmental monitoring. Basically, DNA biosensors convert the Watson–Crick base-pair recognition event into a readable analytical signal. A DNA biosensor is prepared by the immobilization of a single-stranded (ss) oligonucleotide (probe) onto a transducer surface to recognize its complementary (target) DNA sequence via hybridization. The DNA duplex formed on the electrode surface is called a hybrid. This binding event is then converted into an electrochemical signal by the transducer. The most important advantage of utilizing electrochemistry for transducing hybridization is its suitability for microfabrication. Thus compact, user-friendly, and handheld devices can easily be designed by combining electrochemistry and microfabrication.

THE ADVANTAGES OF USING DISPOSABLE ELECTROCHEMICAL DNA BIOSENSORS

The miniaturization of electrochemical DNA biosensors[1–3] enabled the mass production of disposable sensor strips. These sensor strips have gradually taken the place of the conventional three-electrode system consisting of bulky working,

DOI: 10.1201/9781003247432-8

reference, and counter electrodes, and beakers with a large amount of buffer and analyte solution volume, mostly ranging between 1 and 2 mL. However, the disposable sensor strips alone can be described as disposable electrochemical cells, which contain the working, reference, and counter electrodes together on a minimal scale and space. A small droplet of the buffer and an analyte solution of about 20 μL would be sufficient for an electrochemical reaction to take place. Especially, the "memory-effects" resulting from the insufficient cleaning of the electrode surface have been completely eliminated with these single-use sensor strips. Thus, more reproducible results could be obtained by using a new electrode for each experiment. Such a significant decrease in the required analyte volume, and increase in reproducibility while shrinking the required space in the laboratory, had a revolutionary impact on the field of electrochemical biosensors, so that the term "on-field analysis" has become possible with the help of the miniaturization of such electrochemical potentiostats and sensor strips, enabling "bedside clinical diagnosis" and "on-site environmental monitoring." Even the diabetes patients themselves have become capable of measuring their own blood glucose levels by using such small handheld electrochemical devices in connection with sensor strips.[4] Nowadays, the biggest challenge facing electrochemists is to develop such a small device, which can perform nucleic acid-based tests with a simple and rapid procedure, as well as high sensitivity and selectivity.

DISPOSABLE ELECTRODE MATERIALS

Nowadays, two important electrode materials are in widespread use for disposable sensor technology: carbon and gold (Au).[5] Especially, carbon is desired because of its rich surface chemistry, low background current, wide potential window, low-cost, and chemical inertness. Disposable carbon electrodes are suitable for various kinds of surface modifications, thus enabling a major number of applications in many fields. However, the electron-transfer rates obtained from carbon electrodes are reported to be slower than those obtained from metal electrodes.[6] Recently, this major drawback has been overcome by the modification of the carbon surface with carbon nanotubes (CNTs). CNTs enabled enhanced electron transfer on electrode surface owing to their small size providing a larger active surface for easy DNA attachment.[7] CNT-based screen-printed electrodes (CNTSPEs) have been fabricated by Wang and Musameh.[8] CNTSPEs with their well-defined electrochemical activity and mechanical stability are promising candidates for DNA-based testing.

Highly oriented pyrolytic graphite electrodes (HOPGE) have also been attractive for electrochemical DNA biosensor research.[9-12] The renewal of HOPGE surface is also simple and rapid. A freshly cleaved surface of HOPGE can easily be prepared by contacting a piece of adhesive tape to the graphite surface, and then removing a thin layer of graphite with the tape. Thus the same electrode can be used for several different measurements. DNA and DNA–drug interaction were examined on thin-film mercury-coated HOPGE by Hason et al.[9] Voltammetric microanalysis of DNA adducts with osmium tetroxide-2,2′-bipyridine (Os, bipy) using a HOPGE provided the detection of 140 pg of DNA-Os, bipy after a 5 min accumulation period.[10] Anodic voltammetry and atomic force microscopy (AFM) imaging were performed for the detection of adriamycin and DNA interaction on HOPGE surfaces.[11] AFM surface characterization of the effect of pH and applied potential on the adsorption of DNA on HOPGE has recently been reported by Oliveira-Brett and Chiorcea.[12]

The use of a rigid carbon–polymer composite material as an electrochemical transducer in hybridization genosensors has recently been reported by Alegret and coworkers.[13–15] Graphite–epoxy composites (GEC) have an uneven surface suitable for strong DNA adsorption. Especially, ssDNA was reported to bind strongly to GEC in a way that prevents the strands from self-associating, while permitting hybridization with complementary DNA. Hybridization was detected by monitoring guanine oxidation signal[15] and also through biotin–streptavidin interaction using a streptavidin conjugated to horseradish peroxidase without nonspecific adsorption onto GEC, even when the surface was treated by blocking reagents.[14] Thus, screen-printed GEC-based electrodes are also promising candidates for highly specific DNA hybridization detection.

Au is the noble metal of choice for screen printing of disposable sensor strips. Au strips offer a very favorable electron-transfer kinetics and a wide anodic potential range. The main disadvantage would be the limitation of the cathodic potential window.

The disposable electrode designs can be classified into two groups. The most generally used one is the screen-printed (thick-film) electrode (SPE). The second one is the recently developed pencil graphite electrode (PGE). Screen-printed electrode technology has developed rapidly with the help of the recent advancements in microfabrication; however, the PGEs still need to be improved to eliminate the use of external reference and counter electrodes. Whereas SPE is sufficient to perform an electrochemical analysis, a PGE still requires a beaker-type system. The important advantage of using PGE would be that the preparation does not require sophisticated instruments, because commercially available carbon graphite leads constitute the main source of PGEs. A carbon graphite lead with 1 mm i.d. provides a more comfortable electrode system in a beaker as well as a significantly bigger electrode surface with its cylindrical shape in comparison with the planar surface of the traditional Teflon-encased electrodes. A simple mechanical extrusion enables the renewal of the surface of PGE-based biosensors, hence obviating the need for an additional regeneration step and erasing memory effects.

DISPOSABLE DNA BIOSENSORS BASED ON PENCIL GRAPHITE ELECTRODES

Thus, these low-cost and easy-to-prepare PGEs have been the ideal choice for several electrochemical DNA biosensor reports. An electrochemical DNA biosensor for the detection of the Factor V Leiden single-nucleotide polymorphisms (SNPs) from polymerase chain reaction (PCR) amplicons using the oxidation signal of colloidal gold (Au) was described by Ozsoz et al.[16] A PGE modified with target DNA (Figure 8.1a), when hybridized with complementary probes conjugated to Au nanoparticles, responded with the appearance of Au oxide wave at ~+1.20 V (Figure 8.1b).

(a)

Covalent attachment of target DNA
onto PGE surface

(b)

Hybridization with Au nanoparticle-modified probe

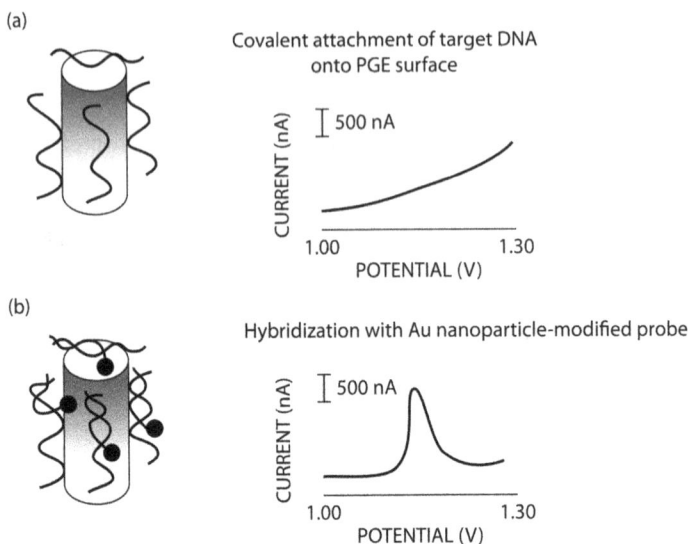

FIGURE 8.1 Detection of hybridization using Au nanoparticle-modified probes. (a) Covalent attachment of target DNA onto pencil graphite electrode (PGE); (b) Au oxide signal appears after hybridization with the complementary Au nanoparticle-conjugated probe.

The discrimination against homozygous and heterozygous genotypes has become possible by using specific oligonucleotides, which are full complementary to either wild type (WT) or mutant (MT) real samples. When the WT probe, which had a complementary DNA base sequence to the WT target, was immobilized on Au nanoparticles, a high Au oxidation signal (Figure 8.2a) showed that the PCR product contained WT DNA. When the MT probe was hybridized with the WT target, about one-half of the signal was obtained indicating the presence of SNPs (Figure 8.2b). Nearly a similar electrochemical response was obtained from the attachment of Au nanoparticle-modified WT probe onto bare PGE (Figure 8.2c). No electrochemical signals could be obtained when the noncomplementary target was immobilized on the PGE surface (Figure 8.2d).

The intrinsic redox activity of DNA was employed for detecting the duplex formation on a PGE-based biosensor by Wang et al.[17,18] Inosine-substituted probes and monitoring of the electrochemical guanine oxidation signal for hybridization events offer several advantages over the common use of external indicators including the appearance of a new peak, a flat background, and simplicity. Inosine is a hypoxanthine ribonucleoside, one of the basic compounds comprising cells, and also a precursor to adenosine and uric acid. In guanine signal-based electrochemical DNA biosensor schemes, inosine-substituted probes have been intensively used,[19,20] because they are electrochemically inactive. When an inosine-substituted probe-modified electrode is exposed to the target DNA, the guanine signal appears after hybridization. The selectivity of the device of Wang et al.[17] was demonstrated for the detection of a SNP in the BRCA1 breast cancer gene. Such low-cost, renewable

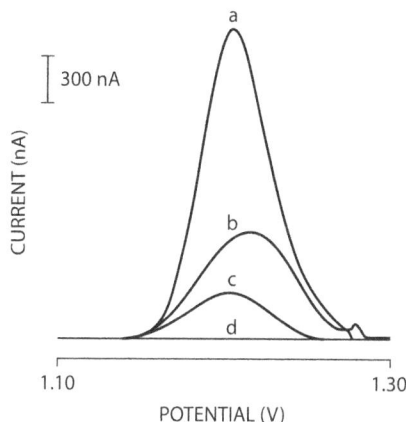

FIGURE 8.2 Differential pulse voltammograms for the oxidation signal of Au at PGE in TBS at (a) wild type (WT) probe-modified Au nanoparticle after hybridization with WT target; (b) mutant (MT) probe-modified Au nanoparticle after hybridization with WT target; (c) WT probe-modified Au nanoparticle after hybridization with no target on PGE surface; and (d) WT probe-modified Au nanoparticle after hybridization with noncomplementary target.

graphite transducers provided an attractive alternative to conventional carbon electrodes used for transducing DNA hybridization. The device offered a greatly simplified operation and held promise for decentralized genetic testing by eliminating the regeneration step and an external electroactive indicator.

DISPOSABLE DNA BIOSENSORS BASED ON SCREEN-PRINTED ELECTRODES

Progress in screen-printing technology with PC-controlled, fully automatic, and sophisticated instruments enabled the large-scale fabrication of the SPEs in various electrode layout designs. The printing patterns of conductors and insulators onto the surface of plastic or ceramic planar substrates can be designed and controlled by the software of these instruments. The most suitable substrate material as well as the conducting and insulating ink materials should be chosen for the purpose of the electrochemical test. The most commonly applied conducting inks are the carbon- and Au-based ones. There are mainly four steps involved in the screen-printing process. First, the conducting ink suspension is deposited onto the screen, followed by the loading of the screen mesh with the conducting ink, then the ink is forced through the screen with the help of a squeegee. Finally, the printed pattern is dried and cured. The printing of the insulating layer and the other cover layers, which can vary according to the electrochemical test, is also performed as described above onto the conducting ink-printed surface. However, the overall performance of the SPEs is greatly influenced by the chemical composition of the ink materials and the printing conditions such as applied pressure and temperature.

Screen-printed electrode is the commonly used electrode type in reports not only for clinical but also environmental and food monitoring purposes. As for the clinical applications, Lucarelli et al.[21] reported carbon SPEs for the detection of SNPs related to apolipoprotein E from PCR-amplified samples. The duplex formation was detected by measuring the guanine oxidation signal. The biosensor format involved the immobilization of an inosine-substituted probe onto an SPE and the voltammetric detection of the duplex formation with the appearance of the guanine signal.

Gold electrode-based indicator-free DNA hybridization detection has been reported by using a conventional-type gold electrode.[22] Figure 8.3 shows the oxidation signal of guanine at about ~+0.73 V by using differential pulse voltammetry on self-assembled L-cysteine monolayer (SAM)-modified screen-printed gold electrodes (AuSPE). The electrochemical determination of hybridization between an inosine-substituted probe (Figure 8.3b) and native target DNA (Figure 8.3a) was accomplished. The indicator-free detection of hybridization on AuSPE is greatly advantageous over the existing carbon-based materials, because of its potential applicability to microfabrication techniques.

Indicator-free, disposable electrochemical DNA-modified carbon SPEs have also been successfully applied for environmental monitoring. The device of Chiti et al.[23] relied on the intercalative or electrostatic collection of toxic aromatic amines onto an immobilized double-stranded DNA (dsDNA) or ssDNA layer in connection with chronopotentiometric analysis. The anodic signal of the guanine bases of DNA-coated SPEs was affected by structural or conformational modifications of the DNA layer due to the DNA–analyte association. The changes in the guanine oxidation signal were taken as an index for the affinity of the analyte

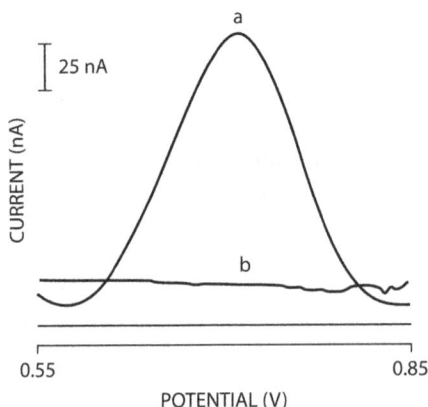

FIGURE 8.3 Differential pulse voltammograms for the oxidation signal of guanine (a) after and (b) before hybridization with target DNA at inosine-substituted probe-modified AuSPE.

toward nucleic acids. Submicromolar detection limits were obtained for molecules with more than two aromatic rings after a 2 min accumulation.[23]

Indicator-based disposable electrochemical DNA biosensors were also described by using carbon SPEs in connection with a well-described intercalator, daunomycin, as an indicator of the hybridization reaction by Marrazza et al.[24,25] Synthetic probes have been immobilized onto carbon SPEs using adsorption at a controlled potential. The hybrids formed on the electrode surface were evaluated by chronopotentiometric stripping analysis of daunomycin. The DNA biosensor was able to detect 0.2 mg L^{-1} of a 21-base target sequence. The determination of low-molecular weight toxic compounds with affinity for DNA, such as polychlorinated biphenyls (PCBs) and aflatoxin B1, was carried out by monitoring their effect on the guanine oxidation signal.[25] River water samples were also used as real matrices for monitoring of these compounds by using the disposable DNA-modified SPEs.[25]

CONCLUSION

The attractive performance of disposable PGEs and SPEs for the biosensing of DNA hybridization has been reviewed. Such performance compares favorably with that of conventional beaker-type electrodes commonly used for transducing DNA hybridization. The simple and fast utility of these test strips provides the basis of the day-to-day practicality of the biosensor and holds great promise for decentralized genetic testing. An electrochemical device based on such single-use electrodes, thus, is a strong alternative to other optical, gravimetric, or surface plasmon resonance (SPR)-based DNA biosensors, when rapid and cost-effective hybridization assays are concerned. Various electrochemical protocols, such as indicator-free or intercalator-based ones, can be readily combined with a disposable DNA biosensor. However, the integration of a battery-operated electronic microprocessor, as well as the reference and counter electrodes, is crucial for the improvement of PGEs. The excellent performance of the electrochemical self-care devices for blood glucose in connection with disposable electrodes provides an encouraging example for electrochemists to develop a similar device for genetic testing. We are looking forward to the entrance of an electrochemical self-care DNA-based device into the booming biotechnology market.

ACKNOWLEDGMENTS

K. Kerman acknowledges the Monbukagakusho scholarship for research students from the Japan Ministry of Education, Culture, Sports, Science and Technology (MEXT).

REFERENCES

1. Millan, K.M.; Mikkelsen, S.R. Sequence-selective biosensor for DNA based on electroactive hybridization indicators. Anal. Chem. **1993**, *65*, 2317–2323.

2. Palecek, E.; Fojta, M. DNA hybridization and damage. Anal. Chem. **2001**, *73*, 74A–83A.
3. Kerman, K.; Kobayashi, M.; Tamiya, E. Recent trends in electrochemical DNA biosensor technology. Meas. Sci. Technol. **2004**, *15*, 1R–11R.
4. Meric, B.; Kilicaslan, N.; Kerman, K.; Ozkan, D.; Kurun, U.; Aksu, N.; Ozsoz, M. Performance of precision G blood glucose analyzer with a new test strip G2B on neonatal samples. Clin. Chem. **2002**, *48*, 179–180.
5. Lucarelli, F.; Marrazza, G.; Turner, A.P.F.; Mascini, M. Carbon and gold electrodes as electrochemical transducers for DNA hybridisation sensors. Biosens. Bioelectron. **2004**, *19*, 515–530.
6. McCreery, R.L. Carbon electrodes: structural effects on electron transfer kinetics. In Electroanalytical Chemistry; Bard, A.J., Ed.; Marcel Dekker: New York, NY, **1991**; Vol. 18.
7. Wang, J.; Kawde, A.-N.; Musameh, M. Carbon-nanotube-modified glassy carbon electrodes for amplified label-free electrochemical detection of DNA hybridization. Analyst **2003**, *128*, 912–916.
8. Wang, J.; Musameh, M. Carbon nanotube screen-printed electrochemical sensors. Analyst **2004**, *129*, 1–2.
9. Hason, S.; Dvorák, J.; Jelen, F.; Vetterl, V. Impedance analysis of DNA and DNA–drug interactions on thin mercury film electrodes. Crit. Rev. Anal. Chem. **2002**, *32*, 167–179.
10. Fojta, M.; Havran, L.; Kizek, R.; Billová, S. Voltammetric microanalysis of DNA adducts with osmium tetroxide,2,2′-bipyridine using a pyrolytic graphite electrode. Talanta **2002**, *56*, 867–874.
11. Oliveira-Brett, A.M.; Piedade, J.A.P.; Chiorcea, A.-M. Anodic voltammetry and AFM imaging of picomoles of adriamycin adsorbed onto carbon surfaces. J. Electroanal. Chem. **2002**, *267*, 538–539.
12. Oliveira-Brett, A.M.; Chiorcea, A.-M. Effect of pH and applied potential on the adsorption of DNA on highly oriented pyrolytic graphite electrodes. Atomic force microscopy surface characterisation. Electrochem. Commun. **2003**, *5*, 178–183.
13. Pividori, M.I.; Merkoçi, A.; Barbé, J.; Alegret, S. PCR-genosensor rapid test for detecting *Salmonella*. Electroanalysis **2003**, *15*, 1815–1823.
14. Pividori, M.I.; Merkoçi, A.; Alegret, S. Graphite–epoxy composites as a new transducing material for electrochemical genosensing. Biosens. Bioelectron. **2003**, *19*, 473–484.
15. Erdem, A.; Pividori, M.I.; del Valle, M.; Alegret, S. Rigid carbon composites: a new transducing material for label-free electrochemical genosensing. J. Electroanal. Chem. **2004**, *567*, 29–37.
16. Ozsoz, M.; Erdem, A.; Kerman, K.; Ozkan, D.; Tugrul, B.; Topcuoglu, N.; Ekren, H.; Taylan, M. Electrochemical genosensor based on colloidal gold nanoparticles for the detection of Factor V Leiden mutation using disposable pencil graphite electrodes. Anal. Chem. **2003**, *75*, 2181–2187.
17. Wang, J.; Kawde, A.-N.; Sahlin, E. Renewable pencil electrodes for highly sensitive stripping potentiometric measurements of DNA and RNA. Analyst **2000**, *125*, 5–7.
18. Wang, J.; Kawde, A.-N. Pencil-based renewable biosensor for label-free electrochemical detection of DNA hybridization. Anal. Chim. Acta **2001**, *431*, 219–224.
19. Ozkan, D.; Erdem, A.; Kara, P.; Kerman, K.; Meric, B.; Hassmann, J.; Ozsoz, M. Allele-specific genotype detection of Factor V Leiden mutation from polymerase chain reaction amplicons based on label-free electrochemical genosensor. Anal. Chem. **2002**, *74*, 5931–5936.

20. Kara, P.; Ozkan, D.; Erdem, A.; Kerman, K.; Pehlivan, S.; Ozkinay, F.; Unuvar, D.; Itirli, G.; Ozsoz, M. Detection of achondroplasia G380R mutation from PCR amplicons by using inosine modified carbon electrodes based on electrochemical DNA chip technology. Clin. Chim. Acta **2003**, *336*, 57–64.

21. Lucarelli, F.; Marrazza, G.; Palchetti, I.; Cesaretti, S.; Mascini, M. Coupling of an indicator-free electrochemical DNA biosensor with polymerase chain reaction for the detection of DNA sequences related to the apolipoprotein E. Anal. Chim. Acta **2002**, *469*, 93–99.

22. Kerman, K.; Morita, Y.; Takamura, Y.; Tamiya, E. Label-free electrochemical detection of DNA hybridization on gold electrode. Electrochem. Commun. **2003**, *5*, 887–891.

23. Chiti, G.; Marrazza, G.; Mascini, M. Electrochemical DNA biosensor for environmental monitoring. Anal. Chim. Acta **2001**, *427*, 155–164.

24. Marrazza, G.; Chiti, G.; Mascini, M.; Anichini, M. Detection of human apolipoprotein E genotypes by DNA electrochemical biosensor coupled with PCR. Clin. Chem. **2000**, *46*, 31–37.

25. Marrazza, G.; Chianella, I.; Mascini, M. Disposable DNA electrochemical biosensors for environmental monitoring. Anal. Chim. Acta **1999**, *387*, 297–307.

9 Handheld Nucleic Acid Analyzer

James A. Higgins
United States Department of Agriculture–Agricultural
Research Service, Beltsville, Maryland, U.S.A.

CONTENTS

INTRODUCTION

Since the advent of the polymerase chain reaction (PCR) in the late 1980s, investigators have sought to expand the potential of this technology to that of on-site, "bedside" diagnostics, in which clinical parameters can be determined in real time, without the need for transport of samples to a remote testing facility. Discoveries of clandestine bioweapons programs in Iraq and the former Soviet Union in the early 1990s led the U.S. Department of Defense to sponsor development of nucleic acid-based instruments suitable for detecting a variety of pathogens at both frontline and rear-echelon locations. Both of these research initiatives have begun to yield commercial handheld instruments, capable of performing PCR and reverse transcription-PCR (RT-PCR) in real time; future devices promise to offer direct detection of target nucleic acids without the need for amplification. Other instrument packages look to provide the end user with both sample preparation/nucleic acid extraction and detection in one portable instrument. This entry reviews existing handheld nucleic acid detection instrumentation, discusses their advantages and disadvantages, and previews newer devices and platforms expected to continue to make this area of medical diagnostics particularly innovative and exciting.

THREAT AGENT DETECTION: IMPETUS FOR HANDHELD NUCLEIC ACID ANALYZER DEVELOPMENT

Successful performance of PCR obviously depends on the ability of the thermal cycling platform to maintain accurate gradations of temperature in the manifold or

DOI: 10.1201/9781003247432-9

block containing the reaction tubes. The heating and cooling apparatus necessary to provide these temperature changes places constraints on the size and weight of the thermal cycler. "First-generation" thermal cyclers, such as the Perkin Elmer 480 model instrument, while reasonably light in weight and small in footprint, were not designed with field use in mind. However, during operations in the Persian Gulf in 1990–1991, field laboratories for detection of threat agents were deployed and operated by both the U.S. Army and the U.S. Navy. PCR on conventional thermal cyclers, with agarose gel electrophoresis used to confirm amplification products, was used on a regular basis by these laboratories with a high degree of success, despite the trying conditions associated with performing molecular assays in a desert environment. By the mid-1990s the U.S. Army adopted PCR-based assays as a critical component of the field-deployable 520th Theatre Area Medical Laboratory (TAML), which was designed to provide rapid threat agent detection in battlefield conditions.

Just as Western defense and public health personnel were coming to grips with the disclosure of the size and scope of the Iraqi bioweapons program, it was revealed that the Soviet Union had a clandestinely operated sizeable, well-funded program of its own. Among the pathogens selected for weaponization were smallpox, plague, tularemia, and anthrax. The likelihood that U.S. and NATO troops may encounter such agents heightened the importance of on-site diagnostic techniques and, consequently, research in this area was prioritized by the US Department of Defense (DoD), primarily under the auspices of the Defense Advanced Research Projects Agency (DARPA).

Research on portable nucleic acid-based detection devices was considerably aided by the discovery in 1991 that the 5′–3′ exonuclease activity of *Taq* polymerase, the enzyme mediating PCR, could be harnessed to provide real-time monitoring of the reaction via the hybridization of a fluorogenic oligonucleotide probe.[1] In 1995, researchers at Genentech, Inc., and Applied Biosystems reported the first use of a commercial probe expressly designed for real-time PCR, the "TaqMan®" probe.[2] With this assay system, fluorescence generated by the probe could be detected by the appropriate instrumentation and the user provided with a graphical depiction of the accumulation of the PCR amplicons over the length of the reaction. The TaqMan assay was notable for being conducted in the same reaction tube from start to finish; there was no need for postreaction manipulations. Probe design and synthesis was relatively straightforward and a TaqMan-based protocol for the detection of the foodborne bacterium *Listeria monocytogenes* was published in 1995.[3] These results gave impetus to the design and evaluation of TaqMan probes against a variety of bacterial and viral threat agents, such as orthopoxviruses.[4]

The implementation of fluorogenic probes freed the user from the need to conduct electrophoresis to determine whether or not a PCR assay was successful, and had obvious implications for reducing dependence on a laboratory infrastructure. However, the need for power sources and optics capable of monitoring the fluorescence produced by a real-time PCR assay was a major impediment to the design of portable thermal cyclers. For example, the first commercial platform dedicated to real-time PCR, the Perkin Elmer/Applied Biosystems 7700 model Prism® instrument, appeared in 1996–1997, but was too heavy (140 kg) for field use. U.S. Department of Defense-sponsored research sought to address this issue, and in 1997 researchers at the Lawrence Livermore National Laboratory (Livermore, CA) succeeded in fabricating a battery-powered, suitcase-sized real-time thermal cycler capable of using TaqMan probes: the MATCI, or Miniature Analytical Thermal Cycler Instrument.[5] Central to the operation of the device was its use of silicon "thermal cycler" units to mediate the heating and cooling of the associated 25 μl volume plastic reaction tube; this allowed for a one cycle per minute cycling speed. The MATCI provided sufficient temperature fidelity to allow for differentiation between sequences, with a high degree of homogeneity (such as poxvirus hemagglutinin gene), on the basis of a single nucleotide substitution.[6]

The MATCI could only accommodate one reaction tube at a time; the next iteration of the technology, the Advanced Nucleic Acid Analyzer (ANAA), which appeared in 1998, could perform 10 reactions simultaneously, and detection of as few as 500 cells of *Erwinia* was accomplished after only 7 min of thermal cycling.[7] The ANAA was a suitcase-sized instrument and researchers at the Lawrence Livermore National Laboratory were interested in advancing the miniaturization of the system to a handheld format; this led to the fabrication of the Handheld Advanced Nucleic Acid Analyzer (HANAA), arguably the world's first handheld, real-time thermal cycler instrument. The HANAA could perform four reactions simultaneously, with the use of two light emitting diodes (at 490 and 525 nm, with a combined emission of over 1 mW of power) allowing for monitoring of two dyes (e.g., fluorescein amidite [FAM] in one channel and carbosy-4',5'-dichloro-2',7'-dimethoxyfluorescein [JOE] in the other).[8] The HANAA weighed less than 1 kg, and several sequential reactions could be performed, powered off a 12 V/3.5 A battery pack.[9] Reactions of up to 25 μl in volume were conducted in customized plastic tubes, and an improved silicon/platinum thermal cycler unit (22×6.6×2 mm) permitted "traditional" cycling parameters (e.g., 95°C for 0 s, 55°C for 10 s, and 72°C for 15 s for 40 cycles) to be conducted in less than 20 min; strongly positive samples could generate positive results by 13 min of assay time.[10]

Prototype HANAAs were distributed to a variety of testers and feedback was favorable enough to result in licensing of the technology to Smiths Detection, Edgewood, MD; the result was production of the BioSeeq™ handheld real-time thermal cycler, available for purchase in 2003 (Figures 9.1 and 9.2). The BioSeeq can accommodate up to six independently programmable reactions, each with a volume of 25 μl. In their evaluation of the BioSeeq during an 8-month beta testing phase, Emanuel et al.[11] used TaqMan probes directed against the *Francisella tularensis tul4* and *fopA* genes. Detection limits for the BioSeeq were 200 and 300 fg, respectively; these are approximately 100 and 150 genome equivalents of *F. tularensis*. In comparison, the detection limit of the assay when performed on the ABI 7900 real-time thermal cycler instrument was 50 fg, approximately 25 genome equivalents. Assay specificity, determined by testing a panel of 27 bacterial species and strains, including other possible threat agents, was the same for both targets on both instruments. When DNA extracted from tissues of mice infected with the *F. tularensis* live vaccine strain (LVS) strain via aerosol exposure was assayed on the ABI 7900 and the BioSeeq, results were comparable, indicating that the handheld instrument was successful in analyzing the DNA in reaction volumes half the size of those of the ABI 7900 (which uses a volume of 50 μl). Custom reagents for a variety of threat agents are currently available.

FIGURE 9.1 The Smiths Detection BioSeeq™ instrument weighs 6.5 lb (2.9 kg) and measures 12×8×2 in. (30.4×20.3×5.0 cm) and can run for up to 1 hr on the power provided by eight alkaline "C" batteries. There are six independently programmable reaction modules and data are displayed to the user via a small inset screen; the instrument can also be linked to a laptop computer for more involved assay programming and data display.

Source: Photo courtesy of P. Emanuel, U.S. Army, Edgewood, MD.

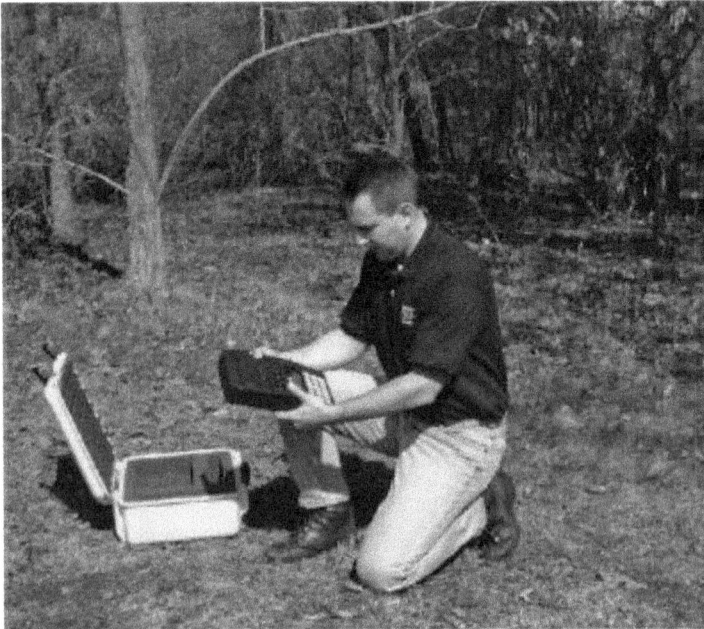

FIGURE 9.2 Demonstration of the use of the BioSeeq™ instrument in the field.

Source: Photo courtesy of P. Emanuel, U.S. Army, Edgewood, MD.

OTHER DOD-SPONSORED THREAT AGENT DETECTION PLATFORMS

In 2004, another real-time, handheld nucleic acid-based detection platform became available for use by U.S. government agencies: the Razor, manufactured by Idaho Technology, Inc. (Salt Lake City, UT; Figure 9.3). The Razor builds on the experience of Idaho Technology's suitcase thermal cycler, the Ruggedized Advanced Pathogen Identification System (RAPID), which was designed in collaboration with personnel from the U.S. Air Force. The RAPID is essentially a "combat" version of the LightCycler® (which is licensed by Roche, Inc., for sale to non-U.S.-government laboratories) and has been used extensively for the detection of threat agents in the United States, Europe, and the Middle East. The RAPID is capable of using prepackaged, freeze-dried real-time PCR and RT-PCR reagents, and this convenience has been carried over to the Razor, in which reactions are performed in plastic pouches containing the freeze-dried reagents of interest. The end user simply uses a conventional syringe to inject sterile water into the pouch, reconstitutes the reagents, injects template, and proceeds with the assay; up to 12 reactions can be done in one pouch. The Razor, which weighs 8 lb (3.6 kg) and can analyze the entire 12-sample

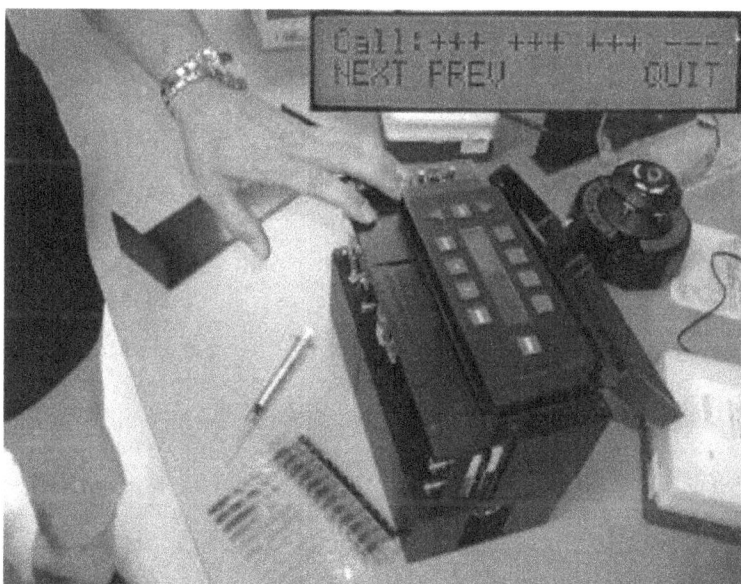

FIGURE 9.3 The Idaho Technology, Inc., Razor instrument is an 8 lb (3.6 kg) handheld real-time thermal cycler that performs the reactions in specially formatted plastic pouches (foreground) that contain freeze-dried reagents for up to 12 samples. The reagents are reconstituted via injection of sterile water through syringe ports on either end of the pouch. Inset: results of an assay are displayed to the user as plus/minus scoring on an LCD display panel.

Source: Photo: James Higgins; inset: Matt Scullion, Idaho Technology, Inc., Salt Lake City, UT.

pouch in approximately 22 min, is designed for field use off a battery power source. The instrument's software is preloaded with thermal cycling protocols for detection of a variety of threat agents, and results are presented to the user either via plus/ minus scoring on the integral LCD display (Figure 9.3), or via a more comprehensive graphical display on an attached laptop computer.

HANDHELD NUCLEIC ACID ANALYZERS FOR CLINICAL USE

Simultaneously with DoD-sponsored research, a number of private companies have pursued the development and fabrication of handheld nucleic acid analyzers with an eye toward capturing what is expected to be a very lucrative and burgeoning market in "point-of-care" (POC) diagnostics.

Convincing physicians and clinical laboratory directors that POC nucleic acid analyzers, much less molecular methods per se, are worth embracing has traditionally been difficult.[12] Sample processing, particularly of matrices with a high concentration of PCR inhibitors (such as blood), continues to be a major hurdle. Also, firms hoping to market POC diagnostics must convince physicians and laboratory directors that the sometimes higher costs of their assays can be justified by improved sensitivity and specificity over existing methods. Nonetheless, according to *Small Times* magazine author Marlene Bourne, more than 21 companies are working on molecular diagnostic devices involving analysis of nucleic acids; at least 12 appear to be interested in the POC market ("Bantamweights vs. giants: Is a lab-on-a-chip brawl a mismatch?," *Small Times*, October 31, 2003). While an exhaustive recounting of these projects is beyond the scope of this entry, some of the more interesting platforms are worthy of mention.

Clinical Microsensors (CMS), a division of Motorola, was founded in 1995 to exploit technology, developed by researchers at the California Institute of Technology, for "bioelectronic" detection of nucleic acid hybridization events. In late 1998 CMS had produced a prototype handheld instrument, powered by a 9 V battery, that could accommodate a "DNA chip" fabricated to contain 14 gold electrodes with DNA probes attached to their surface. The device promised a degree of sensitivity sufficient to detect DNA in wastewater, sludge, and blood, all very challenging sample matrices.[13] In 2000, CMS unveiled a desktop instrument, the 4800 eSensor™, which can assay from 1 to 48 eSensor chips at one time. According to a December 2003 conversation with Motorola scientist Tim Tiemann, development of a handheld analyzer that would provide onboard sample processing and nucleic acid purification, as well as incorporating eSensor-based detection technology, is actively ongoing.

One company produces a handheld nucleic acid analyzer suitable for use in a clinical diagnostic setting; this is the Verigene™ ID instrument (Figure 9.4) which exploits the nanoparticle detection method developed in the laboratory of Chad Mirkin at Northwestern University, Illinois.[14] The initial Verigene ID assay is for detection of SNPs associated with coagulation disorders; future assays are planned for infectious agents and other clinical parameters.

Other firms developing POC diagnostic platforms based on array-mediated detection of nucleic acids include Nanogen, Affymetrix, and Caliper Technologies;

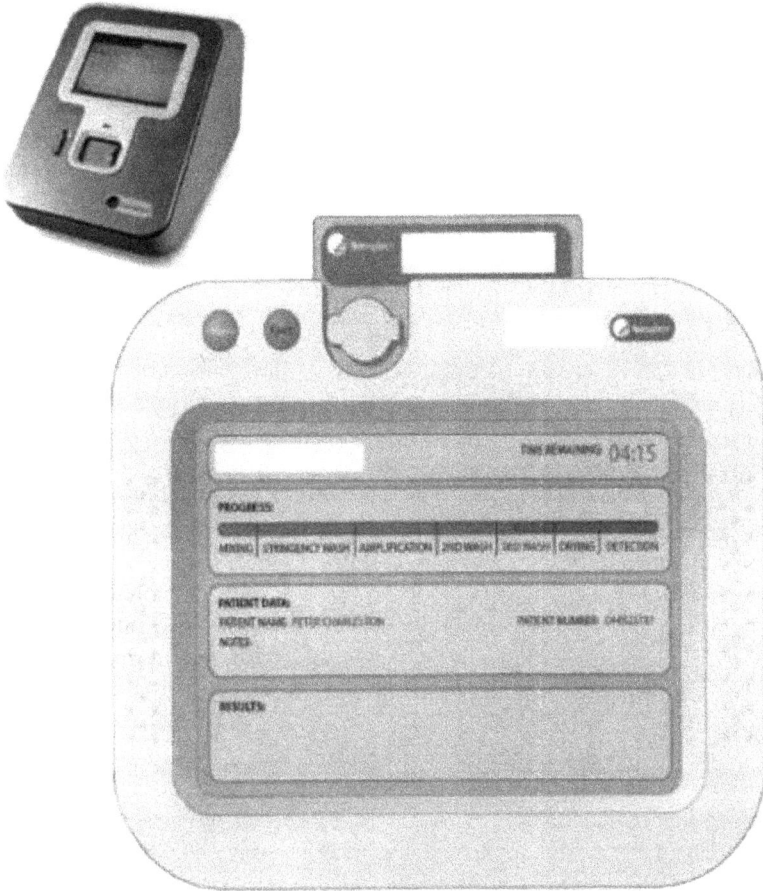

FIGURE 9.4 Image of a prototype of Nanosphere's Verigene Mobile device, which will allow the user to perform sample processing and nucleic acid detection via hybridization assay in one handheld instrument suitable for the clinical diagnostic market. Inset: the company currently markets the portable Verigene ID system, which provides detection of nucleic acids via a proprietary, nanoparticle-mediated hybridization chemistry.

Source: Photo courtesy of Jamie Abrams, Nanosphere, Northbrook, IL.

all of which have experience in the fabrication of large, benchtop instruments. In addition to clinical tests, it is anticipated that handheld devices will also be useful for other situations in which molecular detection is particularly helpful, such as in monitoring crops and agricultural products for the presence of genetically modified organisms (GMOs); testing of environmental samples for the presence of viral, bacterial, fungal, and protozoal pathogens; and analysis of food for the presence of both pathogenic and spoilage-associated microorganisms.

For the devices mentioned above, the chemistry mediating detection of the nucleic acids of interest may or may not involve PCR and nucleic acid hybridization; for example, NuGen Technologies, Inc. (San Carlos, CA) is relying on a proprietary method called exponential single primer isothermal amplification that may allow for detection of very low quantities of target nucleic acids and can be adapted for use in microarray or microfluidic formats. Another interesting amplification chemistry is the loop-mediated isothermal amplification (LAMP), in which *Bst* DNA polymerase is used to mediate autocycling strand displacement DNA synthesis.[15]

COMBINING SAMPLE PREPARATION AND ANALYSIS IN ONE PLATFORM

The ideal handheld nucleic acid-based analysis device would incorporate both sample processing and real-time analysis and detection in one instrument. The technical challenges are formidable but some progress in fabricating such a device is underway. Nanosphere's Verigene Mobile instrument (Figure 9.4) combines microfluidic purification of the sample and subsequent nucleic acid detection via hybridization with probe-bound nanoparticles, all in one handheld device. The Verigene Mobile was designed to be a dedicated POC instrument suitable for use by personnel who do not have intensive training in molecular biology-based assays; the sample is deposited into a cartridge that is in turn inserted into the device for automated processing, with results presented to the user in real time.

CONCLUSION

By the winter of 2004 at least three handheld nucleic acid analyzers—the Smiths Detection BioSeeq, the Idaho Technology, Inc. Razor, and the Nanosphere Verigene—were available for use. The first two instruments will rely on PCR-mediated amplification of the DNA or RNA template, whereas the Verigene relies on its proprietary nanoparticle-mediated hybridization format to detect target nucleic acids. Other devices that promise the same degree of portability will become available and offer the end user the ability to extract DNA or RNA from a variety of sample matrices and volumes. With continuous advances in highly active fields of research such as microfluidic systems, micro-total analysis systems (μTAS), nanotechnology-based "lab on a chip" devices, and microelectromechanical systems (MEMS), the future looks very exciting for the advent of novel handheld molecular biology-based devices suitable for use in the field and at the point of care.

REFERENCES

1. Holland, P.M.; Abramson, R.D.; Watson, R.; Gelfand, D.H. Detection of specific polymerase chain reaction product by utilizing the 5′–3′ exonuclease activity of *Thermus aquaticus* DNA polymerase. Proc. Natl. Acad. Sci. U. S. A. **1991**, *88*, 7276–7280.
2. Livak, K.J.; Flood, S.J.; Marmaro, J.; Giusti, W.; Deetz, K. Oligonucleotides with fluorescent dyes at opposite ends provide a quenched probe system useful for detecting PCR product and nucleic acid hybridization. PCR Methods Appl. **1995**, *4*, 357–362.

3. Bassler, H.A.; Flood, S.J.; Livak, K.J.; Marmaro, J.; Knorr, R.; Batt, C.A. Use of a fluorogenic probe in a PCR-based assay for the detection of *Listeria monocytogenes.* Appl. Environ. Microbiol. **1995**, *61*, 3724–3728.

4. Ibrahim, M.S.; Esposito, J.J.; Jahrling, P.B.; Lofts, R.S. The potential of 5′ nuclease PCR for detecting a singlebase polymorphism in *Orthopoxvirus.* Mol. Cell. Probes **1997**, *11*, 143–147.

5. Belgrader, P.; Smith, J.K.; Weedn, V.W.; Northrup, M.A. Rapid PCR for identity testing using a battery-powered miniature thermal cycler. J. Forensic Sci. **1998**, *43*, 315–319.

6. Ibrahim, M.S.; Lofts, R.S.; Jahrling, P.B.; Henchal, E.A.; Weedn, V.W.; Northrup, M.A.; Belgrader, P. Real-time microchip PCR for detecting single-base differences in viral and human DNA. Anal. Chem. **1998**, *70*, 2013–2017.

7. Belgrader, P.; Benett, W.; Haldey, D.; Richards, J.; Stratton, P.; Mariella, R. Jr.; Milanovich, F. PCR detection of bacteria in seven minutes. Science **1999**, *284*, 449–450.

8. Richards, J.B.; Benett, W.J.; Stratton, P.; Hadley, D.R.; Nasarabadi, S.L.; Milanovich, F.P. Miniaturized Detection System for Handheld PCR Assays. Proceedings of the Society of Photo-Optical Instrumentation Engineers International Symposium on Environmental and Industrial Sensing, Boston, MA, Nov. 5–8, 2000, Bellingham, WA: Society of Photo-Optical Instrumentation Engineers, **2001**; 4200–4212.

9. Benett, W.J.; Richards, J.B.; Stratton, P.; Hadley, D.R.; Bodtker, B.H.; Nasarabadi, S.L.; Milanovich, F.P.; Mariella, R.P. Jr.; Koopman, R.P. Handheld Advanced Nucleic Acid Analyzer. Proceedings of the Society of Photo-Optical Instrumentation Engineers International Symposium on Environmental and Industrial Sensing, Boston, MA, Nov. 5–8, 2000, Bellingham, WA: Society of Photo-Optical Instrumentation Engineers, **2001**; 4200–4211.

10. Higgins, J.A.; Nasarabadi, S.; Karns, J.S.; Shelton, D.R.; Cooper, M.; Gbakima, A.; Koopman, R.P. A handheld real time thermal cycler for bacterial pathogen detection. Biosen. Bioelectron. **2003**, *18*, 1115–1123.

11. Emanuel, P.A.; Bell, R.; Dang, J.L.; McClanahan, R.; David, J.C.; Burgess, R.J.; Thompson, J.; Collins, L.; Hadfield, T. Detection of *Francisella tularensis* within infected mouse tissues by using a handheld PCR thermocycler. J. Clin. Microbiol. **2003**, *41*, 689–693.

12. Check, W. Nucleic acid tests move slowly into clinical labs. Am. Soc. Microbiol. News **2001**, *67*, 560–565.

13. Licking, E. The great DNA chip derby. Business Wkly. **1999**, *3652*, 90–91.

14. Park, S.J.; Taton, T.A.; Mirkin, C.A. Array-based electrical detection of DNA using nanoparticle probes. Science **2002**, *295*, 1503–1506.

15. Kuboki, N.; Inoue, N.; Sakurai, T.; Di Cello, F.; Grab, D.J.; Suzuki, H.; Sugimoto, C.; Igarashi, I. Loop-mediated isothermal amplification for detection of African trypanosomes. J. Clin. Microbiol. **2003**, *41*, 5517–5524.

10 DNA-Binding Fluorophores

Mikael Kubista, Jonas Karlsson, Martin Bengtsson, Neven Zoric, Gunar Westman

TATAA Biocenter AB, Göteborg, Sweden

CONTENTS

INTRODUCTION

The first cyanine dye, named cyanine because of its blue color, was serendipitously synthesized by C.H.G. Williams as early as 1856. Since the end of the 19th century, the class of cyanine dyes have dominated the field of photography as photosensitizers and are unrivalled even today. The generic cyanine dyes consist of two nitrogen centers, one of which is positively charged and linked to the other center by a conjugated chain of an odd number of carbon atoms (Figure 10.1). Symmetrical cyanine dyes commonly contain two benzazole moieties connected by a polymethine chain, whereas the unsymmetrical cyanines usually consist of a benzazole group and a quinoline or pyridine heterocycle, also connected by a methine bridge. The extensive conjugation and the delocalization of charge lead to long-wavelength absorption maxima and large molar absorptivities. Apart from use as photosensitizers, cyanine dyes have been employed as laser materials, photorefractive materials, antitumor agents, and in optical disks as recording media to name but a few applications.

DNA-BINDING CYANINE DYES

Unsymmetrical cyanine dyes have received considerable interest because of their excellent nucleic acid staining properties.[1,2] In 1986, Lee et al.[3] showed that the unsymmetrical cyanine dye thiazole orange (TO; Figure 10.2) exhibits 3000-fold enhancement in fluorescence intensity upon binding to RNA. Later, Rye et al.[4] showed that TO, and the similar dye, oxazole yellow (YO) with a benzoxazolium moiety instead of benzothiazolium (the BO version of the dye), have a dramatic increase in fluorescence also upon binding to DNA. Both these studies indicated that the dyes bind in a nearest-neighbor exclusion stoichiometry typical of intercalators.

DOI: 10.1201/9781003247432-10

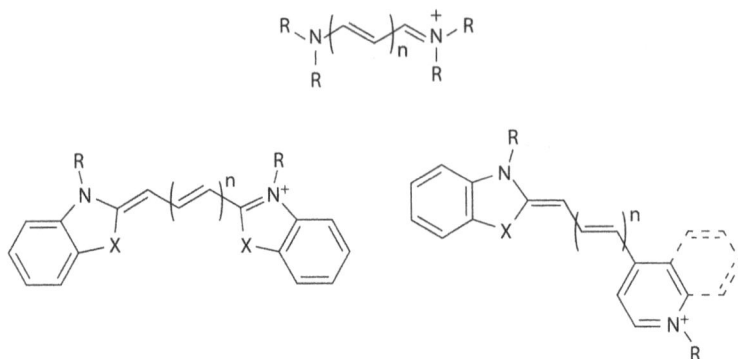

FIGURE 10.1 Generic structure of cyanine dyes (top) and examples of symmetrical (left) and unsymmetrical cyanine dyes (right).

To obtain stains with high affinity for DNA, Rye et al.[5] developed TOTO and YOYO (Figure 10.2), which have two dye moieties connected by a biscationic amine linker. Their affinity for double-stranded DNA is about the square of that of the monomers as a result of bisintercalation and their high positive charge. To enhance the affinity for DNA of monomeric dyes, positively charged substituents (3-propyl trimethyl ammonium bromide) were added in TO-PRO and YO-PRO (Figure 10.3).[6]

The maximum absorption of cyanine dyes depends on the length of the conjugated chain and the nature of the heterocyclic moieties. Today dyes are available

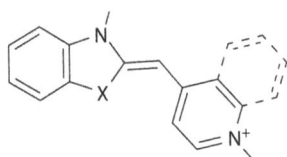

Dye	X	Heterocycle	λ_{max}
BO	S	pyridine	445
TO	S	quinoline	510
PO	O	pyridine	435
YO	O	quinoline	480

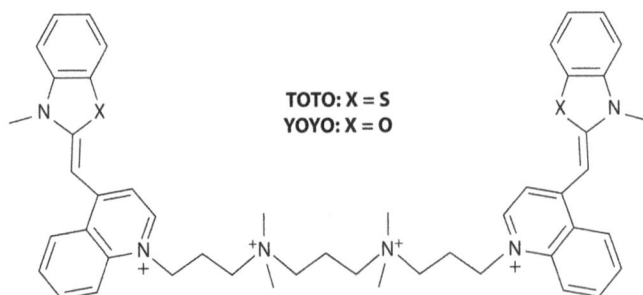

TOTO: X = S
YOYO: X = O

FIGURE 10.2 General structure of the monomeric unsymmetrical cyanine dyes (top), and the homodimeric cyanine dyes TOTO and YOYO (bottom).

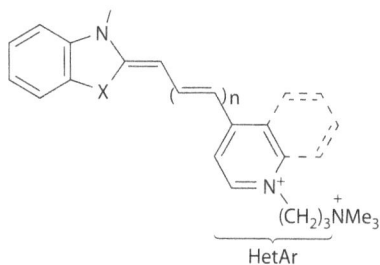

Dye	X	n	HetAr	Ex/Em (nm)
PO-PRO	O	0	pyridine	435/455
BO-PRO	S	0	pyridine	462/481
YO-PRO	O	0	quinoline	491/509
TO-PRO	S	0	quinoline	515/531
PO-PRO-3	O	1	pyridine	539/567
BO-PRO-3	S	1	pyridine	575/599
YO-PRO-3	O	1	quinoline	612/631
TO-PRO-3	S	1	quinoline	642/661

FIGURE 10.3 Divalent unsymmetrical cyanine dyes with different spectral characteristics.

Source: From Ref. [7].

whose absorptions span a broad range of the visible spectrum (Figure 10.3).[6] They typically have several hundredfold enhancements in fluorescence upon binding DNA and a quantum yield in bound state of at least 0.1.

The fluorescence quantum yield of cyanine dyes increases when torsional motion around the methine bridge is restricted, which reduces the probability of nonradiative relaxation from the excited singlet state.[7,8] When the dyes bind DNA, internal rotation is likely to be strongly hindered, which causes the dramatic increase in fluorescence.

The interactions of cyanine dyes with nucleic acids have been thoroughly investigated. Early flow linear dichroism studies indicated that YO-PRO and YOYO intercalate in DNA.[9] Intercalation was also supported by circular dichroism, fluorescence anisotropy, and dye–nucleobase energy transfer measurements. By nuclear magnetic resonance (NMR), the solution structures of TOTO bound to short oligonucleotides have been determined.[10] Frequently, more than one complex was observed. A preferred binding site was bisintercalation in a central CTAG:CTAG binding site with the cationic linker located in the minor groove. The DNA complex is unwound by ~30° and extended ~2 Å per TO moiety, which is typical of intercalation.

Thermodynamic studies of TO binding to nucleic acids of different base compositions showed that binding has little sequence selectivity.[11] At elevated dye:base pair ratios, a secondary binding mode was observed for both monomeric and dimeric dyes.[9,12] In a spectroscopic study, Nygren et al.[13] determined thermodynamic parameters of TO binding to DNA both as monomer and as dimer. With several sequences, the dimer complex was found to have lower fluorescence than the bound monomer. By fluorescence polarization measurements of TOTO–DNA complexes, evidence for external binding was obtained.[14] In hole-burning studies, external binding of TO-PRO-3 to DNA was revealed.[15] Atomic force microscopy and viscometry studies support the notion that TO and in particular TOTO bind in more than one mode to DNA.[16] Clearly, the interaction of cyanine dyes with DNA is complex and yet far from fully understood.

Most cyanine dyes have strong affinity also for single-stranded DNA and obtain high fluorescence upon binding.[13,17] This makes them less useful as specific probes

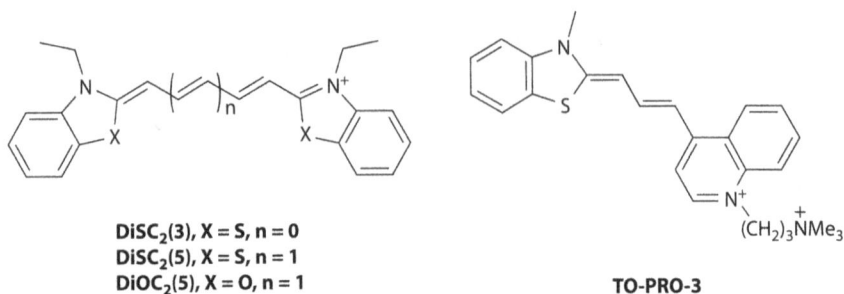

DiSC$_2$(3), X = S, n = 0
DiSC$_2$(5), X = S, n = 1
DiOC$_2$(5), X = O, n = 1

TO-PRO-3

FIGURE 10.4 Symmetrical cyanine dyes (left) and unsymmetrical cyanine dye TO-PRO-3 (right).

for double-stranded DNA. On the one hand similar affinity for single- and double-stranded DNA is typical of intercalators of low complexity that mainly gain binding energy through stacking with bases. Minor groove binders, on the other hand, typically show strong preference for double-stranded molecules. In recent years, some symmetrical cyanine dyes have been found to bind in the minor groove. Seifert et al.[18] showed that the symmetrical pentamethine cyanine dye 3,3′-diethylthiadicarbocyanine (DiSC$_2$[5]) (Figure 10.4) cooperatively binds as head-to-head dimer in the DNA minor groove forming helical H-aggregates. Mikheikin et al.[19] showed that a series of trimethine symmetrical cyanine dyes, e.g., DiSC$_2$,[3] bind in the minor groove mainly as monomers. These studies suggested that the hydrophobicity of the heterocyclic group is a critical determinant for dimerization. Recently, Sovenyhazy et al.[20] showed that the trimethine cyanine dye TO-PRO-3 (Figure 10.4) cooperatively binds in the minor groove of poly(dA–dT)$_2$, although it has previously been considered to mainly intercalate in DNA.

A characteristic of classical minor groove-binding ligands such as Hoechst 33258 and DAPI is its crescent-shaped molecular structure.[21] To improve the preference of cyanine dyes for double-stranded DNA, Karlsson et al.[22,23] designed a series of crescent-shaped unsymmetrical cyanine dyes based on the BO (Benzothiazolium) and TO chromophores (Figure 10.5). Flow linear dichroism, circular dichroism, and electrophoresis unwinding measurements revealed that these dyes prefer minor groove binding in mixed sequence DNA. Like the intercalating dyes, the minor groove binders exhibit large increase in fluorescence upon binding (Table 10.1).

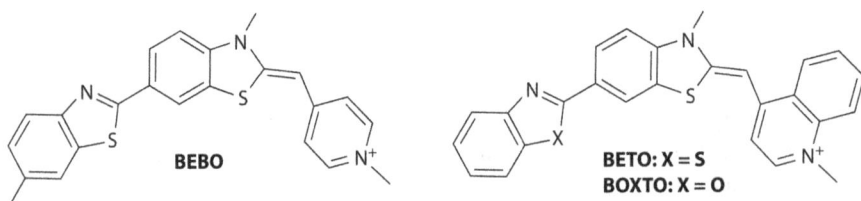

BEBO

BETO: X = S
BOXTO: X = O

FIGURE 10.5 Minor groove-binding unsymmetrical cyanine dyes.

TABLE 10.1

Fluorescence and Absorbance Properties of BETO and BOXTO Compared with Other Minor Groove Binders and with TO

	$\lambda_{abs.max}$ (nm)	$\lambda_{ems.max}$ (nm)	ϕ_F	ϕ_{bound}/ϕ_{free}	F_{bound}/F_{free}	Reference
Free BETO	487	590	0.015			[22]
BETO-ctDNA[b]	516	561	0.21	14	130	[22]
Free BOXTO	482	588	0.011			[22]
BOXTO-ctDNA[b]	515	552	0.52	50	260	[22]
Free BEBO	448	542	0.011			[23]
BEBO-ctDNA	467	492	0.18	16	245	[23]
Free TO	501	–	0.0002			[13]
TO-ctDNA	508	525	0.11	550		[13]
DAPI-ctDNA	349	456	0.34	18		[24]
Hoechst-ctDNA	356	466	0.42	28		[24]

PROBING POLYMERASE CHAIN REACTION (PCR)

In 1992, Higuchi et al.[25] showed that it was possible to follow the amplification of DNA in real time by adding ethidium bromide to the reaction tube and monitor the increase in fluorescence when the dye binds to the accumulating DNA product. A variety of dyes and probes have been developed for the detection of DNA in homogeneous solution since then. To be useful, the dyes should bind DNA with little sequence selectivity in one dominant-binding mode. They should have strong preference for double-stranded DNA and exhibit large fluorescence enhancement upon binding. Binding should be strong enough to give a signal proportional to the amount of DNA present, but without interfering with the amplification reaction. For the same reason, dissociation kinetics should be fast. Although YO-PRO was used in some early assays,[26] the most common dye is SYBR Green I from Molecular Probes.[27] Its chemical structure has not been published in scientific reports, but its spectral characteristics suggest that it is based on the TO-chromophore, probably with a cyclic substituent (U.S. patents 5,436,134 and 5,658,751). Bengtsson et al.[28] tested the minor groove binder BEBO as reporter in real-time polymerase chain reaction (PCR) (Figure 10.6) and found it to compare well with SYBR Green I in all important aspects.

Dyes are excellent reporters in real-time PCR, but they do not distinguish between products. TaqMan,[29] Molecular Beacons,[30] Scorpion Primers,[31] and LightCycler hybridization probes are examples of sequence-specific probes that are based on fluorescence resonance energy transfer (FRET) between two dyes tethered to an oligonucleotide. When these probes bind target DNA, the distance between the dyes changes as a result of either probe degradation or conformational change resulting in fluorescence enhancement. Common dyes in these probes are laser dyes with very high fluorescence quantum yields. An interesting alternative to the two dye probes is the LightUp probe.[32] The LightUp probe has only one dye,

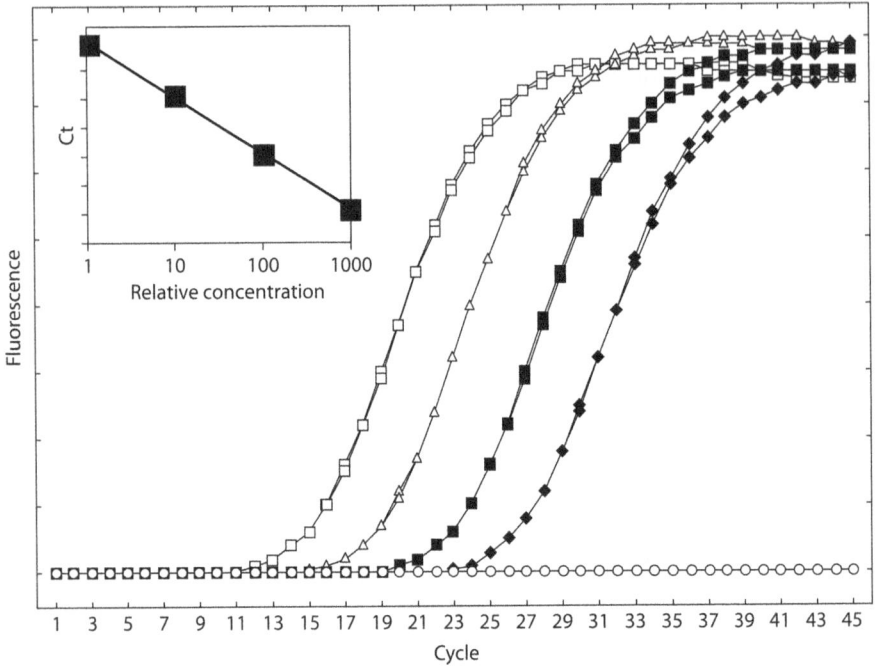

FIGURE 10.6 Real-time PCR amplification with BEBO (0.2 μM) detection. Amplification was performed in a RotorGene with 470 nm excitation and 510 nm detection. Background is subtracted and the amplification curves are normalized. Samples are duplicates of a 1:1, 1:10, 1:100, 1:1000 dilution series of a 222 bp beta tubulin fragment and NTC. Inset: standard curve ($R = 0.99955$, slope −3.8, efficiency 0.83).

which is a cyanine dye, whose fluorescence greatly increases when bound to DNA. Instead of using an oligonucleotide to identify the target sequence, the LightUp probe uses peptide nucleic acid (PNA), for which charged cyanine dyes have negligible affinity. The LightUp probes have been found to be excellent reporters in both real-time PCR and post-PCR applications.[33,34]

CONCLUSION

Fluorescent probes have replaced radioisotopes as labels of nucleic acids. They have important advantages in being nonhazardous and stable upon storage. Moreover, some dyes have different fluorescence in free and bound state, which opens for probing in homogeneous solution, eliminating the need to separate bound and free probes. These dyes were accidentally discovered. Based on detailed structural and thermodynamic investigations of the dyes and their interactions with nucleic acids, we are rapidly learning what governs their properties. This will make it possible to improve the dyes in terms of solubility and sequence nonspecificity, and also to design dyes of different colors. Particularly exciting is the use of the dyes as labels in sequence-specific probes because enhancement in fluorescence can be much more sensitively measured than a change in fluorescence as a result of energy transfer. Also, multiplexing is expected to be much simpler with single-labeled probes.

REFERENCES

1. Mishra, A.; Behera, R.K.; Behera, P.K.; Mishra, B.K.; Behera, G.B. Cyanines during the 1990s: a review. Chem. Rev. **2000**, *100* (6), 1973–2011.
2. Obrien, D.F.; Kelly, T.M.; Costa, L.F. Excited-state properties of some carbocyanine dyes and energy-transfer mechanism of spectral sensitization. Photogr. Sci. Eng. **1974**, *18* (1), 76–84.
3. Lee, L.G.; Chen, C.H.; Chiu, L.A. Thiazole orange—a new dye for reticulocyte analysis. Cytometry **1986**, *7* (6), 508–517.
4. Rye, H.S.; Quesada, M.A.; Peck, K.; Mathies, R.A.; Glazer, A.N. High-sensitivity 2-color detection of double-stranded DNA with a confocal fluorescence gel scanner using ethidium homodimer and thiazole orange. Nucleic Acids Res. **1991**, *19* (2), 327–333.
5. Rye, H.S.; Yue, S.; Wemmer, D.E.; Quesada, M.A.; Haugland, R.P.; Mathies, R.A.; Glazer, A.N. Stable fluorescent complexes of double-stranded DNA with *bis*-intercalating asymmetric cyanine dyes: properties and applications. Nucleic Acids Res. **1992**, *11*, 2803–2812.
6. Haugland, R.P. Nucleic acid detection. In *Handbook of Fluorescent Probes and Research Chemicals*, 9th Ed.; Haugland, R.P., Ed.; Molecular Probes: Eugene, OR, **2002**; 269–286.
7. Khairutdinov, R.F.; Serpone, N. Photophysics of cyanine dyes: subnanosecond relaxation dynamics in monomers, dimers, and H- and J-aggregates in solution. J. Phys. Chem. B **1997**, *101* (14), 2602–2610.
8. Serpone, N.; Sahyun, M.R.V. Photophysics of dithiacarbocyanine dyes—subnanosecond relaxation dynamics of a dithia-2,2′-carbocyanine dye and its 9-methyl-substituted meso analog. J. Phys. Chem. **1994**, *98* (3), 734–737.
9. Larsson, A.; Carlsson, C.; Jonsson, M.; Albinsson, B. Characterization of the binding of the fluorescent dyes YO and YOYO to DNA by polarized light spectroscopy. J. Am. Chem. Soc. **1994**, *116*, 8459–8465.
10. Jacobsen, J.P.; Pedersen, J.B.; Hansen, L.F.; Wemmer, D.E. Site selective *bis*-intercalation of a homodimeric thiazole orange dye in DNA oligonucleotides. Nucleic Acids Res. **1995**, *23*, 753–760.
11. Petty, J.T.; Bordelon, J.A.; Robertson, M.E. Thermodynamic characterization of the association of cyanine dyes with DNA. J. Phys. Chem. B **2000**, *104*, 7221–7227.
12. Netzel, T.L.; Nafisi, K.; Zhao, M.; Lenhard, J.R.; Johnson, I. Base-content dependence of emission enhancements, quantum yields, and lifetimes for cyanine dyes bound to double stranded DNA: photophysical properties of monomeric and bichromomphoric DNA stains. J. Phys. Chem. **1995**, *99*, 17936–17947.
13. Nygren, J.; Svanvik, N.; Kubista, M. The interactions between the fluorescent dye thiazole orange and DNA. Biopolymers **1998**, *46*, 39–51.
14. Schins, J.M.; Agronskaia, A.; deGrooth, B.G.; Greve, J. Orientation of the chromophore dipoles in the TOTO–DNA system. Cytometry **1999**, *37* (3), 230–237.
15. Milanovich, N.; Suh, M.; Jankowiak, R.; Small, G.J.; Hayes, J.M. Binding of TO-PRO-3 and TOTO-3 to DNA: fluorescence and hole-burning studies. J. Phys. Chem. **1996**, *100* (21), 9181–9186.
16. Bordelon, J.A.; Feierabend, K.J.; Siddiqui, S.A.; Wright, L.L.; Petty, J.T. Viscometry and atomic force microscopy studies of the interactions of a dimeric cyanine dye with DNA. J. Phys. Chem. B **2002**, *106* (18), 4838–4843.

17. Rye, H.S.; Glazer, A.N. Interaction of dimeric intercalating dyes with single-stranded DNA. Nucleic Acids Res. **1995**, *23* (7), 1215–1222.

18. Seifert, J.L.; Connor, R.E.; Kushon, S.A.; Wang, M.; Armitage, B.A. Spontaneous assembly of helical cyanine dye aggregates on DNA nanotemplates. J. Am. Chem. Soc. **1999**, *121* (13), 2987–2995.

19. Mikheikin, A.L.; Zhuze, A.L.; Zasedatelev, A.S. Binding of symmetrical cyanine dyes into the DNA minor groove. J. Biomol. Struct. Dyn. **2000**, *18* (1), 59–72.

20. Sovenyhazy, K.M.; Bordelon, J.A.; Petty, J.T. Spectroscopic studies of the multiple binding modes of a trimethine-bridged cyanine dye with DNA. Nucleic Acids Res. **2003**, *31* (10), 2561–2569.

21. Reddy, B.S.P.; Sondhi, S.M.; Lown, J.W. Synthetic DNA minor groove-binding drugs. Pharmacol. Ther. **1999**, *84* (1), 1–111.

22. Karlsson, H.J.; Eriksson, M.; Perzon, E.; Lincoln, P.; Åkerman, B.; Westman, G. Groove-binding unsymmetrical cyanine dyes for staining of DNA: syntheses and characterization of the DNA-binding. Nucleic Acids Res. **2003**, *31* (21), 6227–6234.

23. Karlsson, H.J.; Lincoln, P.; Westman, G. Synthesis and DNA binding studies of a new asymmetric cyanine dye binding in the minor groove of [poly(dA–dT)]₂. Bioorg. Med. Chem. **2003**, *11* (6), 1035–1040.

24. Cosa, G.; Focsaneanu, K.S.; McLean, J.R.N.; McNamee, J.P.; Scaiano, J.C. Photophysical properties of fluorescent DNA-dyes bound to single- and double-stranded DNA in aqueous buffered solution. Photochem. Photobiol. **2001**, *73* (6), 585–599.

25. Higuchi, R.; Dollinger, G.; Walsh, P.S.; Griffith, R. Simultaneous amplification and detection of specific DNA-sequences. Bio-Technol. **1992**, *10* (4), 413–417.

26. Tseng, S.Y.; Macool, D.; Elliott, V.; Tice, G.; Jackson, R.; Barbour, M.; Amorese, D. An homogeneous fluorescence polymerase chain reaction assay to identify *Salmonella*. Anal. Biochem. **1997**, *245* (2), 207–212.

27. Wittwer, C.T.; Herrmann, M.G.; Moss, A.A.; Rasmussen, R.P. Continuous fluorescence monitoring of rapid cycle DNA amplification. BioTechniques **1997**, *22* (1), 130–131, 134–138.

28. Bengtsson, M.; Karlsson, H.J.; Westman, G.; Kubista, M. A new minor groove binding asymmetric cyanine reporter dye for real-time PCR. Nucleic Acids Res. **2003**, *31* (8), e45.

29. Lee, L.G.; Connell, C.R.; Bloch, W. Allelic discrimination by nick-translation PCR with fluorogenic probes. Nucleic Acids Res. **1993**, *21* (16), 3761–3766.

30. Tyagi, S.; Kramer, F.R. Molecular beacons: probes that fluoresce upon hybridization. Nat. Biotechnol. **1996**, *14* (3), 303–308.

31. Whitcombe, D.; Theaker, J.; Guy, S.P.; Brown, T.; Little, S. Detection of PCR products using self-probing amplicons and fluorescence. Nat. Biotechnol. **1999**, *17* (8), 804–807.

32. Svanvik, N.; Westman, G.; Wang, D.; Kubista, M. Light-up probes: thiazole orange-conjugated peptide nucleic acid for detection of target nucleic acid in homogeneous solution. Anal. Biochem. **2000**, *281*, 26–35.

33. Svanvik, N.; Stahlberg, A.; Sehlstedt, U.; Sjoback, R.; Kubista, M. Detection of PCR products in real time using light-up probes. Anal. Biochem. **2000**, *287* (1), 179–182.

34. Isacsson, J.; Cao, H.; Ohlsson, L.; Nordgren, S.; Svanvik, N.; Westman, G.; Kubista, M.; Sjoback, R.; Sehlstedt, U. Rapid and specific detection of PCR products using light-up probes. Mol. Cell. Probes **2000**, *14* (5), 321–328.

11 Fluorescence Resonance Energy Transfer (FRET)

Dino A. De Angelis
Memorial Sloan-Kettering Cancer Center, New York,
New York, U.S.A.

CONTENTS

INTRODUCTION

The phenomenon of fluorescence resonance energy transfer (FRET) was first described by Theodor Förster in 1948. Since then, FRET has become an incredibly useful tool in biology for three reasons: (1) FRET is really sensitive in the range of 100 Å and below, the scale at which transactions between biological macromolecules and complexes occur; (2) the instrumentation is extremely sensitive and is readily amenable to miniaturization, high throughput, and automation; and (3) cell-permeable and genetically encodable FRET probes enable the real-time quantitation of dynamic cellular processes in live cells. In this chapter, we introduce the reader to the photophysics underlying FRET and present a brief general overview of the types of biological measurements enabled by the technique.

FLUORESCENCE

When photons are absorbed by a fluorescent compound (Figure 11.1A), within femtoseconds it undergoes a transition from a singlet ground state (S_0) to an excited state which can be S_1 (thick line) or a higher vibrational energy level of S_1 (denoted by the thin lines above S_1). The excitation maximum of a particular fluorescent compound is the wavelength at which this process occurs most efficiently. Within picoseconds, molecules in higher vibrational states relax to the S_1 state (thick line).

Most fluorescent molecules remain in the S_1 state for 1–10 ns on average, after which they can return to the S_0 ground state through one of a few different relaxation

DOI: 10.1201/9781003247432-11

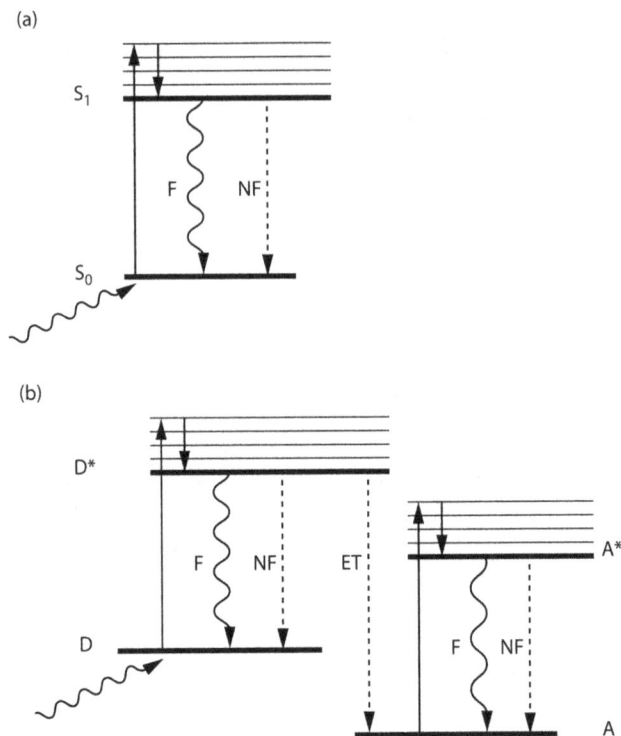

FIGURE 11.1 (a) General mechanisms of excited-state deactivation of a fluorophore. Up arrow represents transition from ground state (S_0) to an S_1 excited state (thick line) or higher (thin lines) after photon absorption (wavy upward arrow). Down arrows represent relaxation, first from higher states down to S_1 (short arrow). Relaxation from S_1 to S_0 involves either fluorescence emission (F, wavy down arrow) or nonfluorescent mechanisms (NF, dashed down arrow). (b) Mechanisms of excited-state deactivation of a donor fluorophore in the presence of an acceptor. Same as A except that energy transfer (ET) is also displayed (longer dashed arrow) which results in less fluorescence emission from the donor excited state (D*) and activation of acceptor molecule from ground state (A) to excited state (A*). Excited state of A can also be depopulated via fluorescent (F) or nonfluorescent (NF) mechanisms.

pathways.[1] The radiative relaxation pathway occurs through the emission of fluorescent light by the excited fluorophore as it returns to the ground state, indicated by the wavy arrow in Figure 11.1A. The light emitted is always of lower energy (longer wavelength) than that of the exciting light.

Several nonradiative pathways compete with fluorescence emission for return to the ground state. These nonfluorescent pathways (labeled NF in Figure 11.1A) include internal conversion (a return to S_0 accompanied by release of heat), intersystem crossing (change in the electron-spin orientation resulting in relaxation to a long-lived triplet excited state before decaying to S_0), and collisional deactivation (involving direct encounters between the dye and neighboring solvent molecules). These deactivation pathways are inherent to fluorescent dyes in solution and are present to various extents.

RESONANCE ENERGY TRANSFER

Resonance energy transfer is also a nonradiative pathway, but it requires a stringent set of conditions, unlike the NF deactivation pathways mentioned above. In resonance energy transfer, the excited dye is a FRET donor (D) which becomes deactivated via transfer of some of its excited-state energy to a compatible acceptor (A) molecule (Figure 11.1B). Because it does not involve the emission of photons from the donor, FRET is more correctly referred to as "Förster resonance energy transfer," or simply RET.[2] Efficient energy transfer from excited D to A requires that (1) the fluorescence emission spectrum of D and the absorption spectrum of A be overlapping (resonant frequencies as shown in Figure 11.2A); (2) the distance between D and A be within 10–100 Å (1–10 nm); (3) the dipole moments of D and A be approximately parallel to one another. If A is also a fluorescent molecule, FRET results in an increase in fluorescence emission from A (Figure 11.1B, indicated by the wavy arrow depopulating A*) in addition to the decrease in fluorescence emission from D. If A is an NF molecule, the net result of FRET is a decrease in fluorescence emission from D. In this case, A is a fluorescence quencher and A* becomes exclusively depopulated by NF pathways (Figure 11.1B, dashed arrow).

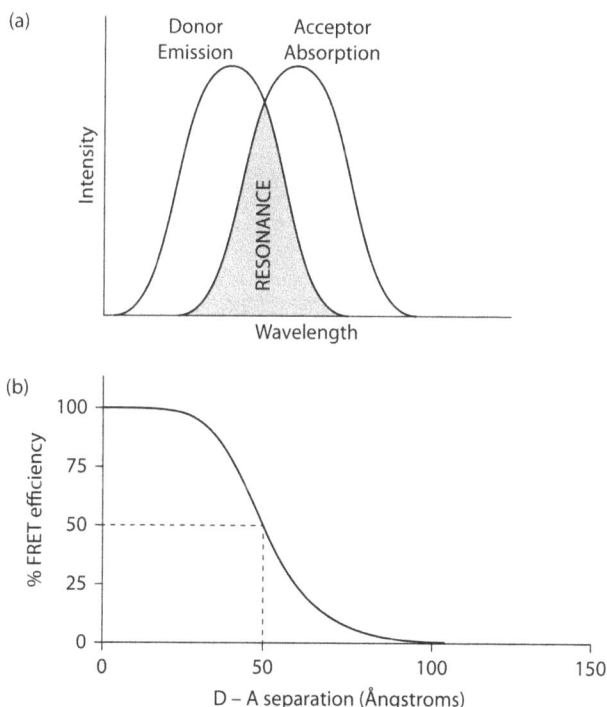

FIGURE 11.2 (a) Fluorescence intensity plotted as a function of donor emission and acceptor absorption wavelengths, displaying a spectral overlap required for resonance energy transfer. (b) FRET efficiency plotted as a function of the distance between a donor and acceptor pair characterized by a R_0 of 50 Å.

DISTANCE DEPENDENCE OF FRET

FRET efficiency is inversely proportional to the sixth power of the distance between a donor and acceptor, and therefore is extremely sensitive to their intermolecular separation. This is illustrated in the equation:

$$E(r) = (R_0)^6 / \left[(R_0)^6 + r^6 \right]$$

where E is FRET efficiency, r is the distance separating D from A, and R_0 is the Förster distance. At the Förster distance, 50% of excited D is deactivated via FRET (ET in Figure 11.1B), and the remainder is deactivated via fluorescence (F) and other NF pathways. The Förster distance varies between different D–A pairs, typically ranging from 25 to 75 Å (Table 11.1). R_0 takes into consideration the amount of spectral overlap between the dyes and assumes that donors and acceptors are randomly oriented. For a donor–acceptor pair with an R_0 of 50 Å, doubling the intermolecular separation from 50 to 100 Å results in a 32-fold decrease in FRET efficiency (Figure 11.2B).

Distance measurements obtained by monitoring FRET efficiency typically represent an average (rather than absolute) distance between D and A, because of the assumption of random orientation in the calculation of R_0, and the fact that biological macromolecules usually fluctuate between thermodynamically equivalent states. Technicalities aside, the exquisite dependence of FRET on distances of the scale of biological macromolecules (Table 11.2) underscores the real usefulness of the technique: detecting biological phenomena involving changes in the proximity of two molecules (or two portions of the same molecule) respectively labeled with D and A.

TABLE 11.1
Förster Distances of Commonly Used FRET Pairs

Donor	Acceptor	R^0 (Å)
Fluorescein	Tetramethylrhodamine	55
IAEDANS	Fluorescein	46
EDANS	Dabcyl	33
Alexa Fluor 488	Alexa Fluor 546	64
Alexa Fluor 488	Alexa Fluor 555	70
Alexa Fluor 488	Alexa Fluor 568	62
Alexa Fluor 488	Alexa Fluor 594	60
Alexa Fluor 488	Alexa Fluor 647	56
Fluorescein	QSY 7 and QSY 9 dyes	61

Source: Haughland, R.P. *Handbook of Fluorescent Probes and Research Products*, Ninth Edition, 2002, Molecular Probes Inc.

TABLE 11.2
Dimensions of Different Biological Macromolecules in Ångstroms

Molecular Feature	Distance in Ångstroms
Carbon–carbon single bond	1.5
1 turn of alpha helix (3.6 amino acid residues)	5.4
Diameter of DNA double helix	22
1 turn of DNA double helix (10 bp)	34
Diameter of a 26-kDa globular protein	40
10 turns of alpha helix (36 amino acid residues)	54
F-Actin filament diameter	60
Eukaryotic plasma membrane thickness	75
Microtubule diameter	250

DETECTION OF FRET

A wide array of phenomena can be measured using FRET, and they tend to fall into either of two main categories: (1) homogeneous in vitro assays performed in a spectrofluorimeter or microplate reader and (2) live-cell FRET imaging performed by fluorescence microscopy. The basic requirements for the detection of FRET in these two domains are the same: (1) a light source for donor excitation; (2) wavelength filters to spectrally separate excitation photons (originating from the light source) from emission photons originating from both D and A; and (3) detectors that can separately register emission photons from D and A as two electrical signals (spectrofluorimeter) or as two digital images (fluorescence microscopy).

IN VITRO: HOMOGENEOUS FRET

Spectrofluorimeters measure the fluorescence properties of homogeneous samples in cuvettes or in multiwell dishes. For a FRET experiment, the light source illuminates at a wavelength intended to excite the donor but not the acceptor. The detectors register photons emitted by both D and A. When D and A are much further apart than the R_0 characterizing the pair, emission from D is much higher than emission from A. The D/A emission ratio is therefore high (excited D deactivates preferentially with the fluorescence emission pathway). In contrast, when D and A are within Förster distance of one another, the D/A emission ratio decreases as D transfers a fraction of its excitation state energy to A, causing less fluorescence emission from D and more fluorescence emission from A. In cases when A is a fluorescence quencher, proximity between D and A simply results in decreased emission from D.

Binding reactions between two molecular entities (receptor–ligand; antibody–antigen; transcription factor–DNA sequence) respectively labeled with a donor and an acceptor can be quantified by FRET.[3] The labeling can be direct (via covalent modification of the molecules of interest) or indirect (i.e., via noncovalent binding of a labeled antibody or other binding moieties). Binding will result in energy transfer from the donor to the acceptor (Figure 11.3A).

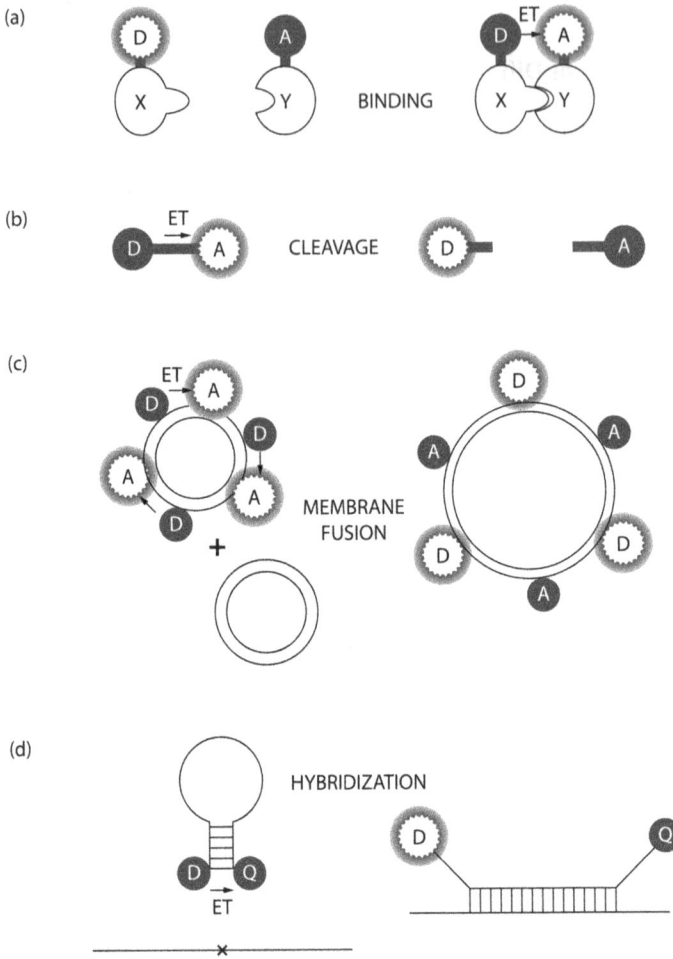

FIGURE 11.3 Types of molecular interactions measured by FRET. (a) Binding of two molecular entities X and Y respectively labeled with donor D and acceptor A results in FRET (ET). (b) Cleavage of D- and A-labeled peptide or nucleic acid sequence via protease or endonuclease results in a decrease in the FRET signal. (c) Fusion of liposomes containing a high concentration of D- and A-labeled phospholipids with unlabeled liposomes results in dilution of D and A, and decrease of FRET in the newly fused membrane. (d) Molecular beacons oligonucleotide probes adopt a stem and loop structure in the absence of a perfectly matching target DNA sequence, resulting in FRET between donor D and nonfluorescent acceptor Q (quencher). Perfect match results in hybridization of loop sequence to the target DNA and increased separation of D from Q.

Peptides containing a specific protease cleavage site, and labeled with D and A at either termini, can serve as FRET sensors.[4] FRET occurs in the intact peptide but incubation with the protease separates D from A causing a decrease in the FRET signal (Figure 11.3B). The activity of endonucleases can similarly be detected using a dual-labeled nucleic acid sequence.

Membrane fusion assays can also be performed using FRET, by incorporating phospholipids labeled with D and A moieties into liposomes at relatively high concentrations.[5] Because of molecular proximity, D and A undergo FRET in this liposome population. Fusion of such liposomes with unlabeled liposomes creates larger liposomes in which FRET is decreased due to the increase in average distance between labeled probes via dilution (Figure 11.3C).

Molecular beacons are FRET probes that can adopt one of two very different conformations depending on the presence of specific target DNA sequences in a sample (see Refs. [6] and [7]). The technique is so sensitive that it can resolve single-base pair mismatches. In the absence of a perfectly matching sequence, the molecular beacon adopts a hairpin-like structure (Figure 11.3D) which brings the donor- and acceptor-labeled ends of the molecule into close proximity. The acceptor is a quencher (Q) in these probes, and therefore very little fluorescence is emitted by the probes in this conformation. In the presence of the exact match, the stem structure opens up, allowing the loop to hybridize to the perfectly matched target and increasing the distance between D and Q.

These are just a few illustrations of diagnostic in vitro assays performed using FRET probes. They are typically easy to set up and can be automated; some are known as "walkaway" assays because no further processing of the sample is required following the initial incubation of reagents. The extent of the reaction (whether binding, cleavage, membrane fusion, or other) can be modulated by adding competitors, agonists, antagonists, or other test compounds, either in purified form or from crude extracts, and their effect on the assay can be measured. The kinetics of any assay can be monitored by collecting the emission from D and A continuously or at timed intervals, without interrupting the reaction. In addition, because D–A pairs can be spectrally quite different from one another, several probes can be monitored simultaneously in the same reaction (multiplexing; see Ref. [7]).

In homogeneous FRET, background noise derived from sample autofluorescence can limit the sensitivity of the technique. Two related techniques extend the sensitivity range of FRET measurements by radically reducing noise in the system. Time-resolved FRET (TR-FRET) uses fluorescence donors with unusually long fluorescence lifetimes (400 ns to 2 ms for chelates of ruthenium and lanthanides compared to 10 ns for typical fluorescent molecules).[8] Time-resolved FRET delays the collection of light by the detector, allowing any fluorescence that is not derived from the long lifetime donors in the sample—including noise—to decay to zero. The slower decaying donor-specific fluorescence (or FRET signal) can therefore be measured in the absence of background.

Another way to reduce background fluorescence in FRET measurements is by omitting the light source altogether. Bioluminescence RET (BRET) uses a luciferase as a donor which produces light only when the substrate (coelenterazine) is added to the reaction. When bioluminescence is produced in close proximity to the acceptor, energy (not photons) is transferred resulting in an increase in the acceptor fluorescence. As there is no extrinsic illumination to activate sample autofluorescence, BRET can increase the sensitivity of donor–acceptor interactions by an order of magnitude compared to FRET.[9]

IN VIVO: LIVE-CELL FRET IMAGING

FRET measurements are not limited to in vitro assays. Cell-permeable FRET probes as well as the green, cyan, and yellow fluorescent proteins (GFP, CFP, and YFP), introduced in cells by DNA transfection, have permitted the study of dynamic events involving changes in the proximity of macromolecules in live cells. Sensors based on GFP variants can be targeted to any compartment for which there is a known signal sequence, enabling local measurement of the activity under study. In FRET imaging, the light required to excite the donor is focused by a microscope lens onto a live-cell sample. Photons emitted from the donor and the acceptors are respectively collected as two digital images (Figure 11.4). The intensity levels of the two images are then ratioed on a pixel-per-pixel basis, yielding a two-dimensional map of the D/A emission ratio for the particular time point. The sample can be imaged over time and thus the spatio-temporal features of the FRET signal can be analyzed. By enabling the analysis of phenomena in the range of Förster distances (1–10 nm), FRET imaging goes beyond the limits resolution of light microscopy (~200 nm).

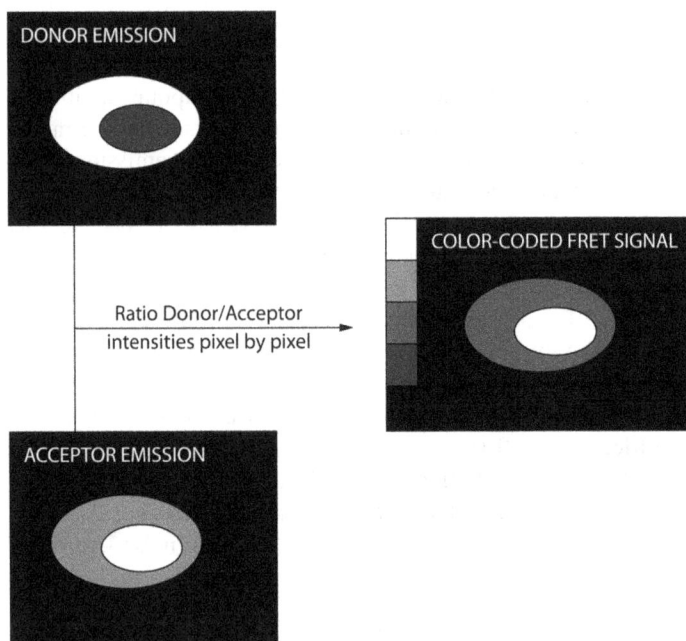

FIGURE 11.4 Fluorescence resonance energy transfer imaging of live cells. A pair of digital images represents the collection of light from the donor emission (top left image) and acceptor emission (bottom left image) following illumination of a sample at the donor excitation wavelength. The intensity of pixels from the donor emission image is divided by the intensity of pixels from the acceptor emission image via software and a color-coded map of the ratios is generated (right). In this diagram, there is a high FRET signal in the nucleus (low donor intensity/high acceptor intensity).

The binding of two different proteins respectively labeled with CFP and YFP (the pair with the best spectral overlap characteristics for FRET) can be measured kinetically by measuring the amount of FRET they undergo. Protease activity can likewise be monitored in live cells by placing YFP and CFP at each extremity of a cleavable peptide. A powerful class of sensors inserts a conformationally active peptide between a CFP–YFP pair. The peptide can adopt different conformations depending on the environment, thus changing the distance and angle between the fluorescent proteins and yielding differential FRET signals. The detection of fluctuations in intracellular Ca^{2+} concentration was the first demonstrated application of this class of sensors, and the list has grown since (see Ref. [10] for a review).

For microscopy measurements, high levels of cellular autofluorescence can be compensated for by using an unlabeled region of the cell as an internal reference; this background is subtracted from the signal in the region of interest prior to calculating FRET. A more serious problem is photobleaching of the donor—basically, a destruction of the fluorophore caused by overillumination which is common in microscopy because the excitation light scans the same sample area repeatedly. Under photobleaching conditions, the efficiency of FRET undergoes an apparent decrease over time as the donor fluorophore is unable to repopulate its excited state. This is donor dependent, but can be corrected for by establishing the rate of photobleaching in a control sample, and normalizing the FRET signal obtained in the test sample to this baseline.[10] BRET measurements can also be performed in live cells; these experiments do not suffer from photobleaching because all of the light is generated via a bioluminescent reaction.[9]

CONCLUSION

This brief survey has only scratched the surface with respect to the breadth and power of FRET for the analysis of binding, dissociation, hybridization, and other phenomena involving changes in the proximity of biological molecules. Development of FRET-based sensors continues unabated, and so is the development of instrumentation designed to quantify FRET—with a trend toward miniaturization and high-throughput operation to enable diagnostic use of the technology on a genomic scale.

REFERENCES

1. Lakowicz, J. Energy transfer. In *Principles of Fluorescence Spectroscopy*, 2nd Ed.; Lakowicz, J.R., Ed.; Kluwer Academic/Plenum Publishers: New York, NY, **1999**; 367–394.
2. Förster, T. Intermolecular energy migration and fluorescence. Ann. Phys. **1948**, *2*, 55–75.
3. Schmid, J.A.; Birbach, A.; Hofer-Warbinek, R.; Pengg, M.; Burner, U.; Furtmuller, P.G.; Binder, B.R.; de Martin, R. Dynamics of NF kappa B and Ikappa Balpha studied with green fluorescent protein (GFP) fusion proteins. Investigation of GFP-p65 binding to DNA by fluorescence resonance energy transfer. J. Biol. Chem. **2000**, *275* (22), 17035–17042.
4. Gulnik, S.V.; Suvorov, L.I.; Majer, P.; Collins, J.; Kane, B.P.; Johnson, D.G.; Erickson, J.W. Design of sensitive fluorogenic substrates for human cathepsin D. FEBS Lett. **1997**, *413*, 379–384.

5. Malinin, V.S.; Haque, M.E.; Lentz, B.R. The rate of lipid transfer during fusion depends on the structure of fluorescent lipid probes: A new chain-labeled lipid transfer probe pair. Biochemistry **2001**, *40* (28), 8292–8299.

6. De Angelis, D.A. Why FRET over genomics? Physiol. Genomics **1999**, *1* (2), 93–99.

7. Tyagi, S.; Bratu, D.P.; Kramer, F.R. Multicolor molecular beacons for allele discrimination. Nat. Biotechnol. **1998**, *16* (1), 49–53.

8. Bazin, H.; Trinquet, E.; Mathis, G.J. Time resolved amplification of cryptate emission: A versatile technology to trace biomolecular interactions. Biotechnology **2002**, *82* (3), 233–250.

9. Boute, N.; Jockers, R.; Issad, T. The use of resonance energy transfer in high-throughput screening: BRET versus FRET. Trends Pharmacol. Sci. **2002**, *23* (8), 351–354.

10. Miyawaki, A.; Tsien, R.Y. Monitoring protein conformations and interactions by fluorescence resonance energy transfer between mutants of green fluorescent protein. Methods Enzymol. **2000**, *327*, 472–500.

12 LightUp® Probes

Mikael Leijon
LightUp Technologies AB, Huddinge, Sweden

CONTENTS

INTRODUCTION

Real-time monitoring of the polymerase chain reaction (PCR) product by fluorescent probes has enabled the homogeneous detection of pathogen deoxyribonucleic acid (DNA) or ribonucleic acid (RNA) in a closed-tube format. The reduced risk of carry-over contamination and the superior sensitivity of PCR-based assays have made real-time PCR the method of choice in many clinical settings. The LightUp® Probe is a novel fluorescent probe where the fluorescent reporter dye is coupled to an oligomer of the uncharged DNA analog peptide nucleic acid (PNA). PNA provides strong, but still specific, binding and serves as the sequence-recognizing element of the probe. The fluorophore is an asymmetric cyanine dye that fluoresces when the probe binds its target DNA, but it is virtually nonfluorescent when the probe is free in solution.

RESULTS

The extraordinary sensitivity of PCR-based assays is an important factor behind the recent tour de force of real-time PCR in molecular biology in general and viral diagnostics in particular.[1,2] Besides the inherent sensitivity of the PCR technique, the real-time fluorescence detection used in modern PCR instrumentation has the added advantage of providing accurate quantification of the initial amount of the amplified nucleic acid. This is achieved by comparing C_t values, i.e., the PCR cycle where the fluorescence signal reaches a certain level, for unknown clinical samples with the C_t values of a dilution series of a reference material with known concentration. This is generally carried out via a standard curve:

$$C_t = \frac{\log(c)}{\log(1+E)^{-1}} + C_t(1)$$

where the parameter E is the PCR efficiency, ideally equal to unity, and $C_t(1)$ is the threshold cycle value where one copy of the standard would appear. Assuming that

DOI: 10.1201/9781003247432-12

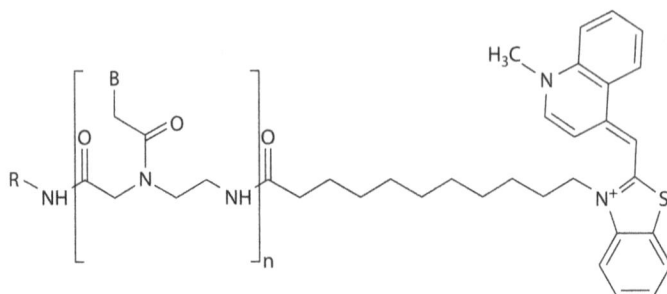

FIGURE 12.1 A schematic view of the LightUp Probe. The dye may be different.

the standard curve parameters also apply to the clinical samples, the concentration of the clinical samples can be determined from the measured C_t values.

DNA-based sequence recognizing fluorescent probes commonly utilize a fluorophore/quencher pair that is, in some manner, separated in the course of the amplification reaction. TaqMan® probes, for example, are degraded by the 5'-nuclease activity of the DNA polymerase, which releases the fluorophore from its sequence-recognizing element, and thereby separates it from the quencher, so that a light signal is emitted upon excitation.[3] Molecular beacons achieve the same goal by a rearrangement of the secondary structure upon hybridization, which causes a separation of the quencher and the fluorophore.[4] Fluorescence resonance energy transfer (FRET) probes work slightly differently. Instead of the acceptor dye functioning as a quencher, it emits the detectable fluorescence when the two probes bind adjacent to each other on the target DNA molecule.[5]

The LightUp® Probe[6] (Figure 12.1) represents a significant departure from this paradigm, because the DNA has been entirely replaced by the homomorphous but uncharged and achiral DNA analog PNA (Figure 12.2). PNA is composed of N-(2-aminoethyl)glycine subunits[7] and forms very stable hybrid complexes with both DNA and RNA with a sequence specificity that is higher than that DNA has for itself.[8] An asymmetric cyanine dye, such as a thiazole orange derivative or 4-[(3-methyl-6-(benzothiazol-2-yl)-2,3-dihydro-(benzo-1,3-thiazole)-2-methylidene)]-1-methyl-pyridinium iodide (BEBO),[9] is generally used as the fluorescent marker (Figure 12.1).

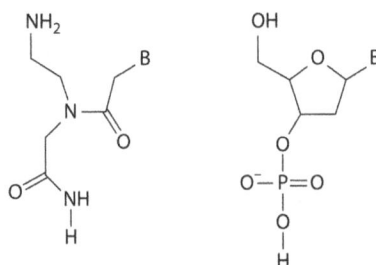

FIGURE 12.2 The homomorphous DNA and PNA subunits.

This type of dye releases excitation energy by internal motion when it is free in solution and is, under these conditions, virtually nonfluorescent. When bound to DNA, however, the dye becomes brightly fluorescent.[10–13] Because the dye is positively charged, a significantly higher background signal would result from internal binding of the dye if DNA is used instead of PNA.

The LightUp Probes have several advantages compared to probes with sequence recognizing elements consisting of DNA. Because it binds during the annealing phase, an optimal three-step PCR cycle can be used. The primer design is also very flexible because the amplicon does not need to be kept short, apart from what is compatible with the maximum sensitivity of the PCR system. A typical amplicon length is generally in the range of 100–500 bp. Because PNA is not degraded by proteases or nucleases, kits utilizing LightUp Probe detection are very stable with typical storage conditions at 4–6°C for at least 1 year. Possibly, the different chemistry by which the dye is attached to the PNA as compared to DNA contributes to this favorable stability. Recently, it has become increasingly apparent that the long (typically 20–40 bases) DNA probes are not suitable in some applications where high specificity is desirable, for instance, in SNP (single nucleotide polymorphism) detections. Obviously, a one-base mismatch out of 40 bases makes a smaller difference than one base out of ten. For this reason, minor groove binding (MGB) probes were designed, where MGB ligands have been covalently attached to the DNA probes to increase stability upon hybridization and enable a shortening of the probe.[14] The uncharged LightUp Probe is intrinsically much more stable than a DNA probe because of the absence of electrostatic backbone–backbone repulsion. Furthermore, the positively charged dye (Figure 12.1) plays a similar role as the minor groove-binding moiety of the MGB probes. Hence, the LightUp Probes are typically 8–12 bases long, about one-third of the length of a typical DNA probe and one-sixth of the length of a FRET pair. Many viral pathogens have rapidly mutating genomes, making it very challenging to find conserved target regions for DNA-based probes/quencher systems. The LightUp technology provides a significant advantage in these applications.[13] Finally, the spectral region occupied by a LightUp Probe is just one-half that of a fluorophore/quencher pair or a FRET pair, because the LightUp Probe contains only a single dye. This should be advantageous in multiplex applications, but because several instrument platforms use a single wavelength excitation source (e.g., the ABI (Applied Biosystems Instruments); real-time PCR instruments and the Roche LightCycler), it is necessary to design asymmetric cyanine dyes with greatly varying Stokes' shift, which may prove difficult, to enable multiplex detection with LightUp Probes. However, the trend is that new real-time PCR instrumentation provides several excitation wavelengths (e.g., the Rotor-Gene™ from Corbett Research, the SmartCycler® from Cepheid Research, and the iCycler from Bio-Rad).

Figure 12.3A shows the real-time amplification curves, in duplicates, for a successive dilution of an Epstein–Barr virus (EBV) DNA solution, obtained by using the LightUp ReSSQ EBV Assay. The initial EBV template concentration ranges from 20 to 200,000 genome copies, as indicated. Prominent Hook effects are observed at all concentrations and the end signal is independent of the initial template amount. This shows that the probe is stable throughout the PCR run. With

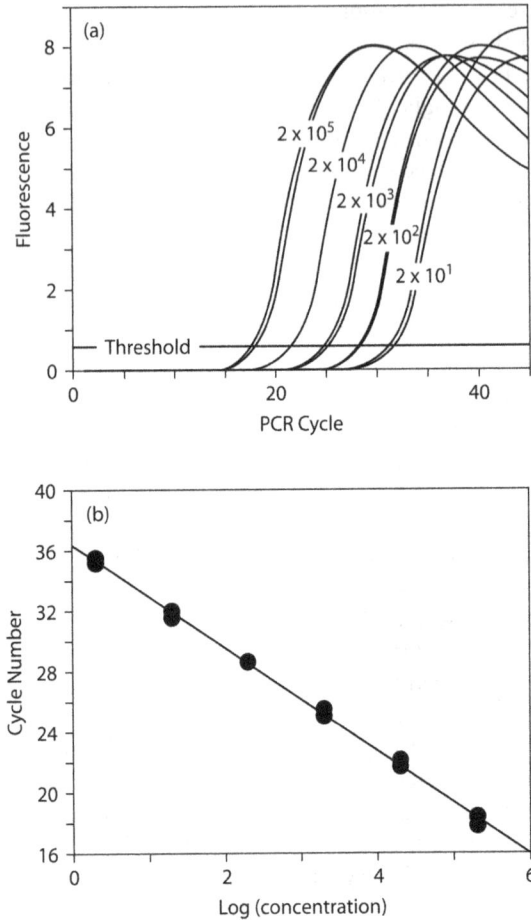

FIGURE 12.3 (a) The LightUp ReSSQ™ Epstein–Barr virus real-time PCR assay. (b) Graph relating number of RT-PCR cycles to the log (concentration) of the Epstein–Barr concentration.

DNA probes, the end signal commonly decreases for low template concentrations, which is likely to be an effect of probe degradation. By fitting a standard curve to the threshold cycle numbers in Figure 12.3A and the logarithm of the known concentrations (Figure 12.3B), it is found that the PCR efficiency is 96% and that the $C_t(1)$ value is 36.4 cycles. The viral load of an unknown clinical sample could easily be determined from its C_t value and this standard curve. As an example, a clinical evaluation study shows that quantification with the LightUp Cytomegalovirus (CMV) ReSSQ Assay is very similar to that of the Roche COBAS Amplicor CMV Monitor, but with higher sensitivity (to be published).

Applications of LightUp Probes also include bacterial DNA diagnostics. One example is Salmonella typing. It has been shown that the *Salmonella bongori* serotype could readily be distinguished from the *Salmonella entertidis* and *Salmonella anatum* serotypes from a single mismatch in the probe target region.[13] In another study, *Yersinia enterocolitica* was quantified over four log units.[15] LightUp Probes have also found application in cancer diagnostics. The relative expression of the immunoglobulin κ and λ genes is a sensitive marker for non-Hodgkin B-cell lymphomas. In a healthy individual, 60% of the B-cells produce κ chains, while the rest produce λ chains. A significant perturbation of this ratio is indicative of B-cell lymphoma. In a recently published study, 28 of 32 clinical samples were correctly diagnosed B-cell lymphoma by cDNA quantification using LightUp Probes.[16]

CONCLUSION

LightUp Probes provide a useful combination of strong affinity and high specificity. The resistance against degradation is an important advantage when storing and manufacturing diagnostic kits. Freeze–thaw cycles are often deleterious to long DNA, such as viral genomes and plasmids. This can be avoided by using LightUp Probe technology, because the probe does not need to be stored frozen. Furthermore, the probes can be made short: 8–12 residues are typical. This is a very useful property in viral diagnostics, where the mutation frequency often is high and long conserved target regions may be rare. This is particularly true for RNA viruses such as HIV, influenza, and severe acute respiratory syndrome (SARS). In fact, several probes could, in principle, be targeting the same amplicon-generating independent signals, thereby making the assay even more robust and less prone to false negatives. The greatest challenge for the future success of the LightUp Probe technology will be to develop and implement fluorescent dyes suitable for multiplex applications.

REFERENCES

1. Foy, C.A.; Parkes, H.C. Emerging homogeneous DNA-based technologies in the clinical laboratory. Clin. Chem. **2001**, *47* (6), 990–1000.
2. Mackay, I.M.; Arden, K.E.; Nitsche, A. Real-time PCR in virology. Nucleic Acids Res. **2002**, *30* (6), 1292–1305.
3. Livak, K.J.; Flood, S.J.; Marmaro, J.; Giusti, W.; Deetz, K. Oligonucleotides with fluorescent dyes at opposite ends provide a quenched probe system useful for detecting PCR product and nucleic acid hybridization. PCR Methods Appl. **1995**, *4* (6), 357–362.
4. Tyagi, S.; Kramer, F.R. Molecular beacons: Probes that fluoresce upon hybridization. Nat. Biotechnol. **1996**, *14* (3), 303–308.
5. Bernard, P.S.; Lay, M.J.; Wittwer, C.T. Integrated amplification and detection of the C677T point mutation in the methylenetetrahydrofolate reductase gene by fluorescence resonance energy transfer and probe melting curves. Anal. Biochem. **1998**, *255* (1), 101–107.
6. Svanvik, N.; Kubista, M. Probe for analysis of nucleic acids. World Patent 9745539, December 4 **1997**.
7. Nielsen, P.E.; Egholm, M.; Berg, R.H.; Buchardt, O. Sequence-selective recognition of DNA by strand displacement with a thymine-substituted polyamide. Science **1991**, *254* (5037), 1497–1500.

8. Egholm, M.; Buchardt, O.; Christensen, L.; Behrens, C.; Freier, S.M.; Driver, D.A.; Berg, R.H.; Kim, S.K.; Norden, B.; Nielsen, P.E. PNA hybridizes to complementary oligonucleotides obeying the Watson–Crick hydrogen-bonding rules. Nature **1993**, *365* (6446), 566–568.

9. Bengtsson, M.; Karlsson, H.J.; Westman, G.; Kubista, M. A new minor groove binding asymmetric cyanine reporter dye for real-time PCR. Nucleic Acids Res. **2003**, *31* (8), e45.

10. Nygren, J.; Svanvik, N.; Kubista, M. The interactions between the fluorescent dye thiazole orange and DNA. Biopolymers **1998**, *46* (1), 39–51.

11. Svanvik, N.; Westman, G.; Wang, D.; Kubista, M. Light-Up probes: Thiazole orange-conjugated peptide nucleic acid for detection of target nucleic acid in homogeneous solution. Anal. Biochem. **2000**, *281* (1), 26–35.

12. Svanvik, N.; Ståhlberg, A.; Sehlstedt, U.; Sjöback, R.; Kubista, M. Detection of PCR products in real time using light-up probes. Anal. Biochem. **2000**, *287* (1), 179–182.

13. Isacsson, J.; Cao, H.; Ohlsson, L.; Nordgren, S.; Svanvik, N.; Westman, G.; Kubista, M.; Sjöback, R.; Sehlstedt, U. Rapid and specific detection of PCR products using light-up probes. Mol. Cell. Probes **2000**, *14* (5), 321–328.

14. Afonina, I.; Zivarts, M.; Kutyavin, I.; Lukhtanov, E.; Gamper, H.; Meyer, R.B. Efficient priming of PCR with short oligonucleotides conjugated to a minor groove binder. Nucleic Acids Res. **1997**, *25* (13), 2657–2660.

15. Wolffs, P.; Knutsson, R.; Sjöback, R.; Rådström, P. PNA-based light-up probes for real-time detection of sequence-specific PCR products. Biotechniques **2001**, *31* (4), 766769–766771.

16. Ståhlberg, A.; Aman, P.; Ridell, B.; Mostad, P.; Kubista, M. Quantitative real-time PCR method for detection of B-lymphocyte monoclonality by comparison of kappa and lambda immunoglobulin light chain expression. Clin. Chem. **2003**, *49* (1), 51–59.

13 Invader® Assay

Stephen Day, Andrea Mast
Third Wave Technologies, Madison, Wisconsin, U.S.A.

CONTENTS

INTRODUCTION

The Invader® platform (Third Wave Technologies Inc., Madison, WI) is a homogeneous, isothermal, DNA probe-based system for highly sensitive, quantitative detection of specific nucleic acid sequences. It is an accurate and specific detection method for single-base changes, insertions, deletions, gene copy number, infectious agents, and gene expression. Invader reactions can be performed directly on genomic DNA- or on RNA-, polymerase chain reaction (PCR-), or RT-PCR-amplified products. A target-specific signal is amplified but not the target itself.

BACKGROUND

The Invader technology is based on the ability of a Cleavase® enzyme to recognize and cleave a specific nucleic acid structure generated by an overlap of two oligonucleotides (oligos)—the Invader oligo and the primary probe—on the nucleic acid target. Cleavase enzymes are a family of naturally occurring as well as engineered thermophilic structure-specific 5′ endonucleases.[1] Cleavase enzymes for use in DNA assays are derived from members of the flap endonuclease (FEN-1) family, typically found in thermophilic archaebacteria.[1] Invader RNA assays use Cleavase enzymes that are engineered from the 5′-exonuclease domain of DNA polymerase I of thermophilic eubacteria.[2] The optimal substrate for these nucleases is composed of distinct upstream, downstream, and template strands, which mimic the replication fork formed during displacement synthesis. The enzymes cleave at the 5′ end of the downstream strand between the first 2 base pairs and remove the single-stranded 5′ arm or flap.[1]

DOI: 10.1201/9781003247432-13

STRUCTURE FORMATION

The primary probe consists of two functionally different regions: a 5'-flap sequence and a 3' target-specific region (TSR). The sequence of the 5'-flap varies in length (typically 1–15 nucleotides) and, because it is independent of the target, can consist of any sequence. The 3' base of the Invader oligo overlaps with the TSR of the primary probe at a base referred to as "position 1," creating a substrate for the Cleavase enzyme. The specificity of the Cleavase enzymes requires that position 1 of the primary probe be complementary to the target for cleavage to occur, providing the ability to discriminate single-base changes. The generation of the proper enzyme substrate is dependent on association between the primary probe, Invader oligo, and target. A noncomplementary base in the target at position 1 results in the formation of a nicked structure, rather than an invasive structure, because position 1 of the primary probe effectively becomes part of the flap. The nicked structure is not a substrate for the Cleavase enzyme, and thus the noncomplementary primary probe does not get cleaved. Discrimination relies on enzymatic recognition of the properly assembled structure in addition to the sequence specificity of oligonucleotides to the target sequence.

DNA DETECTION

The TSR of the primary probe is designed to have a melting temperature close to the reaction temperature, typically 63°C. Its inherent instability allows for the exchange of cleaved and uncleaved probes onto the target. When the specific sequence is present, the Invader oligo and TSR of the primary probe form an invasive structure on the target. The Cleavase enzyme removes the 5'-flap plus position 1 of the TSR. Following cleavage, the primary probe dissociates from the target and is replaced with another uncleaved primary probe, which is present in excess. Thus, numerous 5'-flaps are released for each target molecule present, resulting in a linear amplification of signal, but not target.

The cleaved 5'-flap then acts as an Invader oligo to generate another invasive structure with a synthetic molecule, the fluorescence resonance energy transfer (FRET) oligo, which contains a donor fluorophore on the overlapping base (position 1) and a quencher fluorophore on the other side of the cleavage site. Cleavage of the FRET oligo separates the donor fluorophore (F) from the quencher fluorophore (Q) to generate a fluorescent signal. Using two different 5'-flap sequences and their complementary FRET oligos, each labeled with different fluorophores, two different target sequences may be detected in a single well. In addition to producing a signal, the FRET oligo provides a second level of signal amplification. In the DNA assay, the 5'-flap forms an unstable, invasive duplex with a FRET oligo. After the fluorophore has been released, the 5'-flap dissociates and is able to form an invasive structure with a new, uncleaved FRET oligo. Signal amplification results from the combined effect of having multiple flaps cleaved for each target molecule and from each of those released flaps driving the cleavage of multiple FRET oligos. Both reactions—cleavage of the primary probes and of the FRET probes—occur simultaneously at a single temperature near the melting temperature of the primary probe (Figure 13.1).[3–5]

DNA Invader® assay

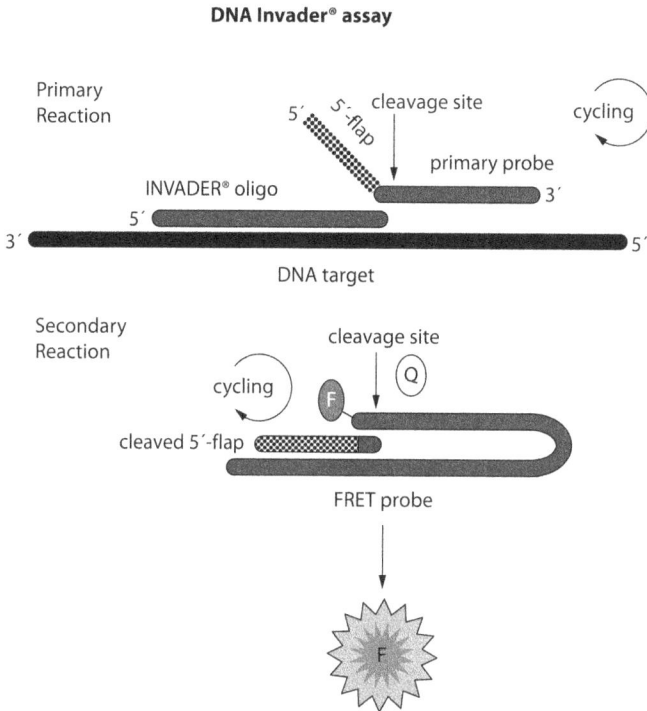

FIGURE 13.1 Schematic of the Invader® DNA assay.

RNA DETECTION

The FEN-type Cleavase enzymes that are used in Invader DNA assays are not able to recognize RNA targets; instead, pol-type Cleavase enzymes that recognize both DNA and RNA targets are used. The altered substrate specificity of the pol-type Cleavase enzymes requires an adaptation to the reaction used to cleave the FRET oligo. RNA assays occur in two sequential steps: the primary reaction and the secondary reaction. During the primary reaction, the Invader oligo and primary probe form an invasive structure on the RNA target. An additional oligo, the stacker oligo, is designed to coaxially stack with the 3' end of the primary probe. The stacker oligo allows the primary probe to cycle on and off the target at a higher temperature and increases the sensitivity of the assay. In the secondary reaction, the cleavage product of the primary reaction (the cleaved 5'-flap plus one base of the TSR) forms a one-base overlap structure with a secondary reaction template (SRT) and a FRET oligo. The enzymatic cleavage of the FRET oligo separates a fluorophore from a quencher molecule, as described for DNA assays, generating a fluorescent signal. In contrast to the DNA format, the released 5'-flap remains bound to the SRT while the FRET oligos cycle on and off the SRT. Consequently, multiple FRET oligos are cleaved per 5'-flap, resulting in a linear accumulation of fluorescence signal. An arrestor oligo, which hybridizes to the TSR plus a portion of the 5'-flap of uncleaved

RNA Invader® assay

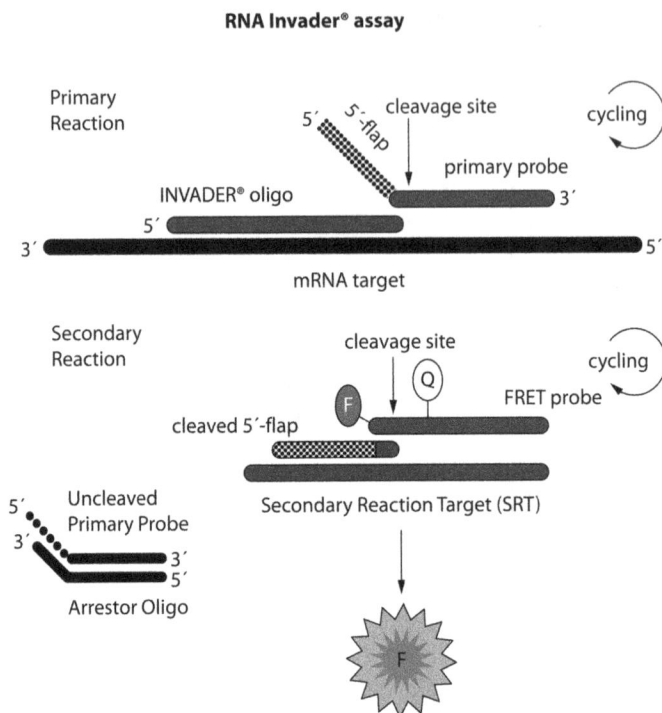

FIGURE 13.2   Schematic of the Invader® RNA assay.

primary probes, is also included in the secondary reaction. The arrestor oligo allows for greater signal generation by sequestering uncleaved primary probes and preventing competitive inhibition of 5′-flap hybridization with the SRT during the secondary reaction (Figure 13.2).[3,4,6]

DETECTION FORMATS

Because its composition is independent of the target being detected, the 5′-flap can be detected in a variety of ways, e.g., by size, sequence, charge, or fluorescence. Based on these properties, the Invader technology can be applied to numerous alternate detection formats including mass spectrometry, capillary electrophoresis, microfluidics, universal array chips, capture, eSensor™, fluorescence polarization, and time-release fluorescence.

ACLARA Biosciences (Mountain View, CA) produces small, electrophoretically distinct, fluorescent molecules called eTag reporter molecules that can be incorporated into primary probes as 5′-flaps. The released eTag flaps are resolved and quantitated using universal separation conditions on standard capillary DNA-sequencing instrumentation. High-level multiplex capabilities are possible with the development of hundreds of eTag reporters. ACLARA has demonstrated the ability to detect 26 Invader RNA assays from a single tube.

SPECIFICITY

Superior specificity has been achieved with the Invader technology by combining sequence identification with enzyme structure recognition. The key to the assay's specificity is the strict requirement for an invasive structure. If the sequence at the cleavage site is not complementary to the intended target, an invasive structure does not form, and the 5'-flap is not released for detection in the secondary reaction. For mutation detection applications, such as cystic fibrosis transmembrane conductance regulator (CFTR) gene screening, this discrimination capability provides high accuracy, as demonstrated by its ability to classify the three possible CFTR ΔF508 genotypes and discriminate between the ΔI507 and ΔF508 mutations (Figure 13.3). In a multisite study to determine the

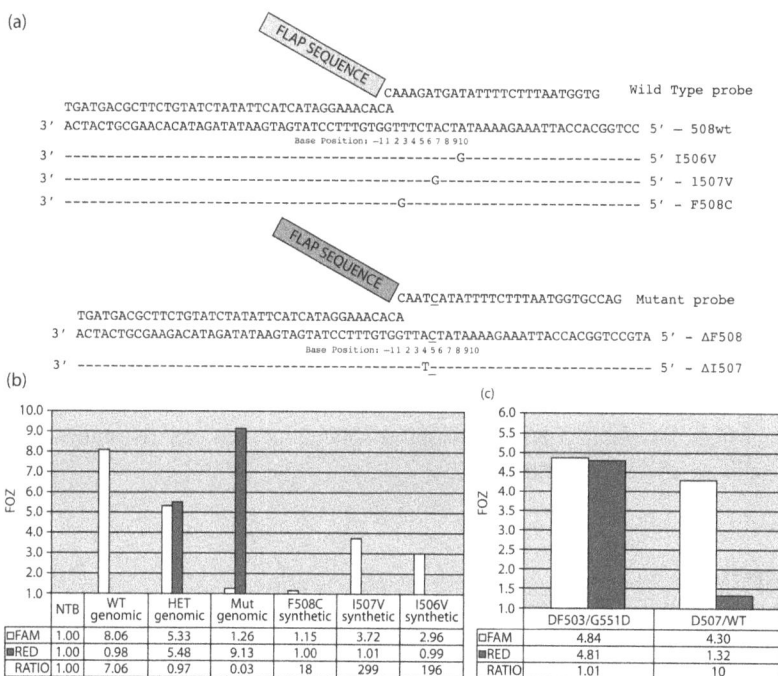

FIGURE 13.3 Discrimination and detection of ΔF508 DNA. (a) ΔF508 assay design and alignment of WT, I506V, I507V, F508C, ΔF508, and ΔI507 DNA sequences. The wild-type (WT) probe is completely homologous to the WT target but mismatches at positions 3, 7, and 10 on the respective targets F508C, I507V, and I506V. The Mutant (Mut) probe has a single mismatch at position 5 (underlined) on the ΔF508 target and two mismatches (positions 4 and 5) on the ΔI507 target. The position 5 mismatch was designed to reduce cross-reactive signal on the ΔI507 target, which differs from ΔF508 only at position 4. (b) The specific signal generated on WT, HET (WT/ΔF508), and Mut (ΔF508/ΔF508) genomic samples. The WT probe generates sufficient signal with the I507V and I506V targets (mismatches at positions 7 and 10, respectively), but not with the F508C target (mismatch at position 3). An additional test is required to determine the F508C genotype. (c) The specificity of the Mut probe: the mismatch at position 5 does not hinder the signal generated from a ΔF508 HET genomic sample and creates additional discrimination of the ΔI507 genomic sample.

accuracy of genotyping five polymorphisms associated with venous thrombosis, the Invader assays correctly genotyped 99.9% of samples compared with 98.5% correct genotypes called by PCR-based methods.[7] Such ability to discriminate single-base changes also makes the Invader platform ideally suited for genotyping highly homologous strains of infectious agents.

The Invader platform is capable of specifically genotyping highly polymorphic regions such as the cytochrome P450 gene family, which contain multiple pseudogenes, gene deletions, and gene duplications. To prevent false-positive results or inflated wild-type signals caused by the association of Invader oligos to the pseudogenes, the region of interest can be amplified using PCR primers specific to the region of interest. The resulting amplicons can be used as templates for Invader assay-based detection. Using this approach, results from 11 Invader CYP2D6 assays provided a clear genotype for 100% of the 171 anonymous donor samples that had a visible PCR product on an agarose gel.[5]

The quantitative aspect of the Invader platform allows for the determination of chromosome and gene copy number directly from genomic DNA. This analysis may be accomplished by comparing the specific signal generated from the gene or chromosome of interest with that of a reference gene such as α-actin, which is not known to be polymorphic for either duplication or deletion.[5] Careful design of the Invader oligos to regions with minimal homology to other regions within the genome enables accurate quantitation of the desired target sequence.

Invader RNA assays have the same ability as the DNA assays to discriminate single-base changes. By positioning the cleavage site of the primary probe at a nonconserved site within the mRNA target, Invader assays are able to differentiate highly homologous RNA sequences. Cross-reactivity is undetectable in the presence of 10,000-fold excess of highly related mRNA.[4] In addition, assays with cleavage sites located at particular exons or splice junctions can distinguish alternatively spliced mRNA variants.[6]

SENSITIVITY

Because of its high level of precision, Invader assays can determine changes in gene expression as low as 1.2 fold.[6] Intraassay coefficients of variation (CVs) are typically less than 10% and range, on average, from 2% to 3%.[4,6] Reading the assays in real-time or at multiple time points can extend the dynamic range of detection, which covers three orders of magnitude with a single endpoint read.[4]

Signal amplification eliminates the need for target amplification prior to analysis, which, in turn, simplifies sample processing. Genomic DNA obtained from small blood volumes can be used directly in genotyping assays. Invader DNA assays are able to statistically detect as little as 10 ng of genomic DNA (approximately 1500 heterozygous genomic copies)[3]; however, a minimum of 100 ng is generally recommended. For RNA assays, sample preparation can be simplified by using unpurified cell lysates directly in the assay. The Invader RNA assays can routinely detect as little as 0.005–0.01 amol RNA target (approximately 3000–6000 copies of RNA transcript or messenger RNA).[6]

Invader assays have the unique ability to tolerate the presence of secondary polymorphisms. Whereas the formation of the proper enzyme substrate is dependent on base pairing at critical positions within the primary probe, other regions may be mismatched. This provides the ability to detect or disregard polymorphisms, which increases the assay detection sensitivity in highly polymorphic genes or strains of infectious agents.

ROBUSTNESS

Ease of use, flexibility, and scalability are some of the advantages that make the Invader platform ideally suited for automation. Assays are readily adaptable from manual to semiautomated to fully automated ultra-high-throughput settings using a variety of standard plate formats. Most automated liquid handling systems can handle the assay setup, which requires only a few pipetting steps. Regardless of format, e.g., 96 or 384 wells, or setup, e.g., manual or automated, the robustness of the platform remains constant.[3]

The Invader platform—a highly accurate, cost-effective, and specific method for the detection of specific nucleic acid sequences—has been applied to the detection of thousands of single nucleotide polymorphisms (SNPs) in genome-wide association studies.[8,9] Simple, fast, reliable, and inexpensive diagnostic tools are essential not only for research purposes but for clinical settings as well. Several analyte-specific reagents (ASRs) have been developed for use in clinical diagnostics, including Factor V Leiden (FVL), apolipoprotein E (ApoE), and cystic fibrosis (CFTR) mutation detection. The current clinical applications of the Invader platform include the detection of coagulopathies and inherited diseases.

CONCLUSION

The Invader platform is a homogeneous, isothermal DNA probe-based system for the quantitative detection of specific nucleic acid sequences. Invader assays can be performed directly on DNA-, RNA-, PCR-, or RT-PCR-amplified products. Based on the ability of the Cleavase enzymes to recognize and cleave-specific nucleic acid structures, the Invader platform is a highly specific method for detecting SNPs, insertions, deletions, and gene copy numbers. Its ability to amplify signals and tolerate secondary polymorphisms makes the Invader platform extremely sensitive as well. Because of the flexibility, ease of use, and robustness of the technology, it can be used in both small and high-throughput laboratories.

REFERENCES

1. Kaiser, M.W.; Lyamicheva, N.; Ma, W.-P.; Miller, C.; Neri, B.; Fors, L.; Lyamichev, V.I. A comparison of eubacterial and archaeal structure-specific 5'-exonucleases. J. Biol. Chem. **1999**, *274* (30), 21387–21394.
2. Lyamichev, V.; Brow, M.A.D.; Dahlberg, J.E. Structure-specific endonucleolytic cleavage of nucleic acids by eubacterial DNA polymerases. Science **1993**, *260*, 778–783.

3. de Arruda, M.; Lyamichev, V.I.; Eis, P.S.; Iszczyszyn, W.; Kwiatkowski, R.W.; Law, S.M.; Olson, M.C.; Rasmussen, E.B. Invader technology for DNA and RNA analysis: Principles and applications. Expert Rev. Mol. Diagn. **2002**, *2* (5), 487–496.

4. Kwiatkowski, R.W.; Lyamichev, V.; de Arruda, M.; Neri, B. Clinical, genetic, and pharmacogenetic applications of the Invader assay. Mol. Diagn. **1999**, *4* (4), 353–364.

5. Neville, M.; Selzer, R.; Aizenstein, B.; Maguire, M.; Hogan, K.; Walton, R.; Welsh, K.; Neri, B.; deArruda, M. Characterization of cytochrome P450 2D6 alleles using the Invader system. BioTechniques **2002**, *32*, S34–S43.

6. Eis, P.S.; Olson, M.C.; Takova, T.; Curtis, M.L.; Olson, S.M.; Vener, T.I.; Ip, H.S.; Vedvik, K.L.; Bartholomay, C.T.; Allawi, H.T.; Ma, W.-P.; Hall, J.G.; Morin, E.D.; Rushmore, T.H.; Lyamichev, V.I.; Kwiatkowski, R.W. An invasive cleavage assay for direct quantitation of specific RNAs. Nat. Biotechnol. **2001**, *19*, 673–676.

7. Patnaik, M.; Dlott, J.S.; Fontaine, R.N.; Subbiah, M.T.; Hessner, M.J.; Joyner, K.A.; Ledford, M.R.; Lau, E.C.; Moehlenkamp, C.; Amos, J.; Zhang, B.; Williams, T.M. Detection of genomic polymorphisms associated with venous thrombosis using the Invader biplex assays. J. Mol. Diagn. **2004**, *6* (2), 137–144.

8. Mein, C.A.; Barratt, B.J.; Dunn, M.G.; Siegmund, T.; Smith, A.N.; Esposito, L.; Nutland, S.; Stevens, H.E.; Wilson, A.J.; Phillips, M.S.; Jarvis, N.; Law, S.; de Arruda, M.; Todd, J.A. Evaluation of single nucleotide polymorphism typing with Invader of PCR amplicons and its automation. Genome Res. **2000**, *10* (3), 330–343.

9. Ohnishi, Y.; Tanaka, T.; Ozaki, K.; Yamada, R.; Suzuki, H.; Nakamura, Y. A high-throughput SNP typing system for genome-wide association studies. J. Hum. Genet. **2001**, *46* (8), 471–477.

14 Laser-Capture Microdissection

Falko Fend
Institute of Pathology and Neuropathology, University
of Tuebingen, Tuebingen, Germany

CONTENTS

INTRODUCTION

The molecular analysis of tissue specimens has given significant insight into the pathogenesis of human disease and led to the identification of a broad range of diagnostic and prognostic biomarkers. High-throughput approaches for the examination of DNA, RNA, and proteins have changed tissue-based diagnostics profoundly and techniques such as next generation sequencing (NGS) complement conventional histological examination. However, the reliability of many tests based on tissue or cell extracts critically depends on the abundance of the cell population in question. Primary tissues contain a variety of cellular elements including specialized, organotypic parenchyma and a large range of stromal and inflammatory cells. The relative percentages of these different cell types vary widely, especially in pathologically altered tissues. This inherent complexity of primary tissues can lead to misinterpretation in molecular studies, both for diagnostics, as well as in research, if bulk tissue extracts are used. Although examinations on the DNA level, e.g. for mutational studies of tumors, are considered robust and are routinely performed on formalin-fixed, paraffin-embedded (FFPE) clinical specimens, the detection level of different methods is critical for obtaining correct results. Whereas the PCR detection threshold for some genetic alterations such as tumor-specific chromosomal translocations are in the range of 1 in 10^4 cells or less carrying the marker, and modern NGS-based mutational analysis has superior sensitivity compared to conventional Sanger sequencing, other tests such as the detection of clonal immunoglobulin or T-cell receptor gene rearrangements with consensus primers, loss of heterozygosity or determination of gene methylation require significantly higher percentages of target cells. In addition,

DOI: 10.1201/9781003247432-14

129

the examination of early or precursor lesions such as carcinoma *in situ* and epithelial dysplasias, which can shed light on the first seminal steps of carcinogenesis, is virtually impossible from bulk tissue extracts. These problems are magnified at the RNA and protein level, as frequently the relative abundances of the targets, rather than their presence or absence, are of relevance. Furthermore, signaling pathways are governed by low-abundance mRNA and protein specimens and critically depend not only on cell type but also on specific neighborhoods and direct cellular interactions, thus requiring topography-based analysis for a deeper understanding. Although bioinformatic tools for deconvolution of gene expression or methylation data derived from complex tissues have been developed, these techniques do not reflect well spatial relationships and may carry significant bias.[1] *In situ* studies such as immunohistochemistry including multiplexing techniques and *in situ* hybridization provide valuable alternatives but are limited in their application range. Other techniques such as spatial transcriptomics or matrix-assisted laser desorption/ionization (MALDI) imaging mass spectrometry require expensive specialized equipment and a complex workflow. Therefore, microdissection strategies, ranging from relatively crude, manual microdissection to micromanipulation of single cells, are used extensively for the isolation of pure cell populations from primary tissues in a broad range of applications and have been instrumental for a range of seminal discoveries.

PRINCIPLES OF LASER-CAPTURE MICRODISSECTION

A variety of laser-assisted microdissection devices making use of the unique properties of laser light have been developed and are commercially available. Two different technical principles can be discerned. One of them, initially termed laser microbeam microdissection (LMM), uses a pulsed ultraviolet (UV) laser with a small beam focus to cut out areas or cells of interest by photoablation of adjacent tissue.[2,3] The tissue or cells of interest are subsequently either catapulted in a recipient cap by a laser pulse or fall into a tube by gravity. Sections can be put onto special membranes to cut out larger pieces of tissue. The second technique is laser-capture microdissection (LCM), in which a near-infrared (IR) laser pulse is used to selectively adhere visually targeted cells and tissue fragments to a thermoplastic membrane placed on the tissue section.[4,5] (Figure 14.1)

Both techniques have advantages and disadvantages, and several studies have performed direct comparisons of UV laser-based and IR laser-based microdissection in terms of feasibility and ease of use, as well as concerning macromolecule recovery, with variable results. Overall, however, both techniques show similar performance, and the choice of method depends on availability, type of tissue and downstream analyses.[6,7]

Recently, systems have been developed which have integrated both IR and UV lasers, combining the advantages of both principles.[8,9] These new systems consist of an inverted microscope, a solid-state near-IR laser diode and a pulsed diode-pumped UV-laser (355 nm) with 20 kHz sub-nano-second pulse rate for cutting, a joystick-controlled microscope stage, computer hardware and software tools for laser control, image archiving and marking areas of interest for automated capture. In addition to conventional bright-field microscopy, epifluorescence can be used, and

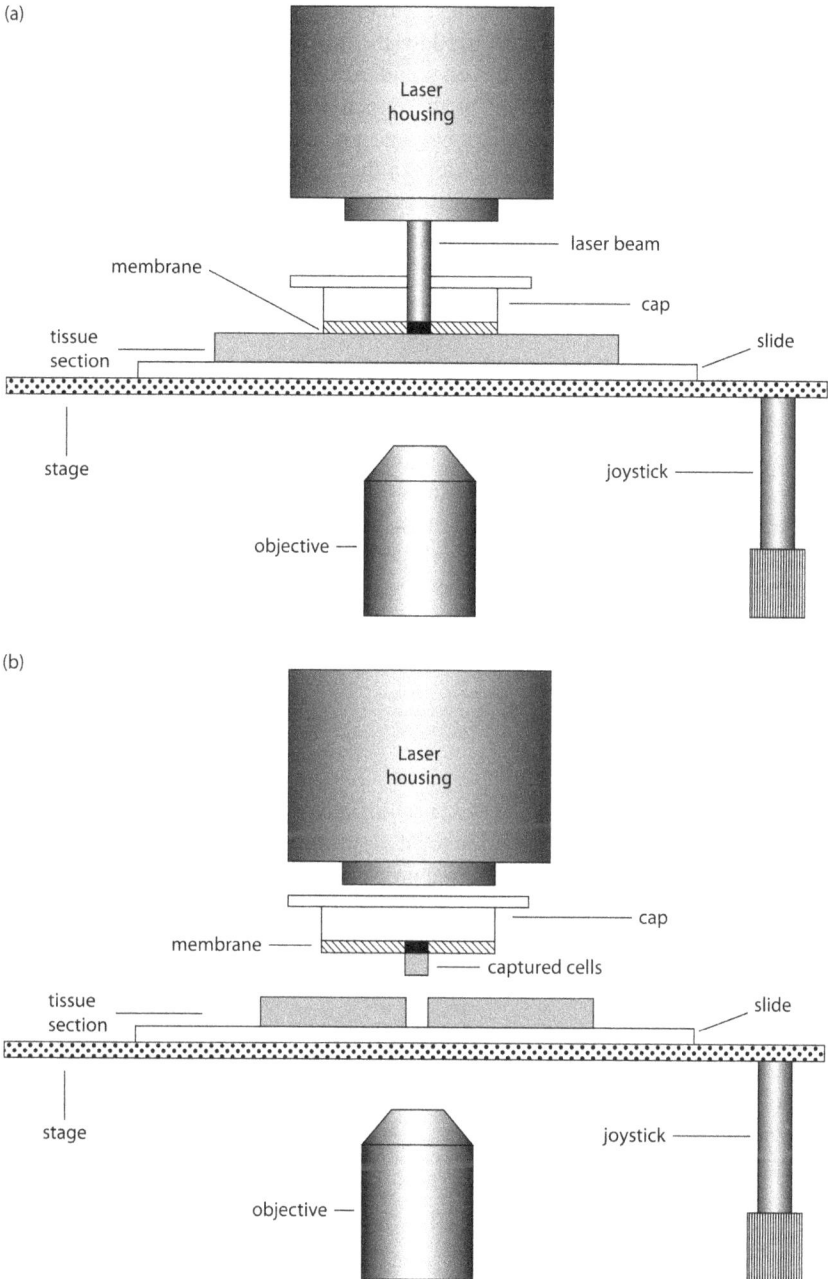

FIGURE 14.1 Schematic representation of laser-capture microdissection using a near-IR laser. A) Activation of the laser leads to focal melting of the thermoplastic membrane attached to the lower portion of the optically transparent cap. B) Lifting of the cap leads to selective detachment of cells adherent to the molten (activated) parts of the membrane.

Source: From Ref. [3].

image analysis software can be used to automatically identify areas of interest/target cells. The thermoplastic membrane used for the transfer of selected cells is mounted on an optically clear cap which fits on standard 0.5 mL microcentrifuge tubes for further processing. After selection of the targeted cells, activation of the IR laser leads to focal melting of the ethylene vinyl acetate (EVA) membrane, which has its absorption maximum near the wavelength of the laser. The melted polymer expands into the section, resolidifies within milliseconds, and forms a composite with the tissue. This allows selective removal of the cells attached to the activated membrane. Laser "shots" can be repeated multiple times across the whole cap surface to collect large numbers of cells.[5,10] Depending on the chosen laser spot size, the architectural features of the tissue, and the desired precision of the microdissection, even thousands of cells can be collected within a few minutes. After dissection, the cap is transferred to a tube containing the buffer solutions required for isolation of the molecules of interest. As most of the energy is absorbed by the membrane, the maximum temperature reached by the tissue upon laser activation is in the range of 90°C for several milliseconds, thus leaving intact biological macromolecules of interest. The precision of LCM can reach the single cell range, but mainly depends on the ability to visually identify the desired cell population on the non-coverslipped slide, and to a lesser degree on the size and configuration of target cells. Microdissection using the pulsed UV laser is performed as described above. Of note, the UV laser can also be used to destroy unwanted tissues by photoablation before capturing the cells of interest.

APPLICATIONS OF LCM

Due to its versatility, virtually every branch of biological and medical sciences has employed laser-based microdissection for their purposes. Major advantages of LCM are its easy integration into routine tissue preparation workflows and the broad range of specimens to which it can be applied, including frozen sections, sections from fixed, paraffin-embedded archival material, certain cytological preparations, cell cultures, chromosomal spreads, microorganisms and forensic materials.[5,11,12] The most important parameters to consider are the type and fixation of the tissue to be examined, the morphologic features suitable for target identification, the stains used for visualization, the desired target biomolecule—DNA, RNA, protein, or others — and the type of downstream analysis. A plethora of techniques for the isolation of macromolecules from captured tissues have been published, and many companies offer specialized kits optimized for extraction from microdissected tissues.

DNA ANALYSIS

DNA is relatively resistant to fixation and staining procedures and provides an ideal target for PCR- or NGS-based analysis. Mutational profiling can be greatly facilitated by collecting highly enriched tumor cell populations with LCM, thus increasing the diagnostic yield and obviating the need for more labor-intensive analytical strategies. This is especially true for analyses which require a high percentage of lesional cells, such as assessment of promoter methylation, loss of heterozygosity (LOH) or numerical aberrations.[13–15] Microdissection is crucial for small lesions such as preneoplastic changes or areas of tumor heterogeneity which are not amenable to conventional examination of bulk tissue extracts and require stringent correlation with morphology or immunophenotypical features. [16–18] The investigation of microdissected precursor lesions helps to correlate genetic alterations with the morphological stages of tumor development, thus testing models of the stepwise acquisition of progressive genetic changes *in vivo* and allowing the construction of genealogical trees starting from the earliest identifiable clonal cell populations.[19,20] In combination with techniques for the random amplification of the whole genome, array comparative genomic hybridization (CGH), e.g. using OncoScan® molecular inversion probe technology, whole genome allelotyping, sequencing or other applications can be performed on a small number or even single cells.[21] On the other hand, optimized protocols allow for whole genome sequencing of small numbers of microdissected cells (100–1000) for mutational analysis without pre-amplification, thus reducing potential amplification artifacts.[22]

Other applications include the analysis of genetic mosaicism and forensic studies for the precise individual assignment of cells in contaminated samples.[11] However, the smaller the amount of template available, the higher is the risk for artifacts such as allelic dropout, biased amplification of certain sequences, or introduction of Taq polymerase errors. Critical evaluation of results and introduction of multiple controls, such as repeat analyses and continuous parallel investigation of adjacent normal tissue, are mandatory.

GENE EXPRESSION ANALYSIS

The analysis of tissue- or cell-specific gene expression is of paramount importance for many fields of biological and biomedical research. However, tissue heterogeneity can make it very difficult to assign gene expression profiles or specific messages to defined cell populations if gross tissue extracts serve as a source for different RNA species, including mRNA, miRNA, and others. Therefore, significant efforts have been put into developing cell isolation strategies which allow cell- or tissue-specific gene expression analysis, with laser-assisted microdissection one of several approaches. RNA is more sensitive to fixation, is quickly degraded by ubiquitous RNases, and requires stringent RNase-free conditions during specimen handling. Keeping these requirements in mind, RNA of good quality can be recovered from microdissected frozen tissue samples down to the single-cell level, suitable for RT-PCR (reverse transcription polymerase chain reaction), quantitative real-time RT-PCR and techniques for global gene expression profiling (GEP), including massively parallel RNA sequencing and alternative techniques such as the Nanostring platform.[3,9,23–27] Rapid, optimized immunohistochemical or immunofluorescence staining protocols allow the combination of phenotypical identification of target cells with mRNA analysis.[28,29] The superior speed of LCM and the fact that capturing is performed from dried slides allow sampling of large numbers of cells without significant RNA degradation during the procurement procedure. If very small numbers of cells serve as template or if large-scale expression profiling using cDNA or oligonucleotide arrays is used, random mRNA amplification protocols can be employed.[30] In initial studies, T7 RNA polymerase-based strategies have been the preferred technique in order to achieve an unbiased mRNA amplification from small amounts of tissue. Array-based expression screening of LCM-captured tissues has been used successfully for delineating expression differences of microdissected small and large neurons from dorsal root ganglia[23] or expression changes during breast cancer progression,[24] to name only a few examples. More recent approaches have applied techniques developed for single cell transcriptomics to laser-captured cells, termed topographic single-cell sequencing.[25,31]

Although frozen tissue is the preferred source for mRNA analysis, the vast amounts of available archival specimens make paraffin-embedded tissues a valuable resource, and RNA derived from formalin-fixed paraffin blocks is suitable for certain types of analysis. A few hundred cells from paraffin sections render enough RNA for several robust RT-PCR assays without the need for preamplification steps, provided that the amplified fragment is below 100–120 bp.[32] Using the TaqMan technology with fluorescence-labeled probes, highly reproducible expression values can be obtained from microdissected archival material over a broad range of starting material and template concentrations. The precision achievable by combining target cell purification by LCM with quantitative RT-PCR and the ease with which these two techniques can be applied to paraffin sections make this approach a valuable tool for advanced molecular diagnostics. Novel techniques to analyze gene expression without amplification such as Nanostring® technology provide a robust platform for GEP of FFPE tissues and in combination with laser-assisted microdissection allow spatial resolution on the cellular level, complementing and validating *in silico* tools for deconvoluting expression data from mixed cell populations.[9]

ANALYSIS OF PROTEINS AND OTHER MACROMOLECULES

Proteins are the effector molecules of the cell and can be modified through a variety of posttranscriptional mechanisms, including phosphorylation and glycosylation. Therefore, a cell-specific analysis of protein patterns may ultimately prove to be more informative than expression profiling of primary tissues. For these reasons, a variety of proteomic techniques have been applied to microdissected tissues, although the inherent limitation of starting material initially presented a major obstacle for protein analyses, given the lack of amplification techniques.[33–35] Current technologies allow application of both bottom-up, unbiased global approaches such as liquid-chromatography mass spectrometry (LC-MS), as well as targeted methods such as reverse-phase protein arrays on microdissected tissues.[36–38] Extraction techniques adapted to FFPE material and optimized proteomic workflows combined with the improvement in "signal-to-noise" ratio offered by LCM dissection render an improved coverage of the proteome and phosphoproteome of specific cell populations and allow a better analysis of stroma-tumor crosstalk and a cell-specific separation of activated signaling pathways.[36,39–42] Beyond research applications, protocols for the analysis of proteins from FFPE tissues have already entered clinical diagnostics, e.g. for subtyping amyloid deposits in clinical biopsies not resolvable by immunohistochemical staining.[43]

Although the topographically informed analysis of nucleic acids and proteins still remain the dominant applications of LCM, other macromolecules such as lipids or cellular metabolites can also be analyzed from microdissected tissues, as long as target stability and dedicated extraction techniques are taken into consideration.[44–46]

CONCLUSION

Laser-capture microdissection and related laser-assisted microdissection techniques have developed into a key technology for the in-depth molecular analysis of tissues and cells for both research and diagnostic purposes. Although recently novel techniques for isolating cells from complex tissues have been added to the technical armamentarium, such as flow sorting or microfluidic devices, the ease of use and versatility of LCM and its closeness to standard histological techniques facilitates its integration into the workflow of any molecular laboratory without much additional effort. In addition, in contrast to other cell isolation techniques, LCM allows parallel observation of morphological features and cellular neighborhoods. Combined with downstream high-throughput techniques, microdissection is a big step toward "molecular morphology."

REFERENCES

1. Avila Cobos, F.; Alquicira-Hernandez, J.; Powell, J.E.; Mestdagh, P.; De Preter, K. Benchmarking of cell type deconvolution pipelines for transcriptomics data. Nat. Commun. **2020**, *11*, 5650. doi: 10.1038/s41467-020-19015-1
2. Böhm, M.; Wieland, I.; Schütze, K.; Rübben, H. Microbeam MOMeNT: Non-contact laser microdissection of membrane-mounted native tissue. Am. J. Pathol. **1997**, *151*, 63–67.
3. Schutze, K.; Lahr, G. Identification of expressed genes by laser-mediated manipulation of single cells. Nat. Biotechnol. **1998**, *16*, 737–742.
4. Emmert-Buck, M.R.; Bonner, R.F.; Smith, P.D.; Chuaqui, R.F.; Zhuang, Z.; Goldstein, S.R.; Weiss, R.A.; Liotta, L.A. Laser capture microdissection. Science **1996**, *274*, 998–1001.
5. Fend, F.; Raffeld, M. Laser capture microdissection in pathology. J. Clin. Pathol. **2000**, *53*, 666–672.
6. Vandewoestyne, M.; Goossens, K.; Burvenich, C.; Van Soom, A.; Peelman, L.; Deforce, D. Laser capture microdissection: should an ultraviolet or infrared laser be used? Anal. Biochem. **2013**, *439*, 88–98. doi: 10.1016/j.ab.2013.04.023
7. Hunt, A.L.; Pierobon, M.; Baldelli, E.; Oliver, J.; Mitchell, D.; Gist, G.; Bateman, N.W.; Larry Maxwell, G.; Petricoin, E.F.; Conrads, T.P. The impact of ultraviolet- and infrared-based laser microdissection technology on phosphoprotein detection in the laser microdissection-reverse phase protein array workflow. Clin Proteomics **2020**, *17*, 9. doi: 10.1186/s12014-020-09272-z
8. Gallagher, R.I.; Blakely, S.R.; Liotta, L.A.; Espina, V. Laser capture microdissection: Arcturus(XT) infrared capture and UV cutting methods. Methods Mol. Biol. **2012**, *823*, 157–178. doi: 10.1007/978-1-60327-216-2_11
9. Golubeva, Y.; Salcedo, R.; Mueller, C.; Liotta, L.A.; Espina, V. Laser capture microdissection for protein and NanoString RNA analysis. Methods Mol. Biol. **2013**, *931*, 213–257. doi: 10.1007/978-1-62703-056-4_12

10. Emmert-Buck, M.R.; Gillespie, J.W.; Paweletz, C.P.; Ornstein, D.K.; Basrur, V.; Appella, E.; Wang, Q.H.; Huang, J.; Hu, N.; Taylor, P.; Petricoin, E.F., 3rd. An approach to proteomic analysis of human tumors. Mol. Carcinog. **2000**, *27*, 158–165.

11. Vandewoestyne, M.; Deforce, D. Laser capture microdissection in forensic research: A review. Int. J. Legal Med. **2010**, *124*, 513–521. doi: 10.1007/s00414-010-0499-4

12. Podgorny, O.V.; Lazarev, V.N. Laser microdissection: A promising tool for exploring microorganisms and their interactions with hosts. J. Microbiol. Methods **2017**, *138*, 82–92. doi: 10.1016/j.mimet.2016.01.001

13. Feltmate, C.M.; Mok, S.C. Whole-genome allelotyping using laser microdissected tissue. Methods Mol. Biol. **2005**, *293*, 69–77. doi: 10.1385/1-59259-853-6:069

14. Eberle, F.C.; Hanson, J.C.; Killian, J.K.; Wei, L.; Ylaya, K.; Hewitt, S.M.; Jaffe, E.S.; Emmert-Buck, M.R.; Rodriguez-Canales, J. Immunoguided laser assisted microdissection techniques for DNA methylation analysis of archival tissue specimens. J. Mol. Diagn. **2010**, *12*, 394–401. doi: 10.2353/jmoldx.010.090200

15. Gagnon, J.F.; Sanschagrin, F.; Jacob, S.; Tremblay, A.A.; Provencher, L.; Robert, J.; Morin, C.; Diorio, C. Quantitative DNA methylation analysis of laser capture microdissected formalin-fixed and paraffin-embedded tissues. Exp. Mol. Pathol. **2010**, *88*, 184–189. doi: 10.1016/j.yexmp.2009.09.020

16. Pretlow, T.P.; Brasitus, T.A.; Fulton, N.C.; Cheyer, C.; Kaplan, E.L. K-ras mutations in putative preneoplastic lesions in human colon. J. Natl. Cancer Inst. **1993**, *85*, 2004–2007. doi: 10.1093/jnci/85.24.2004

17. Cong, P.; Raffeld, M.; Teruya-Feldstein, J.; Sorbara, L.; Pittaluga, S.; Jaffe, E.S. In situ localization of follicular lymphoma: Description and analysis by laser capture microdissection. Blood **2002**, *99*, 3376–3382. doi: 10.1182/blood.v99.9.3376

18. Mo, A.; Jackson, S.; Varma, K.; Carpino, A.; Giardina, C.; Devers, T.J.; Rosenberg, D.W. Distinct transcriptional changes and epithelial-stromal interactions are altered in early-stage colon cancer development. Mol. Cancer Res. **2016**, *14*, 795–804. doi: 10.1158/541-7786.MCR-16-0156

19. Wistuba, I.I.; Behrens, C.; Milchgrub, S.; Bryant, D.; Hung, J.; Minna, J.D.; Gazdar, A.F. Sequential molecular abnormalities are involved in the multistage development of squamous cell lung carcinoma. Oncogene **1999**, *18*, 643–650. doi: 10.1038/sj.onc.1202349

20. Wang, S.; Du, M.; Zhang, J.; Xu, W.; Yuan, Q.; Li, M.; Wang, J.; Zhu, H.; Wang, Y.; Wang, C.; Gong, Y.; Wang, X., et al. Tumor evolutionary trajectories during the acquisition of invasiveness in early stage lung adenocarcinoma. Nat. Commun. **2020**, *11*, 6083. doi: 10.1038/s41467-020-19855-x

21. Casasent, A.K.; Schalck, A.; Gao, R.; Sei, E.; Long, A.; Pangburn, W.; Casasent, T.; Meric-Bernstam, F.; Edgerton, M.E.; Navin, N.E. Multiclonal invasion in breast tumors identified by topographic single cell sequencing. Cell **2018**, *172*, 205–217.e12.

22. Ellis, P.; Moore, L.; Sanders, M.A.; Butler, T.M.; Brunner, S.F.; Lee-Six, H.; Osborne, R.; Farr, B.; Coorens, T.H.H.; Lawson, A.R.J.; Cagan, A.; Stratton, M.R., et al. Reliable detection of somatic mutations in solid tissues by laser-capture microdissection and low-input DNA sequencing. Nat. Protoc. **2021**, *16*, 841–871. doi: 10.1038/s41596-020-00437-6

23. Luo, L.; Salunga, R.C.; Guo, H.; Bittner, A.; Joy, K.C.; Galindo, J.E.; Xiao, H.; Rogers, K.E.; Wan, J.S.; Jackson, M.R.; Erlander, M.G. Gene expression profiles of laser-captured adjacent neuronal subtypes. Nat. Med. **1999**, *5*, 117–122.

24. Ma, X.-J.; Salunga, R.; Tuggle, J.T.; Gaudet, J.; Enright, E.; McQuary, P.; Payette, T.; Pistone, M.; Stecker, K.; Zhang, B.M.; Zhou, Y.-X.; Varnholt, H., et al. Gene expression profiles of human breast cancer progression. Proc. Natl. Acad. Sci. U. S. A. **2003**, *100*, 5974–5979.

25. Nichterwitz, S.; Chen, G.; Aguila Benitez, J.; Yilmaz, M.; Storvall, H.; Cao, M.; Sandberg, R.; Deng, Q.; Hedlund, E. Laser capture microscopy coupled with Smart-seq2 for precise spatial transcriptomic profiling. Nat. Commun. **2016**, *7*, 12139. doi: 10.1038/ncomms12139

26. Peterson, L.A.; Brown, M.R.; Carlisle, A.J.; Kohn, E.C.; Liotta, L.A.; Emmert-Buck, M.R.; Krizman, D.B. An improved method for construction of directionally cloned cDNA libraries from microdissected cells. Cancer Res. **1998**, *58*, 5326–5328.

27. Strell, C.; Hilscher, M.M.; Laxman, N.; Svedlund, J.; Wu, C.; Yokota, C.; Nilsson, M. Placing RNA in context and space – methods for spatially resolved transcriptomics. FEBS J. **2019**, *286*, 1468–1481. doi: 10.111/febs.14435

28. Fend, F.; Emmert-Buck, M.R.; Chuaqui, R.; Cole, K.; Lee, J.; Liotta, L.A.; Raffeld, M. Immuno-LCM: laser capture microdissection of immunostained frozen sections for mRNA analysis. Am. J. Pathol. **1999**, *154*, 61–66.

29. Murakami, H.; Liotta, L.; Star, R.A. IF-LCM: Laser capture microdissection of immunofluorescently defined cells for mRNA analysis rapid communication. Kidney Int. **2000**, *58*, 1346–1353.

30. Luzzi, V.; Mahadevappa, M.; Raja, R.; Warrington, J.A.; Watson, M.A. Accurate and reproducible gene expression profiles from laser capture microdissection, transcript amplification, and high density oligonucleotide microarray analysis. J. Mol. Diagn. **2003**, *5*, 9–14.

31. Nichterwitz, S.; Benitez, J.A.; Hoogstraaten, R.; Deng, Q.; Hedlund, E. LCM-Seq: A method for spatial transcriptomic profiling using laser capture microdissection coupled with PolyA-based RNA sequencing. Methods Mol. Biol. **2018**, *1649*, 95–110. doi: 10.1007/978-1-4939-7213-5_6

32. Specht, K.; Richter, T.; Muller, U.; Walch, A.; Werner, M.; Hofler, H. Quantitative gene expression analysis in microdissected archival formalin-fixed and paraffin-embedded tumor tissue. Am. J. Pathol. **2001**, *158*, 419–429.

33. Banks, R.E.; Dunn, M.J.; Forbes, M.A.; Stanley, A.; Pappin, D.; Naven, T.; Gough, M.; Harnden, P.; Selby, P.J. The potential use of laser capture microdissection to selectively obtain distinct populations of cells for proteomic analysis – Preliminary findings. Electrophoresis **1999**, *20*, 689–700.

34. Ornstein, D.K.; Englert, C.; Gillespie, J.W.; Paweletz, C.P.; Linehan, W.M.; Emmert-Buck, M.R.; Petricoin, E.F., 3rd. Characterization of intracellular prostate-specific antigen from laser capture microdissected benign and malignant prostatic epithelium. Clin. Cancer Res. **2000**, *6*, 353–356.

35. Simone, N.L.; Remaley, A.T.; Charboneau, L.; Petricoin, E.F., 3rd, Glickman, J.W.; Emmert-Buck, M.R.; Fleisher, T.A.; Liotta, L.A. Sensitive immunoassay of tissue cell proteins procured by laser capture microdissection. Am. J. Pathol. **2000**, *156*, 445–452.

36. Buczak, K.; Kirkpatrick, J.M.; Truckenmueller, F.; Santinha, D.; Ferreira, L.; Roessler, S.; Singer, S.; Beck, M.; Ori, A. Spatially resolved analysis of FFPE tissue proteomes by quantitative mass spectrometry. Nat. Protoc. **2020**, *15*, 2956–2979.

37. Mueller, C.; Davis, J.B.; Liotta, L.A. Combining the "Sibling Technologies" of laser capture microdissection and reverse phase protein microarrays. Adv. Exp. Med. Biol. **2019**, *1188*, 95–111. doi: 10.1007/978-981-32-9755-5_6

38. Alexovič, M.; Sabo, J.; Longuespée R. Microproteomic sample preparation. Proteomics **2021**, *6*, 202000318.

39. Bateman, N.W.; Conrads, T.P. Recent advances and opportunities in proteomic analyses of tumour heterogeneity. J. Pathol. **2018**, *244*, 628–637. doi: 10.1002/path.5036

40. Becker, K.F.; Schott, C.; Hipp, S.; Metzger, V.; Porschewski, P.; Beck, R.; Nahrig, J.; Becker, I.; Hofler, H. Quantitative protein analysis from formalin-fixed tissues: implications for translational clinical research and nanoscale molecular diagnosis. J. Pathol. **2007**, *211*, 370–378.

41. Pin, E.; Stratton, S.; Belluco, C.; Liotta, L.; Nagle, R.; Hodge, K.A.; Deng, J.; Dong, T.; Baldelli, E.; Petricoin, E.; Pierobon, M. A pilot study exploring the molecular architecture of the tumor microenvironment in human prostate cancer using laser capture microdissection and reverse phase protein microarray. Mol. Oncol. **2016**, *10*, 1585–1594.

42. Ostasiewicz, P.; Zielinska, D.F.; Mann, M.; Wiśniewski, J.R. Proteome, phosphoproteome, and N-glycoproteome are quantitatively preserved in formalin-fixed paraffin-embedded tissue and analyzable by high-resolution mass spectrometry. J. Proteome. Res. **2010**, *9*, 3688–3700. doi: 10.1021/pr100234w

43. Vrana, J.A.; Gamez, J.D.; Madden, B.J.; Theis, J.D.; Bergen, H.R.; 3rd, Dogan, A. classification of amyloidosis by laser microdissection and mass spectrometry-based proteomic analysis in clinical biopsy specimens. Blood **2009**, *114*, 4957–4959.

44. Cummings, M.; Massey, K.A.; Mappa, G.; Wilkinson, N.; Hutson, R.; Munot, S.; Saidi, S.; Nugent, D.; Broadhead, T.; Wright, A.I.; Barber, S.; Nicolaou, A., et al. Integrated eicosanoid lipidomics and gene expression reveal decreased prostaglandin catabolism and increased 5-lipoxygenase expression in aggressive subtypes of endometrial cancer. J. Pathol. **2019**, *247*, 21–34. doi: 10.1002/path.5160

45. Dilillo, M.; Ait-Belkacem, R.; Esteve, C.; Pellegrini, D.; Nicolardi, S.; Costa, M.; Vannini, E.; Graaf, E.L.; Caleo, M.; McDonnell, L.A. Ultra-high mass resolution MALDI imaging mass spectrometry of proteins and metabolites in a mouse model of glioblastoma. Sci. Rep. **2017**, *7*, 603. doi: 10.1038/s41598-017-00703-w

46. Ikeda, K.; Taguchi, R. Highly sensitive localization analysis of gangliosides and sulfatides including structural isomers in mouse cerebellum sections by combination of laser microdissection and hydrophilic interaction liquid chromatography/electrospray ionization mass spectrometry with theoretically expanded multiple reaction monitoring. Rapid Commun. Mass Spectrom. **2010**, *24*, 2957–2965. doi: 10.1002/rcm.4716

15 Helper-Dependent Adenoviral Vectors

Philip Ng
Baylor College of Medicine, Houston, Texas, U.S.A.

CONTENTS

INTRODUCTION

The adenovirus (Ad) has been extensively exploited as a gene therapy vector because of its ability to efficiently transduce a wide variety of cell types from many different species independent of the cell cycle to direct high-level transgene expression. First-generation Ad (FGAd) vectors typically have the viral early region 1 (E1) replaced by the therapeutic transgene. FGAd vectors are replication deficient and require E1 complementing cells for propagation. However, the majority of the Ad genome remains intact leading to low-level expression of viral genes in the transduced cells. This is directly cytotoxic and also leads to an adaptive cellular immune response against the transduced cells consequently resulting in transient transgene expression and long-term toxicity, thus rendering these vectors unsuitable for many gene therapy applications where long-term transgene expression is desired. In an attempt to further attenuate Ad, vectors have been engineered with deletions or mutations in the viral E2 and E4 genes in addition to deletion of E1. Despite the potential offered by these multiply deleted Ad vectors (also called second- or third-generation Ad vectors), the majority of the viral coding sequences still remain and therefore so does the potential for their expression. The advantages of multiply deleted Ad over FGAd remain controversial as some studies show them to be superior in terms of reduced toxicity and enhanced longevity of transgene expression, whereas others do not. Significant improvement in the safety and efficacy of Ad-based vectors came with the development of helper-dependent adenoviral (HDAd) vectors (also referred to as gutless, gutted, mini, fully deleted, high-capacity, Δ, pseudo) that are

DOI: 10.1201/9781003247432-15

deleted of all viral coding sequences. HDAd vectors retain the advantages of FGAd including high efficiency in vivo transduction and high-level transgene expression. However, owing to the absence of viral gene expression in transduced cells, these HDAds are able to mediate high-level, long-term transgene expression in the absence of chronic toxicity. In addition, because the vector genomes exist episomally in transduced cells, the risks of germ line transmission and insertional mutagenesis leading to oncogenic transformation are negligible. Moreover, the deletion of the viral sequences permits a tremendous cloning capacity of ~37 kb allowing for the delivery of whole genomic loci, multiple transgenes, and large *cis*-acting elements. A summary of the current state of HDAd is presented here and a more comprehensive discussion can be found elsewhere.[1] Discussions regarding Ad and early-generation Ad vectors can be found elsewhere in this volume.

PRODUCTION AND CHARACTERIZATION

Because HDAds are deleted of all viral coding sequences, a helper virus is required for their propagation. The first efficient and currently most widely used method for generating HDAd is the Cre/loxP system developed by Graham and coworkers in 1996[2] (Figure 15.1). In this system the HDAd genome is first constructed in a bacterial plasmid. Minimally, the HDAd genome contains the expression cassette of interest and ~500 bp of *cis*-acting Ad sequences necessary for vector DNA replication (ITRs (inverted terminal repeat)) and packaging (Ψ). Because efficient packaging into the Ad capsid requires a genome size between ~27.7 and ~38 kb, "stuffer" DNA is usually included in HDAd. To rescue the HDAd (to convert the "plasmid form" of the HDAd genome into the "viral form"), the plasmid is first digested with the appropriate restriction enzyme to liberate the HDAd genome from the bacterial plasmid sequences. 293 cells expressing the site-specific Cre recombinase are then transfected with the linearized HDAd genome and subsequently infected with the helper virus. The helper virus bears a packaging signal flanked by loxP sites, the target sequence for Cre, and thus following infection of 293Cre cells, the packaging signal is excised from the helper viral genome by Cre-mediated site-specific recombination between the loxP sites. This renders the helper viral genome unpackagable but still able to undergo DNA replication and thus *trans*-complement the replication and encapsidation of the HDAd genome. The titer of the HDAd is increased by serial coinfection of 293Cre cells with the HDAd and the helper virus and the HDAd is finally purified by CsCl ultracentrifugation. Subsequent improvements of this original system permit efficient and simple large-scale HDAd production with yields of >10,000 vector particles per coinfected producer cell and with helper virus contamination levels of <0.02%.[3] Detailed methodologies for producing HDAd are described in detail elsewhere.[3,4]

IN VIVO APPLICATIONS

LIVER TRANSDUCTION

The liver is an attractive gene therapy target because the fenestrated endothelium permits exposure to intravenously delivered vector, hepatocytes are well suited for

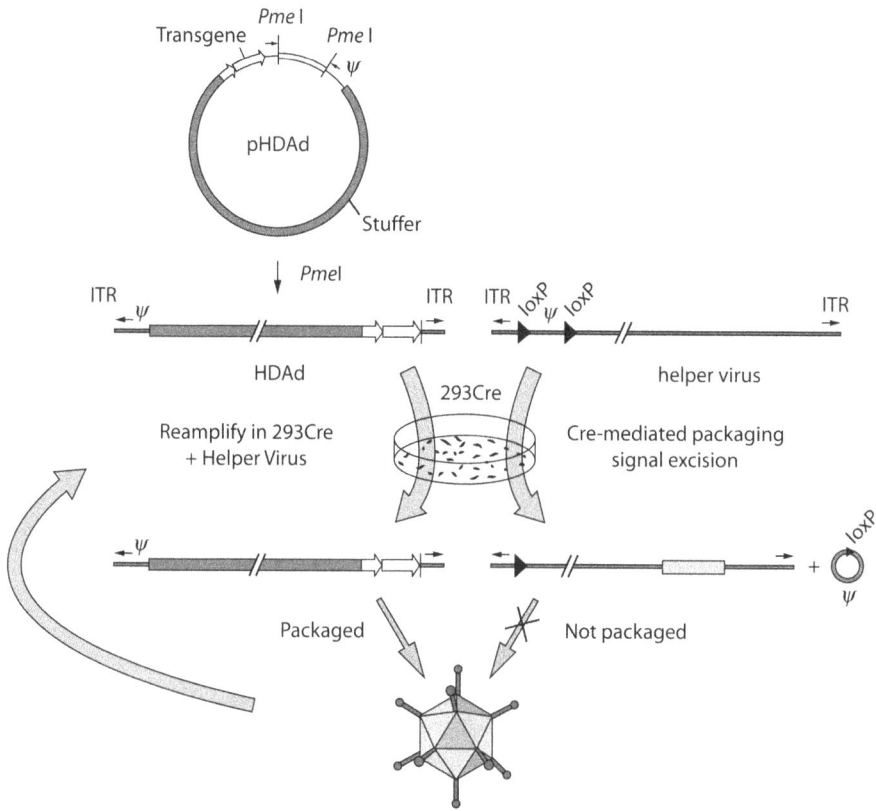

FIGURE 15.1   The Cre/loxP system for generating HD vectors. The HDAd contains only ~500 bp of *cis*-acting Ad sequences required for DNA replication (ITRs) and packaging (ψ), the remainder of the genome consists of the desired transgene and non-Ad "stuffer" sequences. The HDAd genome is constructed as a bacterial plasmid (pHDAd) and is liberated by restriction enzyme digestion (e.g., *Pme*I). To rescue the HDAd, the liberated genome is transfected into 293Cre cells and infected with a helper virus bearing a packaging signal (ψ) flanked by loxP sites. Cre-mediated excision of ψ renders the helper virus genome unpackagable, but still able to replicate and provide all of the necessary *trans*-acting factors for propagation of the HDAd. The titer of the HD vector is increased by serial coinfections of 293Cre cells with the HDAd and the helper virus.

secretion of therapeutic proteins into the circulation for systemic delivery, and it is the affected organ in many genetic disorders. HDAd vectors are particularly attractive vectors for liver-directed gene therapy because of their ability to efficiently transduce hepatocytes following intravenous injection.

To evaluate the utility of liver-directed, HDAd-mediated gene therapy in a large animal model, three baboons were intravenously injected with $3.3–3.9 \times 10^{11}$ vector particles per kilogram of a HDAd expressing human α_1-antitrypsin (hAAT).[5] Human α_1-antitrypsin antagonizes neutrophilic elastase and is abundantly expressed in hepatocytes and at a lower level in macrophages and α_1-antitrypsin-deficient patients have shortened life expectancies because of emphysema. Expression of

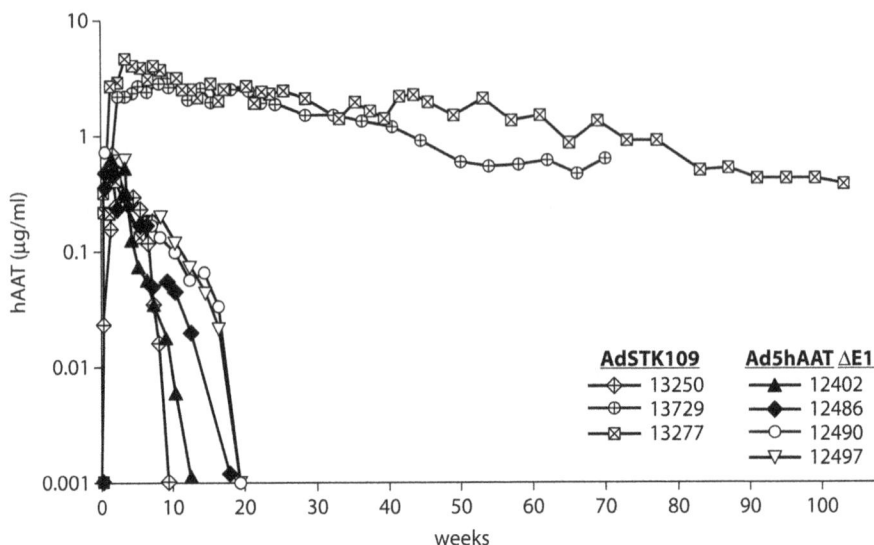

FIGURE 15.2 Serum levels of hAAT in baboons following intravenous administration of the HDAd AdSTK109 or the FGAd AdhAATΔE1. Baboons 12402 and 12486 were injected with 6.2×10^{11} particles/kg of AdhAATΔE1. Baboons 12490 and 12497 were injected with 1.4×10^{12} particles/kg of AdhAATΔE1. Baboons 13250, 13729, and 13277 were injected with 3.3×10^{11}, 3.9×10^{11}, and 3.6×10^{11} particles/kg, respectively, of AdSTK109.

hAAT persisted for more than 1 year in two of the three animals (Figure 15.2). Maximum levels of serum hAAT of 3–4 μg/mL were reached 3–4 weeks postinjection in these two baboons and slowly declined to 8% and 19% of the highest levels after 24 and 16 months, respectively. The slow decline in hAAT expression was attributed to the fact that the baboons were young (7.5 and 9 months old) when injected, and that the decrease in hAAT concentrations was correlative to the growth of the animals. The third baboon had significantly lower levels of serum hAAT that rapidly declined to undetectable levels after 2 months. This baboon had generated anti-hAAT antibodies thus accounting for the low level and rapid loss of serum hAAT. No abnormalities in blood cell counts and chemistries were observed in these three baboons at any time, starting 3 days postinjection. In contrast, hAAT expression lasted only 3–5 months in all four baboons injected with an FGAd expressing hAAT (Figure 15.2). This was shown to be due to the generation of a cellular immune response against viral proteins expressed from the vector backbone of the FGAd resulting in the elimination of transduced hepatocytes. These experiments convincingly demonstrated that HDAds were superior to FGAd with respect to duration of transgene expression and hepatotoxicity in a nonhuman primate.

The potential of liver-directed, HDAd-mediated gene therapy was investigated for the phenotypic correction of hypercholesterolemia in the apolipoprotein E-deficient (apoE$^{-/-}$) mouse model.[6] Apolipoprotein E, a 34-kDa plasma glycoprotein, is a component of all plasma lipoproteins except low-density lipoprotein

(LDL) and plays a major role in lipoprotein catabolism by acting as a ligand for the LDL-receptor (LDLR) and the LDLR-related protein for transport of excess cholesterol from the peripheral tissues to the liver for excretion. The apoE$^{-/-}$ mouse is an excellent model for cardiovascular disease because they develop severe hypercholesterolemia and atherosclerotic lesions similar to those found in humans. Chan and coworkers investigated correction of hypercholesterolemia in apoE$^{-/-}$ mice with either an FGAd or a HDAd expressing apoE.[6] Injection of apoE$^{-/-}$ mice with FGAd resulted in an immediate rise in plasma apoE levels and a concomitant drop in plasma cholesterol levels to within normal range. However, this effect was transient, as apoE levels rapidly declined to pretreatment levels by day 28 and plasma cholesterol levels increased after 28 days, returning to pretreatment levels by 112 days. In contrast, a single injection of HDAd resulted in immediate lowering of plasma cholesterol to subnormal or normal levels for the rest of the natural lifespan of the animal (2.5 years). Plasma apoE levels reached ~200% wild type and remained at supraphysiological levels for >4 months, only to decline slowly to wild-type levels at 1 year and remained at 60%–90% physiological concentrations for the lifetime of the animals (2.5 years). Analysis of total plasma cholesterol revealed normalization of the plasma lipoproteins back to a predominately high-density lipoprotein (HDL) pattern seen in wild-type mice. Aortas in all mice, examined at 2.5 years after treatment with HDAd, were essentially free of atherosclerotic lesions in contrast to saline-injected mice whose aortas were completely covered with lesions. Significantly, this study demonstrated that a single injection of HDAd encoding apoE could confer lifetime protection against aortic atherosclerosis. Toxicity studies revealed that whereas injection of FGAd resulted in significant chronic hepatotoxicity, no such evidence of damage was observed following injection of HDAd. In summary, this study demonstrated the tremendous potential of HDAds for gene therapy; a single injection of HDAd resulted in lifelong expression of a therapeutic transgene and permanent phenotypic correction in a mouse model of a genetic disease without long-term toxicity.

MUSCLE TRANSDUCTION

Duchenne muscular dystrophy (DMD) is a lethal, X-linked, degenerative muscle disease with a frequency of 1 in 3500 male births caused by mutations in the dystrophin gene. Dystrophin is an essential structural component of the skeletal muscle cell membrane, linking intracellular actin filaments with the dystrophin-associated proteins (DAPs) in the sarcolemma. Dystrophin deficiency results in instability of the muscle cell membrane causing muscle fiber degeneration. The length of the dystrophin cDNA (14 kb) precluded its inclusion into most gene therapy viral vectors but following the development of HDAds with large cloning capacity, gene transfer of the full-length dystrophin cDNA became feasible. Gilbert et al.[7] demonstrated that a single injection of a HDAd carrying two copies of the full-length human dystrophin cDNA resulted in transduction of 34% of the fibers of the total tibialis anterior (TA) muscle in neonatal *mdx* mice, a genetic and biochemical mouse model for DMD. The amount of dystrophin produced in these muscles was five times that in normal human muscle. However, only 7% transduction was achieved following injection into the TA muscle of adult *mdx* mice. In these transduced adult fibers, the amount of dystrophin produced was only 10% of the amount in normal humans. However, the high levels of transduction were transient and a humoral immune response was mounted against the foreign human dystrophin protein in the *mdx* mice. Importantly, such a response was not observed in immunodeficient SCID (severe combined immunodeficient) mice suggesting that sustained expression could be achieved in the absence of an immune response to the transgene product. Indeed, injection of an HDAd expressing the full-length murine dystrophin cDNA into the TA muscle of neonatal *mdx* mice resulted in sustained expression for at least 1 year in 52% of the muscle fibers. The treated muscle showed restored dystrophin–glycoprotein complex to the sarcolemma, significant improvement in isometric force production, resistance to high stress muscle contraction damage, and improved muscle histopathology.[8] This study demonstrates the tremendous potential of HDAd-mediated DMD gene therapy.

Brain Transduction

The brain is an attractive target for gene therapy to treat CNS (central nervous system) structural and functional deficits such as aging-related memory loss, Parkinson's disease, Alzheimer's disease, or amyotrophic lateral sclerosis. Zou et al.[9] compared the efficiency, toxicity, and persistence of HDAd and FGAd vector-mediated gene transfer into the hippocampus or lateral ventricle of the brain in 20-month-old rats. Transgene expression peaked 6 days postinjection for both vectors. In the hippocampus, transgene expression from the FGAd decreased rapidly after day 6, being significantly lower than HDAd-injected hippocampus by day 16 and undetectable by day 183. In the ventricle, transgene expression from FGAd was significantly lower than HDAd by day 33 and undetectable by day 66. In contrast, transgene expression from the HDAd remained relatively stable with expression remaining >60% of peak levels on day 183. Overall, expression from HDAd was significantly higher than from FGAd at all-time points after 6 days. FGAd vectors induced substantial inflammatory and immune responses that were significantly higher than those induced by HDAd at all-time points after 3 h. While both vectors induced a rapid increase in the proinflammatory cytokine that peaked at 3 h postinjection, by 3 days, the level was significantly lower in the HDAd-treated brains than the FGAd-treated brains.

Lung Transduction

The lung is another attractive target for gene transfer primarily because of the desire to develop gene therapy for cystic fibrosis (CF). CF is caused by recessive mutations in the CF transmembrane conductance regulatory (CFTR) gene that encodes a membrane chloride channel present in the epithelium of the lung and other organs. Morbidity and mortality in CF patients is due to lung disease characterized by inflammation, obstructive mucus, and persistent infection. HDAd vector efficiently transduces airway epithelial and submucosal cells in mice following intranasal administration, and in contrast to FGAd results in negligible inflammation, with transgene expression lasting for up to 15 weeks.[10] Moreover, intranasal administration of an HDAd bearing the CFTR gene resulted in expression of CFTR in the airway epithelial cells of CFTR-knockout mice and could protect the lung from bacterial challenge.[11] These studies suggest that HDAd-mediated CF gene therapy would benefit CF patients by reducing susceptibility to opportunistic pathogens.

ACUTE TOXICITY

In addition to long-term, chronic toxicity mediated by viral protein expression from the vector backbone, systemic administration of FGAd also results in acute toxicity. This acute toxicity occurs immediately following vector administration and is characterized by high-level inflammatory cytokine production, consistent with activation of an acute inflammatory response.[12–15] While the precise mechanism responsible for this acute response remains to be determined, it appears to be mediated by the viral capsid in a dose-dependent manner with potentially severe and lethal consequences. Indeed, the death of a partial OTC (Ornithine transcarbamylase)-deficient patient, whose clinical course was marked by systemic inflammatory response syndrome (SIRS), disseminated intravascular coagulation, and multi-organ failure, was attributed to acute toxicity from the administration of a second-generation (E1- and E4-deleted) Ad vector.[16] Because the viral capsid of early generation Ad and HDAd are identical, it has been hypothesized that HDAd would also provoke an identical acute response. Indeed, this was confirmed in nonhuman primates in which dose-dependent acute toxicity, consistent with activation of the innate inflammatory immune response, was observed following systemic administration of HDAd.[17] It is important to note that robust activation of the acute inflammatory response is also observed in mice given systemic Ad. However, unlike primates, lethal SIRS does not develop in rodents, even at high doses that are lethal to primates. This may reflect species-to-species differences in the quality of the innate immune response or sensitivities of the end organs to pathologic sequelae. This likely accounts for the plethora of studies reporting negligible toxicity in mice given high-dose HDAd and underscores the importance of safety and toxicity evaluations in larger animals for all gene therapy vectors.

CONCLUSION

HDAd vectors possess many characteristics that make them attractive vectors for gene therapy of a wide variety of genetic and acquired diseases. However, acute toxicity provoked by the viral capsid currently hinders the clinical application of this otherwise promising technology. Clearly, studies to elucidate the mechanism(s) of Ad-mediated activation of the innate inflammatory response are needed. Perhaps with a clearer understanding of this phenomenon strategies can be developed to minimize, if not eliminate, this acute toxic response. If this can be accomplished, then HDAd should be able to provide sustained, high-level transgene expression with no further long-term toxicity.

REFERENCES

1. Ng, P.; Graham, F.L. Helper-dependent adenoviral vectors for gene therapy. In *Gene Therapy: Therapeutic Mechanisms and Strategies*, 2nd Ed.; Templeton, N.S., Ed.; Marcel Dekker Inc.: New York, NY, **2004**; 53–70.

2. Parks, R.J.; Chen, L.; Anton, M.; Sankar, U.; Rudnicki, M.A.; Graham, F.L. A helper-dependent adenovirus vector system: Removal of helper virus by Cre-mediated excision of the viral packaging signal. Proc. Natl. Acad. Sci. U.S.A. **1996**, *93* (24), 13565–13570.

3. Palmer, D.J.; Ng, P. Improved system for helper-dependent adenovirus vector production. Mol. Ther. **2003**, *8* (5), 846–852.

4. Ng, P.; Parks, R.J.; Graham, F.L. Methods for the preparation of helper-dependent adenoviral vectors. Methods Mol. Med. **2001**, *69*, 371–388.

5. Morral, N.; O'Neal, W.; Rice, K.; Leland, M.; Kaplan, J.; Piedra, P.A.; Zhou, H.; Parks, R.J.; Velji, R.; Aguilar-Córdova, E.; Wadsworth, S.; Graham, F.L.; Kochanek, S.; Carey, K.D.; Beaudet, A.L. Administration of helper-dependent adenoviral vectors and sequential delivery of different vector serotype for long-term liver directed gene transfer in baboons. Proc. Natl. Acad. Sci. U.S.A. **1999**, *96* (22), 12816–12821.

6. Kim, I.H.; Jozkowicz, A.; Piedra, P.A.; Oka, K.; Chan, L. Lifetime correction of genetic deficiency in mice with a single injection of helper-dependent adenoviral vector. Proc. Natl. Acad. Sci. U.S.A. **2001**, *98* (23), 12387–13282.

7. Gilbert, R.; Liu, A.; Petrof, B.J.; Nalbantoglu, J.; Karpati, G. Improved performance of a fully gutted adenovirus vector containing two full-length dystrophin cDNAs regulated by a strong promoter. Mol. Ther. **2002**, *6* (4), 509–510.

8. Dudley, R.W.R.; Lu, Y.; Gilbert, R.; Matecki, S.; Nalbantoglu, J.; Petrof, B.J.; Karpati, G. Sustained improvement of muscle function one year after full-length dystrophin gene transfer into *mdx* mice by a gutted helper-dependent adenoviral vector. Hum. Gene Ther. **2004**, *15* (2), 145–156.

9. Zou, L.; Yuan, X.; Zhou, H.; Lu, H.; Yang, K. Helper-dependent adenoviral vector-mediated gene transfer in aged rat brain. Hum. Gene Ther. **2001**, *12* (2), 181–191.

10. Toietta, G.; Koehler, D.R.; Finegold, M.J.; Lee, B.; Hu, J.; Beaudet, A.L. Reduced inflammation and improved airway expression using helper-dependent adenoviral vectors with a K18 promoter. Mol. Ther. **2003**, *7* (5), 649–658.

11. Koehler, D.R.; Sajjan, U.; Chow, Y.H.; Martin, B.; Kent, G.; Tanswell, A.K.; McKerlie, C.; Forstner, J.F.; Hu, J. Protection of Cftr knockout mice from acute lung infection by a helper-dependent adenoviral vector expressing Cftr in airway epithelia. Proc. Natl. Acad. Sci. U.S.A. **2003**, *100* (26), 15364–15369.

12. Muruve, D.A.; Barnes, M.J.; Stillman, I.E.; Libermann, T.A. Adenoviral gene therapy leads to rapid induction of multiple cytokines and acute neutrophil-dependent haptic injury in vivo. Hum. Gene Ther. **1999**, *10* (6), 965–976.

13. Schnell, M.A.; Zhang, Y.; Tazelaar, J.; Gao, G.; Yu, Q.C.; Qian, R.; Chen, S.; Varnavski, A.N.; LeClair, C.; Raper, S.; Wilson, J.M. Activation of innate immunity in nonhuman primates following intraportal administration of adenoviral vectors. Mol. Ther. **2001**, *3* (5 Pt. 1), 708–722.

14. Zhang, Y.; Chirmule, N.; Gao, G.; Qian, R.; Croyle, M.; Joshi, B.; Tazelaar, J.; Wilson, J.M. Acute cytokine response to systemic adenoviral vectors in mice is mediated by dendritic cells and macrophages. Mol. Ther. **2001**, *3* (5 Pt. 1), 697–707.

15. Liu, Q.; Muruve, D.A. Molecular basis of the inflammatory response to adenovirus vector. Gene Ther. **2003**, *10* (11), 935–940.

16. Raper, S.E.; Chirmule, N.; Lee, F.S.; Wivel, N.A.; Bagg, A.; Gao, G.P.; Wilson, J.M.; Batshaw, M.L. Fatal systemic inflammatory response syndrome in a ornithine transcarbamylase deficient patient following adenoviral gene transfer. Mol. Genet. Metab. **2003**, *80* (1–2), 148–158.

17. Brunetti-Pierri, N.; Palmer, D.J.; Beaudet, A.L.; Carey, D.; Finegold, M.; Ng, P. Acute toxicity following high-dose systemic injection of helper-dependent adenoviral vectors into non human primates. Hum. Gene Ther. **2004**, *15* (1), 35–46.

16 Retroviral Vectors

Erin L. Weber, Paula M. Cannon
Childrens Hospital Los Angeles, University of Southern
California Keck School of Medicine, Los Angeles,
California, U.S.A.

CONTENTS

INTRODUCTION

The goal of gene therapy is to treat disease by the introduction of new genetic material into affected or susceptible cells of the body. Gene therapy has been proposed as a treatment for a variety of disorders. Monogenic diseases may be treated simply by replacement of the defective gene with a functional copy. Alternatively, the new genetic material may selectively eliminate tumor cells, provide protection against viral infection, or stimulate the immune system against a specific antigen.

Initiated in 1990, the first gene therapy clinical trial aimed to introduce functional copies of the adenosine deaminase gene into deficient T cells using a retroviral vector.[1] Today, more than 600 clinical trials have been approved, a third of which employs retroviral vectors. The website for the *Journal of Gene Medicine* (http://www.wiley.co.uk/genetherapy/clinical/) provides helpful statistical summaries of approved gene therapy clinical trials to date. The majority of proposed gene therapy treatments are still being evaluated as phase I trials; only one phase III clinical trial has been conducted using retroviral vectors.[2] This trial involved injection of glioblastoma multiforme tumors with murine producer cells to yield in situ production of retroviral vectors expressing the thymidine kinase suicide gene. Unfortunately, therapeutic effectiveness was not improved over standard treatments of surgical resection and radiotherapy.

In general, despite major improvements in vectors since their first clinical use, the therapeutic effectiveness of retroviral vectors remains limited. Advantages and limitations of retroviral vectors, as well as current areas of research, will be discussed in the following review.

DOI: 10.1201/9781003247432-16

OVERVIEW

RETROVIRAL LIFE CYCLE

Retroviruses are RNA viruses containing two copies of a single-stranded RNA genome encapsidated by a protein core and outer lipid layer or envelope. The retroviral genome codes for three basic polyproteins. The *gag* gene codes for structural proteins of the viral core, whereas the *pol* gene generates viral protease, reverse transcriptase, and integrase. The third gene, *env*, codes for a glycoprotein that, following incorporation into the outer lipid layer, directs the retrovirus to bind to receptors on target cells, triggering fusion of the viral lipid envelope with the cell membrane. Following entry of the retroviral core into the cell, the viral RNA genome is copied into a double-stranded DNA provirus by reverse transcriptase. The proviral DNA is then inserted into the host genome by the integrase enzyme, and viral proteins are synthesized from the integrated provirus using cellular machinery (Figure 16.1).

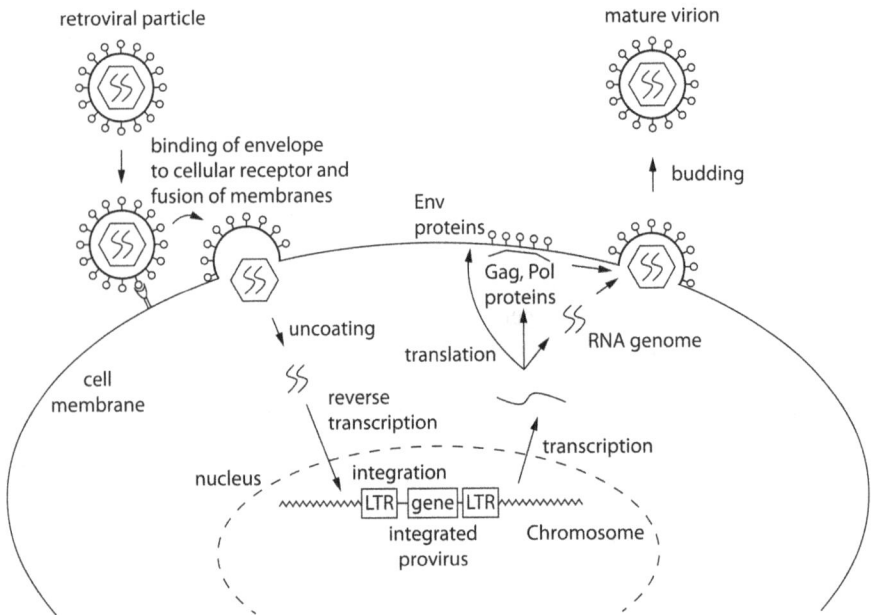

FIGURE 16.1 The retroviral life cycle begins with interaction of the viral envelope with its host receptor, followed by fusion of viral and cellular membranes and entry of the viral core. The viral core disassembles (uncoating) and the viral genome is reverse transcribed into DNA, enters the nucleus, and is integrated into the host genome. The integrated provirus is transcribed by cellular transcriptional machinery, and Gag, Pol, and Env proteins are synthesized. Viral proteins and the newly transcribed viral genome colocalize at the cellular membrane to form viral particles, which bud into the extracellular space.

Retroviral Vectors

Stable integration of the viral genome into host chromosomes allows the possibility of long-term gene expression, making retrovirus-based vectors prime candidates for correction of a variety of deficiencies. The most common retroviral vectors used for gene therapy clinical trials are based on the murine leukemia virus (MLV). The most current vector system is composed of three components: a vector genome that contains only those viral sequences necessary for packaging, reverse transcription, and integration, and which can accept approximately 6–8 kb of exogenous DNA; a packaging construct that provides Gag and Pol proteins in trans but is not packaged due to a deletion in the packaging signal (ψ); and an envelope construct that codes for the Env protein (Figure 16.2).[3] Removal of viral genes from the vector genome and their provision in trans generates replication-defective viral particles, capable of delivering genes to a cell but incapable of subsequent rounds of replication.

LIMITATIONS OF RETROVIRAL VECTORS

For retroviral vectors to be maximally effective, several limitations need to be addressed. First, there is a limited vector size that can be efficiently packaged within the viral core. Additionally, the presence of two copies of the vector genome and the nature of the inserted sequences themselves may contribute to vector instability and

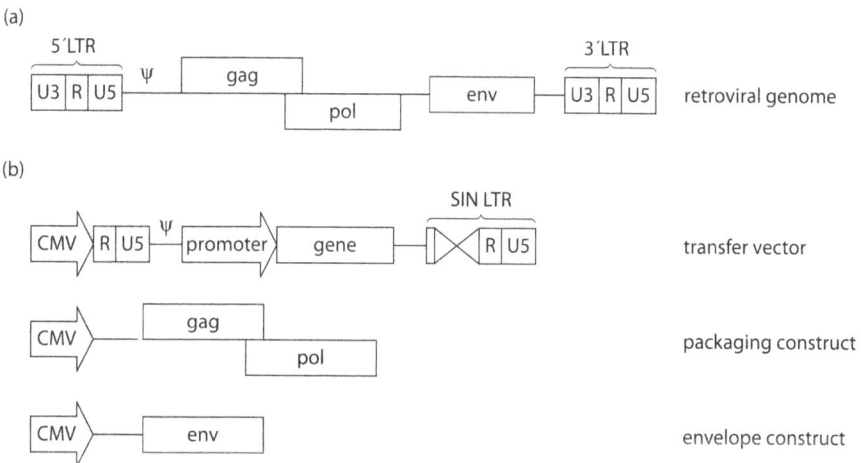

FIGURE 16.2 (a) The retroviral genome is composed of three genes, *gag, pol,* and *env,* flanked by long terminal repeats (LTR) which contain sequences necessary for initiation and termination of transcription, as well as signals for integration into chromosomal DNA. The packing signal, ψ, is required for packing of the retroviral genome into virions. (b) Production of retroviral vector particles is achieved by cotransfection of three plasmids and expression of the viral proteins in trans. The transfer vector is the only sequence packaged into the vector particle as the packaging signal has been removed from the other constructs. Safety of the transfer vector has been improved by deletion of viral enhancer and promoter sequences from the 3′LTR, creating a self-inactivating (SIN) vector with no active viral promoters.

rearrangements during reverse transcription. Beyond these intrinsic limitations of viral structure, improvements are required in three major areas: transduction efficiency, gene expression, and safety.

IMPROVING TRANSDUCTION EFFICIENCIES

The number of cells transduced, or infected, by retroviral vectors is often insufficient to produce a therapeutic effect. MLV vectors require breakdown of the nuclear membrane during cell division for entry into the nucleus and integration into the genome, thus limiting the number of cells that can be transduced. The traditional approach for enhancement of MLV vector transduction efficiencies has been stimulation of cell proliferation by the addition of various growth factors ex vivo.[4] Alternatively, lentiviruses, a subclass of retroviruses, possess nucleophilic signals that permit entry into the nuclei of nondividing cells. However, lentiviral-based vectors are not yet widely used in clinical trials.

The entry of a retrovirus into a cell is also largely determined by binding of the envelope glycoprotein to its cellular receptor, and the distribution of the receptor on various cell types will influence the vector's host range. Envelope proteins from heterologous retroviruses can be used to create pseudotyped vectors with broadened host ranges. For example, pseudotyping with the VSV-G rhabdovirus envelope protein is a common technique used to improve stability and extend the host range of retroviral vectors.[5]

Although pseudotyping and stimulation of cell proliferation have been successful in improving transduction efficiencies of particular cell types ex vivo, such expansion of viral host range may not provide the cell specificity required for gene delivery in vivo. To improve transduction efficiencies in vivo, several methods involving modification of natural retroviral envelopes have been explored to target vectors to a specific cell type (Figure 16.3). The retroviral Env protein is composed of two subunits, the surface (SU) protein containing the receptor recognition domain and the transmembrane (TM) protein, which anchors the complex within the viral lipid envelope. Binding of SU to its receptor is thought to trigger conformational changes, which result in exposure of a fusion peptide on the TM subunit and subsequent fusion of viral and cellular membranes. Most attempts to reengineer the receptor binding site of the SU subunit have involved replacement of the natural receptor-binding domain with a ligand or single-chain antibody with tropism for an alternative cell-surface molecule and have resulted in a lack of virus–cell fusion.[6] Some strategies have resulted in limited success, such as the replacement of the MLV Env receptor-binding surface of SU with the peptide ligand, SDF-1α, which resulted in transduction of human cells via the SDF-1α receptor, CXCR4.[7] However, transduction efficiencies were low.

A second approach to Env targeting, called tethering, concentrates vectors on the surface of specific cell types by the addition of a second binding moiety to the natural Env protein. The native receptor-binding domain remains functional and mediates virus–cell fusion. The insertion of a von Willebrand factor-derived collagen-binding sequence at the N-terminus of the MLV envelope has resulted in improved gene

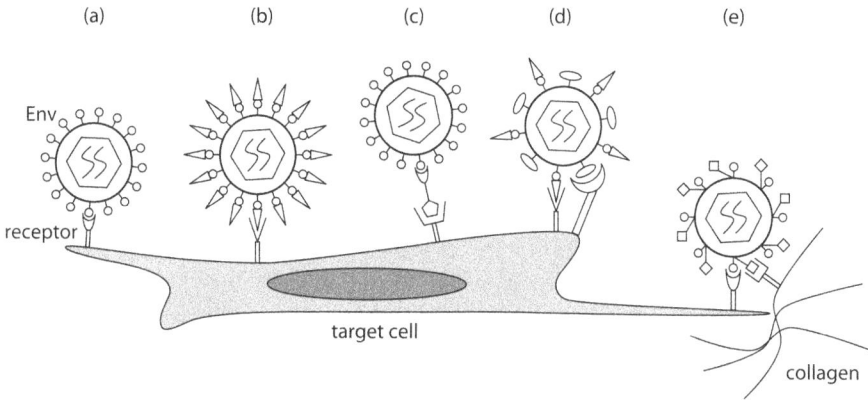

FIGURE 16.3 Efforts to improve transduction of distinct cell populations have involved modification of viral envelope proteins. (a) Retroviral infection begins with interaction of the native retroviral Env with a susceptible cellular receptor. (b) Replacement of the native Env receptor-binding domain with a peptide ligand may direct retroviral binding to a specific receptor. (c) Adaptor proteins contain ligand binding sites from natural Env receptors linked to alternative ligands in an effort to direct retroviral binding to specific receptors. (d) Trans-complementation of binding-defective influenza hemagglutinin proteins and fusion-defective retroviral Env proteins expands the retroviral host range while circumventing fusion defects often seen with retroviral envelopes possessing an alternative ligand at the receptor bind-ing site. (e) Attachment of a peptide ligand to the retroviral envelope protein outside of the receptor-binding domain allows concentration of retroviral particles in the vicinity of target cells by interaction with receptors on nearby structures, e.g., collagen.

delivery in vivo to sites of exposed extracellular matrix within tumor vasculature following systemic administration in mice.[8]

An alternate approach, the use of adaptor proteins, which function as a bridge between the native MLV Env and cell surface receptors, has generally failed due to a lack of virus–cell fusion. However, greater success has been observed with the avian leukosis virus (ALV) Env protein. Use of an adaptor protein consisting of the single-chain antibody, MR1, which binds to a tumor-specific form of the EGF receptor, joined to the extracellular domain of the natural ALV receptor resulted in successful targeting of ALV-pseudotyped MLV vectors to cells expressing the tumor-specific EGF receptor.[9]

Finally, *trans*-complementation represents a strategy for circumventing the fusion defect. Orthomyxoviruses and paramyxoviruses possess two distinct envelope pro-teins which mediate binding and fusion functions separately, suggesting that seg-regation of the two functions may be possible for retroviral vectors as well. Lin et al. have demonstrated improved transduction efficiency into cells expressing Flt-3 by coexpression of a binding-defective influenza hemagglutinin protein and a fusion-defective MLV Env protein possessing the Flt-3 ligand at the receptor-binding domain on an MLV vector.[10]

As an alternative method to improve transduction of tumors in vivo, the use of replication-competent retroviral (RCR) vectors has recently become of interest. As the majority of cells in the body are quiescent, transduction by MLV-based retroviral vectors should be selective for tumor cells. Following the initial transduction event, RCR vectors would replicate, essentially turning each transduced cell into a virus producer cell, and perpetuating gene delivery throughout the tumor. Importantly, RCR vectors typically contain a suicide gene, which is toxic to the transduced cell following prodrug administration, thus eliminating tumor cells and minimizing the spread of virus to noncancerous cells. Injection of human glioma cell tumors in rats with MLV-based RCR vectors delivering the cytosine deaminase suicide gene showed 70–90% transduction rates, compared to 1% transduction with replication-defective vectors, and 100% survival of the rats for at least 60 days.[11] The RCR vectors were not detected in other tissues.

IMPROVING GENE EXPRESSION

Variable levels of gene expression are observed following integration of current retroviral vectors into target cells. These can be attributed to chromosomal position effects at the site of integration, as well as limitations of vector function. Depending on the site of integration, vector genes may be silenced by a closed chromatin environment or subject to inappropriate regulation by host elements. Expression of retroviral vectors is restricted in several cell types, including embryonic stem cells and murine fibroblasts, as a result of the association of repressive factors with sequences in or near the 5′LTR and methylation of viral promoter elements. Additionally, host cells have a tendency to recognize foreign promoters, especially strong viral promoters such as SV40 and CMV typically used to drive expression of the gene of interest, and inactivate them by various mechanisms, including methylation. Even if gene expression remains active, transduced cells often lose viability over time because of an immune response elicited by the product of the delivered gene.

Replacement of repressive viral sequences with similar sequences from heterologous retroviruses and the minimization of CpG dinucleotides within viral promoters has lessened, but not eliminated, silencing of transgene expression.[12] Silencing of viral promoters has sparked an interest in the use of gene-specific regulatory sequences to direct transgene expression. Logg et al. have reported prostate-specific gene expression, which persists over time by incorporation of the prostate-specific probasin promoter into RCR vectors, demonstrating the benefits of improved gene expression and tissue-specificity achieved with an authentic human promoter.[13] Incorporation of gene-specific locus control regions (LCR) into retroviral vectors confers integration site-independent and tissue-specific vector gene expression.[14] Additionally, the use of authentic genomic elements, such as scaffold or matrix attachment regions, and insulator elements has been reported to improve transgene expression when included in a retroviral vector.[15]

IMPROVING VECTOR SAFETY

As retroviral transduction efficiencies and gene expression levels improve, so will the potential for serious adverse effects including retroviremia, generation of replication-competent revertants, germline transmission, and insertional mutagenesis. The recent demonstration that retroviral vectors favor integration sites within transcriptionally active loci has increased the probability of insertional mutagenesis caused by vector integration.[16] Additionally, the development of leukemia-like disease in two French SCID-X1 gene therapy patients as a result of retroviral vector integration in the vicinity of a known oncogene, LMO2, has underscored the necessity for improved safety mechanisms within gene therapy vectors.

Many of the vector modifications described above to improve transduction efficiency and gene expression may also improve vector safety. Elimination of dispensable viral sequences from the vector, separation of necessary viral genes (*gag, pol, env*) onto individual packaging vectors, and replacement of viral sequences with heterologous retroviral components or exogenous *cis*-acting regulatory elements all combine to reduce the risk of homologous recombination between vectors and endogenous viruses and the development of chimeric replication-competent vectors. Additionally, transcriptional targeting, or the use of gene-specific regulatory elements to limit gene expression to desired cell types, will help prevent expression in inappropriate cell types. Targeting of vectors to specific cell types or tissues by reengineering of envelope proteins will also limit gene expression to appropriate tissues. The use of insulator elements and matrix attachment regions within vectors may prevent inappropriate activation of cellular oncogenes by confining the action of vector regulatory elements to vector transgenes. Finally, attempts to target integration to specific genomic locations by modification of the integrase enzyme may eventually direct vector integration to benign sites within the genome. Tan et al. have shown that fusion of the synthetic polydactyl zinc finger protein, E2C, to the HIV-1 integrase results in preferential vector integration near the E2C binding site in vitro. The E2C binding sequence is a unique site within the human genome; however, preferential integration has not yet been demonstrated in vivo.

RETROVIRAL VECTORS AS PHARMACEUTICALS

The development of gene therapy vectors for large-scale therapeutic use will also require improvements in several areas. As viral vectors can only be made by living cells, large-scale manufacturing of gene therapy vectors will require the establishment of producer cell lines, which consistently yield stable, unrearranged viral vectors. Additionally, purification of vectors from producer cell supernatant is a cumbersome procedure and all processes must be carried out using Good Manufacturing Practices and the highest levels of quality control. Safety of viral vectors, namely, the potential to generate replication-competent vectors by recombination among vector components and endogenous retroviruses and insertional mutagenesis, is of significant concern in the development of vector therapy for use in humans. Finally, with in vivo vector therapy as a future goal, innovations are required which will minimize innate and humoral immune responses to systemically delivered vectors, especially inactivation of vectors by human complement and the possible development of anti-vector antibodies.

CONCLUSION

The therapeutic potential of retroviral vectors was first recognized over a decade ago. Due mainly to their ability to integrate into the host genome and the potential for long-term gene expression, retroviral vectors have now become one of the most widely used vehicles for gene therapy. However, despite many beneficial modifications, retroviral vectors are limited by the inability to transduce nondividing cells and inconsistent levels of gene expression. Additionally, with the recent development of leukemia-like disease due to insertional mutagenesis in two patients enrolled in a retroviral vector clinical trial, the safety of retroviral vectors is now a significant concern.

REFERENCES

1. Blaese, R.M.; Culver, K.W.; Miller, A.D.; Carter, C.S.; Fleisher, T.; Clerici, M.; Shearer, G.; Chang, L.; Chiang, Y.; Tostoshev, P.; Greenblatt, J.J.; Rosenberg, S.A.; Klein, H.; Berger, M.; Mullen, C.A.; Ramsey, W.J.; Muul, L.; Morgan, R.A.; Anderson, W.F. T lymphocyte-directed gene therapy for ADA-SCID: Initial trial results after 4 years. Science **1995**, *270* (5235), 475–480.
2. Rainov, N.G. A phase III clinical evaluation of herpes simplex virus type 1 thymidine kinase and ganciclovir gene therapy as an adjuvant to surgical resection and radiation in adults with previously untreated glioblastoma multiforme. Hum. Gene Ther. **2000**, *11* (17), 2389–2401.
3. Soneoka, Y.; Cannon, P.M.; Ramsdale, E.E.; Griffiths, J.C.; Romano, G.; Kingsman, S.M.; Kingsman, A.J. A transient three-plasmid expression system for the production of high titer retroviral vectors. Nucleic Acids Res. **1995**, *25* (4), 628–633.
4. Hacein-Bey, S.; Gross, F.; Nusbaum, P.; Hue, C.; Hamel, Y.; Fischer, A.; Cavazzana-Calvo, M. Optimization of retroviral gene transfer protocol to maintain the lymphoid potential of progenitor cells. Hum. Gene Ther. **2001**, *12* (3), 291–301.
5. Yee, J.K.; Friedmann, T.; Burns, J.C. Generation of high-titer pseudotyped retroviral vectors with very broad host range. Methods Cell Biol. **1994**, *43* (Pt A), 99–112.
6. Zhao, Y.; Zhu, L.; Lee, S.; Li, L.; Chang, E.; Soong, N.W.; Douer, D.; Anderson, W.F. Identification of the block in targeted retroviral-mediated gene transfer. Proc. Natl. Acad. Sci. U.S.A. **1999**, *96* (7), 4005–4010.
7. Katane, M.; Takao, E.; Kubo, Y.; Fujita, R.; Amanuma, H. Factors affecting the direct targeting of murine leukemia virus vectors containing peptide ligands in the envelope protein. EMBO Rep. **2002**, *3* (9), 899–904.
8. Gordon, E.M.; Chen, Z.H.; Liu, L.; Whitley, M.; Wei, D.; Groshen, S.; Hinton, D.R.; Anderson, W.F.; Beart, R.W.; Hall, F.L. Systemic administration of a matrix-targeted retroviral vector is efficacious for cancer gene therapy in mice. Hum. Gene Ther. **2001**, *12* (2), 193–204.
9. Snitkovsky, S.; Young, J.A. Cell-specific viral targeting mediated by a soluble retroviral receptor–ligand fusion protein. Proc. Natl. Acad. Sci. U.S.A. **1998**, *95* (12), 7036–7038.
10. Lin, A.H.; Kasahara, N.; Wu, W.; Stripecke, R.; Empig, C.L.; Anderson, W.F.; Cannon, P.M. Receptor-specific targeting mediated by the coexpression of a targeted murine leukemia virus envelope protein and a binding-defective influenza hemagglutinin protein. Hum. Gene Ther. **2001**, *12* (4), 323–332.
11. Wang, W.J.; Tai, C.K.; Kasahara, N.; Chen, T.C. Highly efficient and tumor-restricted gene transfer to malignant gliomas by replication-competent retroviral vectors. Hum. Gene Ther. **2003**, *14* (2), 117–127.

12. Challita, P.M.; Skelton, D.; el-Khoueiry, A.; Yu, X.J.; Weinberg, K.; Kohn, D.B. Multiple modifications in cis elements of the long terminal repeat of retroviral vectors lead to increased expression and decreased DNA methylation in embryonic carcinoma cells. J. Virol. **1995**, *69* (2), 748–755.

13. Logg, C.R.; Logg, A.; Matusik, R.J.; Bochner, B.H.; Kasahara, N. Tissue-specific transcriptional targeting of a replication-competent retroviral vector. J. Virol. **2002**, *76* (24), 12783–12791.

14. Li, Q.; Peterson, K.R.; Fang, X.; Stamatoyannopoulos, G. Locus control regions. Blood **2002**, *100* (9), 3077–3086.

15. West, A.G.; Gaszner, M.; Felsenfeld, G. Insulators: Many functions, many mechanisms. Genes Dev. **2002**, *16* (3), 271–288.

16. Wu, X.; Li, Y.; Crise, B.; Burgess, S.M. Transcription start regions in the human genome are favored targets for MLV integration. Science **2003**, *300* (5626), 1749–1751.

17 Liposomal Nonviral Delivery Vehicles

Nancy Smyth Templeton
Baylor College of Medicine, Houston, Texas, U.S.A.

CONTENTS

INTRODUCTION

Optimization of cationic liposomal complexes for in vivo applications is complex, involving many diverse components, including nucleic acid purification, plasmid design, delivery vehicle formulation, administration route and schedule, dosing, and detection of gene expression. Optimizing all components of the delivery system is pivotal and will allow broad use of liposomal complexes to treat human diseases or disorders. This chapter will highlight the features of liposomes that contribute to successful delivery, gene expression, and efficacy.

Delivery of nucleic acids using liposomes is promising as a safe and nonimmunogenic approach to gene therapy and would overcome the numerous disadvantages of viral vectors. Furthermore, gene therapeutics composed of artificial reagents can be standardized and regulated as drugs rather than as biologics. Cationic lipids have been used for the efficient delivery of nucleic acids to cells in tissue culture for several years.[1] Much effort has also been directed toward developing cationic liposomes for the efficient delivery of nucleic acids in animals and humans. Most frequently, the formulations that are best to use for transfection of a broad range of cell types in culture are not optimal for achieving efficacy in animals or people. Functional properties defined in vitro do not assess the stability of the complexes in plasma, pharmacokinetics, biodistribution, and colloidal properties that are essential for optimal activity in vivo. Nucleic acid–liposome complexes must also traverse tight barriers in vivo and penetrate throughout the target tissue to produce efficacy for the treatment of many diseases. These are not issues for achieving efficient transfection of cells in culture with the exception of polarized tissue culture cells. Therefore, optimized in vivo liposomes may differ from those used for efficient tissue culture transfection.

Different types of nucleic acids of unlimited size can be delivered using liposomes. Recent advances have dramatically improved transfection efficiencies and efficacy of liposomal vectors.[2–6] We demonstrated broad efficacy of a robust

DOI: 10.1201/9781003247432-17

liposomal delivery system in small and large animal models for lung,[3] breast,[5] head and neck (Hung and Templeton, unpublished data), and pancreatic cancers,[4] and for hepatitis B and C (Clawson and Templeton, unpublished data). This liposomal delivery system is currently used in a clinical trial to treat non-small-cell lung cancer and will be used in upcoming clinical trials to treat other cancers and cardiovascular diseases. Reviews of other in vivo delivery systems and improvements using cationic liposomes have been published recently.[7,8]

OPTIMIZATION OF LIPOSOMAL DELIVERY

Efficient in vivo nucleic acid–liposome complexes have unique features, including their morphology, mechanisms for cell and nuclear entry, targeted delivery, and ability to penetrate across tight barriers and throughout target tissues. Liposomes have different morphologies based on their composition and the formulation method that contribute to their ability to deliver nucleic acids in vivo. Formulations frequently used for the delivery of nucleic acids are lamellar structures, including small unilamellar vesicles (SUVs), multilamellar vesicles (MLVs), or bilamellar invaginated vesicles (BIVs) recently developed in our laboratory (Figure 17.1). Hexagonal structures

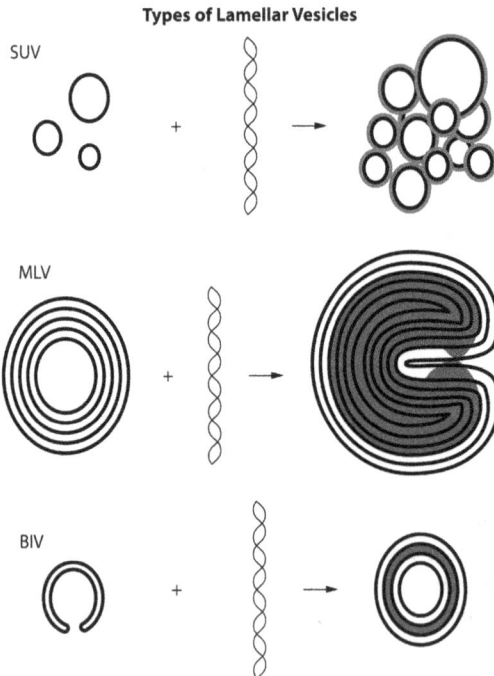

Types of Lamellar Vesicles

FIGURE 17.1   Diagrams drawn from cryoelectron micrographs of lamellar liposomes and complexes. SUVs, "spaghetti and meatballs" structures, condense nucleic acids on the surface. MLV complexes appear as "Swiss rolls." BIVs efficiently encapsulate nucleic acids between two bilamellar invaginated structures.

Source: From Ref. [2].

have demonstrated efficiency primarily for the transfection of some cell types in culture and not for in vivo delivery. SUVs condense nucleic acids on the surface and form "spaghetti and meatballs" structures.[9] SUV complexes produce little or no gene expression upon systemic delivery, although these complexes transfect numerous cell types efficiently in vitro.[1] Furthermore, SUV liposome–DNA complexes cannot be targeted efficiently.

SUV complexes also have a short half-life in circulation, about 5–10 min. Polyethylene glycol (PEG) has been added to liposome formulations to extend their half-life[10–12]; however, PEGylation created other unresolved problems. PEG also hinders delivery of cationic complexes into cells because of its sterically hindering ionic interactions and interferes with optimal condensation of nucleic acids into complexes. Furthermore, extremely long half-life in the circulation, e.g., several days, caused problems for patients because the bulk of the PEGylated liposomal formulation, doxil, which encapsulates the cytotoxic agent, doxorubicin, accumulates in the skin, hands, and feet. Patients contract mucositis and hand and foot syndrome[13,14] that cause extreme discomfort to the patient. Adding ligands to doxil for delivery to specific cell surface receptors did not result in much cell-specific delivery, and most of the injected targeted formulation still accumulated in the skin, hands, and feet. Addition of PEG in BIVs also caused steric hindrance that prevented efficient encapsulation of nucleic acids, and gene expression was substantially diminished. MLV complexes appear as "Swiss rolls" when viewing cross sections by cryoelectron microscopy.[15] These complexes can become too large for systemic administration or deliver nucleic acids inefficiently into cells because of inability to "unravel" at the cell surface. Addition of ligands onto MLV liposome–DNA complexes further aggravates these problems.

BIVs efficiently encapsulate large amounts of nucleic acids of any size between two BIVs.[2] We created these unique structures using 1,2-bis(oleoyloxy)-3-(trimethylammino)propane (DOTAP) and synthetic cholesterol (Chol) and a novel formulation procedure. Addition of other DNA condensing agents, including polymers, is not necessary, and encapsulation of nucleic acids by BIVs alone is spontaneous and immediate. The BIV complexes are also large enough so that they are not cleared rapidly by Kupffer cells in the liver and yet extravasate and penetrate across tight barriers. BIVs penetrated barriers, including the endothelial cell barrier in normal mice and the posterior blood retinal barrier in adult mouse eyes, and diffused throughout large tumors[3] and several layers of smooth muscle cells in pig arteries. We demonstrated efficacy for the treatment of non-small-cell lung cancer[3] using BIV DOTAP:Chol-p53 DNA–liposome complexes and not by using SUV DOTAP:Chol-p53 DNA–liposome complexes. Therefore, morphology of the complexes is essential.

A common belief is that artificial vehicles must be 100 nm or smaller to be effective for systemic delivery. However, this belief is most likely true only for large, inflexible delivery vehicles. Blood cells are several micrometers (up to 7000 nm) in size and yet have no difficulty circulating in the blood, including through the smallest capillaries. However, sickle cell blood cells, which are rigid, do have problems in the circulation. Therefore, we believe that flexibility is a more important issue than size. In fact, BIV DNA–liposome complexes in the size range of 200–450 nm produced the highest levels of gene expression in all tissues after intravenous injection.[2]

Kupffer cells in the liver quickly clear delivery vehicles (including nonviral vectors and viruses) that are not PEGylated and are smaller than 200 nm. Therefore, increased size of liposomal complexes could extend their circulation time particularly when combined with the injection of high colloidal suspensions. BIVs encapsulate nucleic acids and viruses apparently because of the presence of cholesterol in the bilayer, whereas formulations, including L-alpha dioleoyl phosphatidylethanolamine (DOPE), instead of cholesterol could not assemble nucleic acids by a "wrapping type" of mechanism and produce little gene expression in the lungs and no expression in other tissues after intravenous injections. The DOTAP:Chol BIV complexes are flexible and not rigid, are stable in high concentrations of serum, have extended half-life, and circulate efficiently in the bloodstream.

Colloidal properties of nucleic acid–liposome complexes also determine the levels of gene expression produced after in vivo delivery.[2,16] These properties include the DNA:lipid ratio, the overall charge density of the complexes, complex size and shape, lipid composition, formulation, and encapsulation efficiency and the colloidal suspension monitored by turbidity measurements at OD400. Our data showed that transfection efficiency in all tissues corresponded to the OD400 of complexes measured prior to intravenous injection. Colloidal properties affect serum stability, protection from nuclease degradation, blood circulation time, and biodistribution of the complexes.

The BIV complexes efficiently deliver DNA into cells by fusion with the cell membrane and avoid the endocytic pathway (Figure 17.2). Cells are negatively charged on the surface, and cationic complexes have nonspecific ionic charge interactions with cell surfaces that, in part, contribute to efficient transfection. Therefore, we create targeted delivery of our complexes in vivo without the use of PEG to retain predominant entry into cells by direct fusion. These ligand-coated complexes reexpose the overall positive charge by shedding the "mask" as they approach the target cells, using a reversible masking technology to bypass nontarget organs and tissues. Through ionic interactions or covalent attachments, we have added monoclonal antibodies, Fab fragments, proteins, partial proteins, peptides, peptide mimetics,

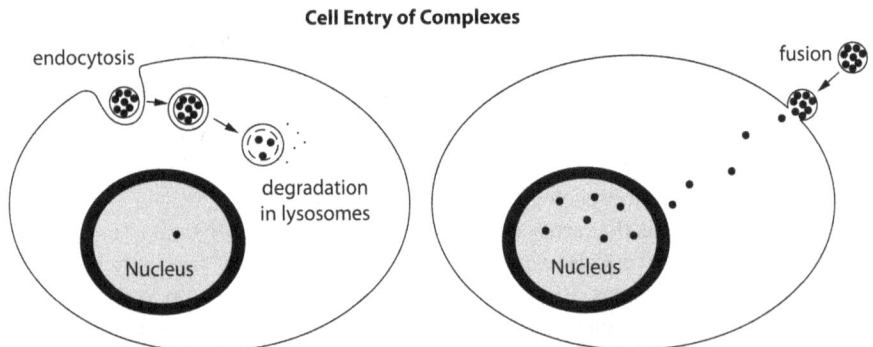

Cell Entry of Complexes

FIGURE 17.2 Mechanisms for cell entry of complexes by endocytosis or direct fusion with the cell membrane. Cell fusion allows the delivery of more nucleic acids to the nucleus because the bulk of the nucleic acids does not enter endosomes.

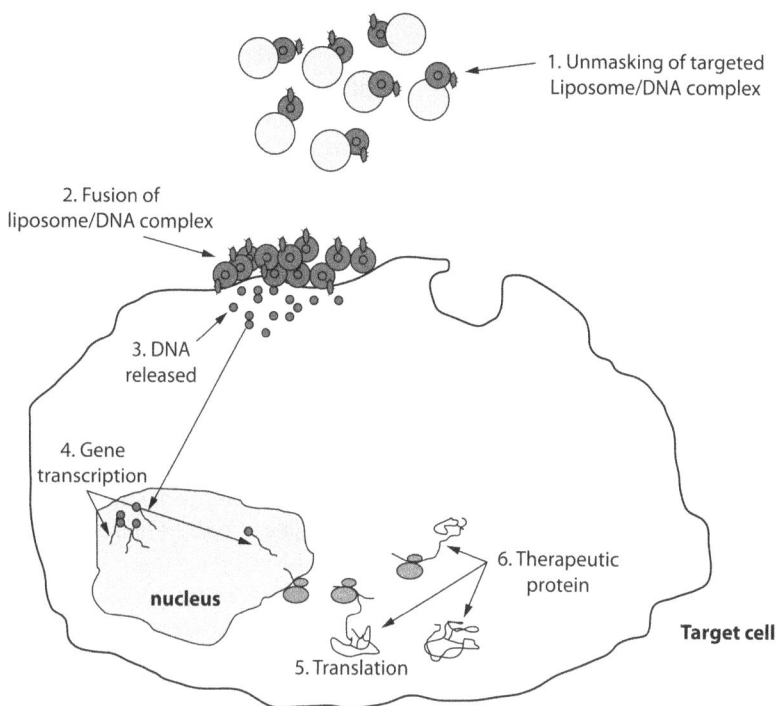

FIGURE 17.3 Optimized strategy for delivery and gene expression in the target cell, including targeted delivery, deshielding, fusion with the cell membrane, entry of nucleic acids into the cell and to the nucleus, and production of gene expression of a cDNA cloned in a plasmid.

small molecules, and drugs to the surface of BIV complexes after mixing. Using novel methods for the addition of these ligands to the complexes results in further increased gene expression in the target cells after transfection. Figure 17.3 shows our optimized strategy to achieve targeted delivery, deshielding, cell fusion, cell and nuclear entry of nucleic acids, and production of gene expression.

Stability of BIV complexes was studied at 37°C out to 24 hr at concentrations of serum ranging from 0% to 100%. The results showed serum stability at the highest concentrations of serum, about 70%–100%, which are physiological concentrations of serum found in the bloodstream, and at no or low concentrations of serum. The complexes were unstable at 10%–50% serum, perhaps because of salt bridging. Therefore, in vitro optimization of serum stability for formulations of cationic complexes must be performed over a broad range of serum concentration to be useful for applications in vivo.

Delivery of DNA to the nucleus and subsequent gene expression may be poorly correlated.[6,17] The following issues should be considered independent of the delivery formulation, including suboptimal promoter-enhancers in the plasmid, poor preparation of plasmid DNA, and insensitive detection of gene expression. Plasmid expression cassettes typically have not been optimized for animal studies. For example, many plasmids lack a full-length CMV promoter–enhancer, and some variations produce greatly reduced or no gene expression in certain cell types.[6] Ideally, investigators design custom promoter–enhancer chimeras that produce the highest levels of gene expression in their target cells. For example, we designed a systematic approach for customizing plasmids used for breast cancer gene therapy by using expression profiling.[6] To increase long-term expression, the use of replication-competent plasmids increases gene expression over time post transfection. Plasmids can also be engineered to provide for specific or long-term gene expression, replication, or integration.

The transfection quality of plasmid DNA is dependent on the preparation protocol and training of the person preparing the DNA. We have also identified large amounts of contaminants in laboratory- and clinical-grade preparations of plasmid DNA and developed three proprietary methods for their detection. These contaminants copurify with DNA by anion exchange chromatography and cesium chloride density gradient centrifugation. Endotoxin removal does not remove these contaminants, and high-performance liquid chromatography cannot detect them. To provide the greatest efficacy, safety, and gene expression, these contaminants must be assessed and removed from plasmid DNA preparations. These contaminants belong to a class of molecules known to inhibit both DNA and RNA polymerase activities. Our group and other investigators have shown that intravenous injections of high doses of improved liposomes alone cause no adverse effects. Plasmids with most of CpG sequences removed apparently reduced toxicity after intravenous injections of cationic complexes.[18] However, only low doses containing up to 16.5 µg of DNA per injection into each mouse reduced toxicity, and no significant dose response to CpG motifs in plasmid DNA was demonstrated. Therefore, we believe that removal of the other contaminants in current DNA preparations is the major block to safe intravenous injection of high doses of DNA–liposome complexes.

Choosing the most sensitive detection method for gene expression is also essential. For example, detection of β-galactosidase (β-gal) expression is far more sensitive than that for the green fluorescent protein (GFP). Specifically, 500 molecules of β-gal per cell are required for detection using X-gal staining, whereas about 1 million molecules of GFP per cell are required for direct detection, and its detection may be impossible if the fluorescence background of the target cell or tissue is too high. Detection of chloramphenicol acetyltransferase (CAT) is extremely sensitive with little or no background detected in untransfected cells. Few molecules of luciferase in a cell can be detected by luminescence assays of cell or tissue extracts post transfection, and the sensitivity is highly dependent on the type of instrument used to measure luminescence. Luciferase data may not predict the therapeutic potential of a nonviral delivery system if several hundred or thousand molecules per cell of a therapeutic protein are required to produce efficacy for a certain disease. Furthermore, noninvasive detection of luciferase expression in vivo is not as sensitive as luminescence assays of cell or tissue extracts post transfection. My colleagues detected luciferase expression in mice by cooled charge-coupled device (CCD) imaging after intravenous injections of BIV DOTAP:Chol-luciferase DNA–liposome complexes and detection of HSV-TK gene expression using microPET imaging.[19]

To establish the maximal efficacy for the treatment of certain diseases or vaccine production, administration of the nonviral gene therapeutic, etc. via different routes may be required. The optimal dose and administration schedule should be determined because administering the highest tolerable dose most frequently may not necessarily produce maximal efficacy. Loss of the therapeutic gene product will vary with the half-life of the protein produced. Therefore, if a therapeutic protein has a longer half-life, then the gene therapy could perhaps be administered less frequently.

CONCLUSION

Some hurdles remain in the broad application of nonviral delivery; however, we are confident that we will successfully overcome these challenges. We predict that eventually the majority of gene therapies will utilize artificial reagents that can be standardized and regulated as drugs rather than biologics. We will also continue to incorporate the molecular mechanisms of viral delivery that produce the efficient delivery to cells into artificial systems. Therefore, the artificial systems, including liposomal delivery vehicles, will be further engineered to mimic the most beneficial parts of the viral delivery systems while circumventing their limitations. We will also maintain the numerous benefits of the liposomal delivery systems discussed in this chapter.

ACKNOWLEDGMENT

I thank Dr. David D. Roberts at the National Cancer Institute, NIH, Bethesda, MD, for the preparation of the figures.

REFERENCES

1. Felgner, P.L.; Gadek, T.R.; Holm, M.; Roman, R.; Chan, H.W.; Wenz, M.; Northrop, J.P.; Ringold, G.M.; Danielson, H. Lipofection: A highly efficient lipid-mediated DNA transfection procedure. Proc. Natl. Acad. Sci. U. S. A. **1987**, *84*, 7413–7417.
2. Templeton, N.S.; Lasic, D.D.; Frederik, P.M.; Strey, H.H.; Roberts, D.D.; Pavlakis, G.N. Improved DNA: Liposome complexes for increased systemic delivery and gene expression. Nat. Biotechnol. **1997**, *15*, 647–652.
3. Ramesh, R.; Saeki, T.; Templeton, N.S.; Ji, L.; Stephens, L.C.; Ito, I.; Wilson, D.R.; Wu, Z.; Branch, C.D.; Minna, J.D.; Roth, J.A. Successful treatment of primary and disseminated human lung cancers by systemic delivery of tumor suppressor genes using an improved liposome vector. Mol. Ther. **2001**, *3*, 337–350.
4. Tirone, T.A.; Fagan, S.P.; Templeton, N.S.; Wang, X.P.; Brunicardi, F.C. Insulinoma induced hypoglycemic death in mice is prevented with beta cell specific gene therapy. Ann. Surg. **2001**, *233*, 603–611.
5. Shi, H.Y.; Liang, R.; Templeton, N.S.; Zhang, M. Inhibition of breast tumor progression by systemic delivery of the maspin gene in a syngeneic tumor model. Mol. Ther. **2002**, *5*, 755–761.
6. Lu, H.; Zhang, Y.; Roberts, D.D.; Osborne, C.K.; Templeton, N.S. Enhanced gene expression in breast cancer cells in vitro and tumors in vivo. Mol. Ther. **2002**, *6*, 783–792.
7. Li, S.; Ma, Z.; Tan, Y.; Liu, F.; Dileo, J.; Huang, L. Targeted delivery via lipidic vectors. In *Vector Targeting for Therapeutic Gene Delivery*; Curiel, D.T., Douglas, J.T., Eds.; Wiley-Liss: Hoboken, NJ, **2002**; 17–32.

8. Pirollo, K.F.; Xu, L.; Chang, E.H. Immunoliposomes: A targeted delivery tool for cancer treatment. In *Vector Targeting for Therapeutic Gene Delivery*; Curiel, D.T., Douglas, J.T., Eds.; Wiley-Liss: Hoboken, NJ, **2002**; 33–62.

9. Sternberg, B. Morphology of cationic liposome/DNA complexes in relation to their chemical composition. J. Liposome Res. **1996**, *6*, 515–533.

10. Senior, J.; Delgado, C.; Fisher, D.; Tilcock, C.; Gregoriadis, G. Influence of surface hydrophilicity of liposomes on their interaction with plasma protein and clearance from the circulation: Studies with poly(ethylene glycol)-coated vesicles. Biochim. Biophys. Acta **1991**, *1062*, 77–82.

11. Papahadjopoulos, D.; Allen, T.M.; Gabizon, A.; Mayhew, E.; Matthay, K.; Huang, S.K.; Lee, K.; Woodle, M.C.; Lasic, D.D.; Redemann, C.; Martin, F.J. Sterically stabilized liposomes: Improvements in pharmacokinetics and antitumor therapeutic efficacy. Proc. Natl. Acad. Sci. U. S. A. **1991**, *88*, 11460–11464.

12. Gabizon, A.; Catane, R.; Uziely, B.; Kaufman, B.; Safra, T.; Cohen, R.; Martin, F.; Huang, A.; Barenholz, Y. Prolonged circulation time and enhanced accumulation in malignant exudates of doxorubicin encapsulated in polyethylene-glycol coated liposomes. Cancer Res. **1994**, *54*, 987–992.

13. Gordon, K.B.; Tajuddin, A.; Guitart, J.; Kuzel, T.M.; Eramo, L.R.; VonRoenn, J. Hand–foot syndrome associated with liposome-encapsulated doxorubicin therapy. Cancer **1995**, *75*, 2169–2173.

14. Uziely, B.; Jeffers, S.; Isacson, R.; Kutsch, K.; Wei-Tsao, D.; Yehoshua, Z.; Libson, E.; Muggia, F.M.; Gabizon, A. Liposomal doxorubicin: Antitumor activity and unique toxicities during two complementary phase I studies. J. Clin. Oncol. **1995**, *13*, 1777–1785.

15. Gustafsson, J.; Arvidson, G.; Karlsson, G.; Almgren, M. Complexes between cationic liposomes and DNA visualized by cryo-TEM. Biochim. Biophys. Acta **1995**, *1235*, 305–312.

16. Templeton, N.S.; Lasic, D.D. New directions in liposome gene delivery. Mol. Biotechnol. **1999**, *11*, 175–180.

17. Handumrongkul, C.; Zhong, W.; Debs, R.J. Distinct sets of cellular genes control the expression of transfected, nuclear-localized genes. Mol. Ther. **2002**, *5*, 186–194.

18. Yew, N.S.; Zhao, H.; Przybylska, M.; Wu, I.-H.; Tousignant, J.D.; Scheule, R.K.; Cheng, S.H. CpG depleted plasmid DNA vectors with enhanced safety and long-term gene expression in vivo. Mol. Ther. **2002**, *5*, 731–738.

19. Iyer, M.; Berenji, M.; Templeton, N.S.; Gambhir, S.S. Noninvasive imaging of cationic lipid-mediated delivery of optical and PET reporter genes in living mice. Mol. Ther. **2002**, *6*, 555–562.

18 Polymer-Based Nonviral Nucleic Acid Delivery and Genome Editing

Janin Germer, Ernst Wagner
Ludwig-Maximilians-Universität, Munich, Germany

CONTENTS

INTRODUCTION

Polyplexes are based on the condensation of negatively charged nucleic acids by electrostatic attraction with polycationic condensing compounds.[1,2] The resulting compact particles protect the nucleic acid and also improve the uptake into the cells. Numerous polycations have been used for formulating nucleic acids into complexes. Ideally, the cationic polymer will carry out multiple tasks (Figure 18.1) which include: compacting nucleic acids into particles that can migrate to the target tissue, shielding the particles against degradation and undesired interactions, and enhancing cell binding and intracellular delivery into cytoplasm and the nucleus. Initially focusing on plasmid DNA (pDNA)-based gene therapy, a series of novel therapeutic nucleic acid entities such as stabilized messenger RNA (mRNA), chemically modified small interfering RNA (siRNA) and micro RNA (miRNA), splice-modifying oligonucleotides, or CRISPR Cas9/single guide RNA (sgRNA) for genome modification have entered research and clinical evaluation. Recent experience indicates that different macromolecular cargos may require different tailor-made carriers based on their different biophysical characteristics, distinct intracellular target sites and mode of action.[3–6] In practical terms, a classical polymer itself is unable to carry out all these tasks. Additional functional domains have to be integrated into the

DOI: 10.1201/9781003247432-18

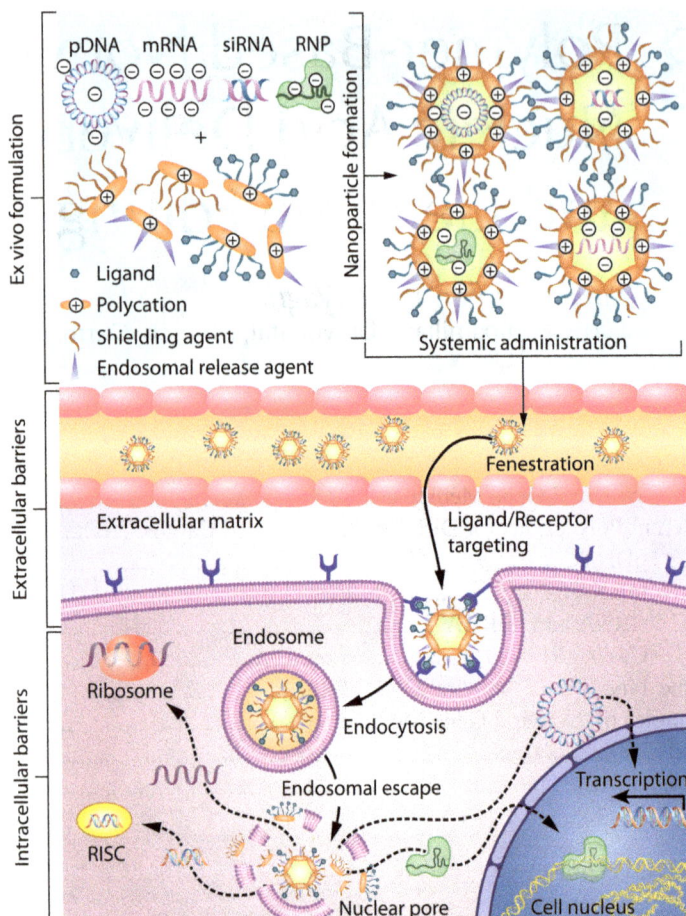

FIGURE 18.1 Nonviral, receptor-mediated polymer-based nucleic acid delivery in vivo. Polymers compacting nucleic acid cargos into nanoparticles, shielding the particle against degradation and undesired interactions, enabling migration to the target tissue, enhancing cell binding and uptake to the cytoplasm. Different intracellular pathways for productive delivery of pDNA, mRNA, siRNA, and Cas9 RNP are represented.

formulation. Advantageously, polymers can be chemically linked to molecules such as cell targeting ligands, including proteins (antibodies, growth factors) and small molecules (carbohydrates, peptides, vitamins). Various polymer-ligand gene delivery systems have been demonstrated to facilitate receptor-cell mediated delivery into cultured cells. Targeted delivery to the lung, the liver, or tumors has been achieved in experimental trials, either by localized or systemic application. Further chemical evolution was driven by the design of sequence-defined polymers, as well as library screening of combinatorial polymers and polyplexes.[6]

By biological evolution, viruses have optimized their strategies to meet dynamic and bioresponsive requirements for successful nucleic acid delivery into the host cell.

Representing potent delivery vehicles, they inspired models for virus-like synthetic carriers containing different domains for mimicking the efficient, dynamic delivery process of viral infection.[7–10] Importantly, 30 years of continuous development yielded numerous improved carriers resulting in the first polymer-containing nucleic acid nanomedicines on market.[11–13]

POLYMERS FOR NONVIRAL GENE MEDICINES

DIFFERENT POLYCATIONS FOR DIFFERENT NUCLEIC ACID CARGOS

Nucleic acid binding polycations include synthetic polymers such as polylysine, polyethylenimine (PEI), cationic dendrimers, carbohydrate-based polymers such as modified chitosan or dextran, or lipid-modified polymers. The characteristics of these polymers and their use in nucleic acid delivery have been reported extensively.[14–16] Numerous studies revealed that different nucleic acid cargos require different polycationic carriers (Table 18.1).

TABLE 18.1

Properties of Cargo Nucleic Acids, Their Sites of Action, and Selected Delivery Carriers

Cargo Nucleic Acid	Properties	Intracellular Target site	Examples of Polymers for Delivery
pDNA	double-stranded (ds) circular DNA large size (~5–15 kbp)	Nucleus	PEI; VIPER; sequence-defined OAAs
mRNA	single-stranded (ss) RNA to be transcribed in the cytoplasm medium size (~1–4 kb)	Cytoplasm	EPE carriers; PEG-PGBA; charge-altering releasable transporters; ionizable amino-polyester (APEs)
siRNA, miRNA	ds RNAs for RNA interference, cleaving (siRNA) or blocking (miRNA) the target mRNA small (21–23 bp)	Cytoplasm	succinylated PEI; T-shaped lipo-OAAs; reducible pBAVE siRNA conjugates LNPs (DLin-MC3-DMA)
sgRNA	ss single-guide RNA for guiding Cas9 to the genomic target site for DNA double strand cleavage medium small size (~100 bases)	Nucleus	ZALs for Cas9 mRNA/sgRNA delivery; PBA modified PAMAM dendrimers, or hydroxystearylated OAA for Cas9 protein/sgRNA delivery

Abbreviations: EPE—ethylenimine-propyleneimine-ethylenimine; LNP—lipid nanoparticle; PAMAM—polyamidoamine; PBA—phenylboronic acid; pBAVE—poly(butyl and amino vinyl ether); PEG-PGBA—polyethylene glycol-polyglycidyl butylamine; PEI—polyethylenimine; OAA—oligo amino amides; VIPER—virus-inspired polymer for endosomal release; ZALs—zwitterionic amino lipids.

---

pDNA

Of the 'first-generation' cationic polymeric carriers evaluated, PEI, has the highest transfection efficiency for pDNA. This can be explained by its intrinsic ability to facilitate endosomal release. The polymer acts as a 'proton sponge', containing protonable amines that after endocytosis slow down endosomal acidification, triggering enhanced endosomal chloride accumulation followed by osmotic swelling and breaking up of endosomes.[17] The active aminoethylene motif of PEI has also been introduced into a series of polymer-based carriers, including the virus-inspired polymer for endosomal release (VIPER).[9,18] Also artificial oligoamino acids were used to design new sequence defined peptide-like pDNA nanocarriers in various distinct structure topologies.[19] The inclusion of histidines enabled fine-tuning of the endosomal pH sensitivity and resulted in enhanced pDNA delivery in vitro and in vivo.[20]

mRNA

Successful intracellular delivery of mRNA enables the cytosolic expression of essentially any desired therapeutic protein (e.g., restoration of function for monogenic diseases, growth factors, therapeutic antibodies, antigens in vaccines), preserving the post-translational modifications of the encoded proteins innate to the host cell. It presents an alternative for the challenging formulation and delivery of protein-based drugs. In contrast to pDNA-based approaches, the delivery into the nucleus is not required, thus avoiding the hazard of insertional mutagenesis and reducing the cellular barriers for delivery. Ideally, the design of mRNA is optimized in sequence for effective translation of the therapeutic protein. Nucleotide modifications can be incorporated to obtain a reduced immunogenicity and prolonged stability of the mRNA. Thus, depending on the application, production of encoded protein may be transient or also persistent for a prolonged period of time.

Efficacy is strongly dependent on intracellular delivery. The pDNA carrier PEI has only moderate efficacy in mRNA transfer. However, small changes (introduction of a methylene group) to the PEI backbone resulted in cationic ethylenimine-propylenimine-ethylenimine (EPE) carriers for efficient mRNA delivery in vitro and in vivo, such as to the liver or the lungs after systemic or aerosol administration, respectively.[21] Functional bioavailability of mRNA for protein translation is a key requirement. One solution presents the design of amphiphilic charge-altering releasable transporters that first protect and deliver mRNA and then degrade by a charge-neutralizing process, resulting in an efficient release of mRNA into the cytoplasm and translation into protein.[22,23] Another solution presents the use of RNA particles formed with a more flexible polyethylene glycol-polyglycidyl butylamine (PEG-PGBA) polymer backbone.[24] Pursuing tissue-selective delivery of mRNA, it was formulated with a diverse selection of ionizable amino-polyester (APEs) into lipid nanoparticles (LNPs), leading to the identification of APE-LNPs predominantly expressing mRNA in lung endothelium, liver hepatocytes, or splenic antigen presenting cells.[25]

siRNA and miRNA

siRNA and miRNA both operate by the RNA interference (RNAi) mechanism. Forming an RNA-induced silencing (RISC) complex with nucleotide sequences complementary to the target mRNA, they prevent the expression of targeted genes by cleavage of the mRNA (for siRNA) or repression of translation (for miRNA). Consequently, specific inhibition of gene expression at the posttranscriptional level without risk of insertional mutagenesis to the host genome can be achieved, presenting a safe alternative for gene therapeutic approaches. Consisting of only 21–23 bp siRNA provides far less negative charges for electrostatic polyplex stabilization. To overcome this hurdle several strategies have been under investigation. Covalent conjugation of the siRNA to the polymer carrier through a reversible disulfide linkage resulted in increased siRNA-mediated knockdown of targeted genes in vivo.[26] siRNA conjugation with PEG for enhanced blood circulation and N-acetylgalactosamine (GalNac) trimer as receptor targeting ligand resulted in potent gene silencing in the liver.[27]

Polymer carriers like PEI that proved efficacy for pDNA delivery were found inefficient for applications with siRNA. Therefore, different modifications to polymer carriers like PEI were made in order to convert them into a suitable siRNA carrier. Succinylation of PEI reduced the polymer's toxicity considerably, consequently allowing formulations of siRNA with higher PEI concentration which mediated effective gene silencing.[28] Hydrophobic modifications with tyrosine lead to self-aggregation of the molecule benefiting the stability of siRNA polyplexes.[29] Peptides and artificial sequence-defined peptide-like carriers were developed for siRNA delivery.[30] For example, T-shaped lipo-oligo amino amides including a central lipid domain and terminal tyrosine tripeptide sequences can form stable siRNA polyplexes.[31] The composition of the lipophilic tail groups are of great relevance for the activity of the carriers. Unsaturated oleic acid residues appeared to be beneficial for the delivery of siRNA[32,33] but not necessarily for other cargos. For phosphorodiamidate morpholino oligomers (PMOs), analogous linolenic acid modified carriers worked best,[34] and for functional delivery of Cas9/sgRNA ribonucleoprotein (RNP), carriers containing hydroxystearic acids displayed highest activity.[35]

LNPs present suitable system for siRNA delivery to the liver. Over three decades LNPs were developed for drug delivery based on nanosized formulation of phospholipids like distearoylphosphatidylcholine (DSPC), cholesterol, and PEG polymer for improved stability and circulation in blood. In siRNA-LNPs, a cationic component binding and neutralizing the siRNA had to be included. A breakthrough was achieved by a formulation that contains a combination of the cationic lipid DLin-MC3-DMA, cholesterol, and DSPC, together with rapidly dissociating PEG C_{14}-lipid.[36] This formulation mediates gene silencing in hepatocytes in vivo at very low doses and presents the basis for Patisiran, the first siRNA-drug that has been approved by the FDA and EMA.[11] A series of other cationic components have been used in LNPs for siRNA delivery to the liver, including a new class of polycationic materials termed 'lipidoids',[37–39] with advantageous physical properties of cationic polymers and lipids.

Cas9/sgRNA

A new era for genetic manipulation has been initiated by revolutionary advances in precise genome modifying nucleases such as zinc finger nucleases (ZFN), transcription activator-like effector nucleases (TALEN), or CRISPR-associated protein 9 (Cas9)/sgRNA. Especially the simple design of the CRISPR Cas9/sgRNA system makes it a most promising genome editing strategy. It is of major priority to develop efficient and safe CRISPR/Cas9 delivery technologies to enable efficient genome engineering.[6,40]

In contrast to standard gene therapy approaches for genetic diseases that require lifelong expression of the therapeutic transgene, a short-term transient expression of genome-modifying nucleases is beneficial. Therefore, nonviral delivery systems provide a secure alternative to viral delivery systems, avoiding undesired complications such as off target effects caused by prolonged expression of the nucleases. Constructs encoding the nuclease based on pDNA or mRNA, or direct delivery of the recombinant Cas9 protein with sgRNA as RNP formulation have been utilized.

An example of efficient in vivo co-delivery of Cas9 mRNA and sgRNA presents the use of zwitterionic amino lipid nanoparticles (ZNPs).[41] Polymers have also been investigated for direct delivery of the Cas9 protein/sgRNA RNP complex. For example, cationic polyamidoamine (PAMAM) dendrimers with phenylboronic acid (PBA) residues enable binding with either negative or positive charged proteins. Successful combined delivery and gene editing of Cas9 protein/sgRNA and various other efficient protein cargos were demonstrated.[42] Lipid-containing oligo amino amides (lipo-OAAs) were identified as efficacious carriers for intracellular Cas9 protein/sgRNA RNP delivery and gene disruption wherein the type of contained fatty acid, hydroxystearic acid, exhibited a crucial effect on the knock out efficiency.[35] To enable effective asialoglycoprotein receptor (ASGPR)-mediated endocytosis into hepatocytes for liver-specific genome editing, Doudna and colleagues designed CRISPR-Cas9 protein conjugates, containing GalNac trimers as targeting ligands. When co-applied with an endosomolytic peptide, effective genome editing in HepG2 cells was demonstrated.[43]

TISSUE TARGETING OF NONVIRAL GENE MEDICINES

For targeted delivery into a distant organ (compare Figure 18.1), the following factors have to be taken into account ideally: (a) nucleic acid polyplexes are stable and inert in blood; (b) they must be able to reach their target tissue and therefore to cross different biologicals barriers, including vascular endothelium, extracellular matrix, and others; (c) once reaching the target cell they should effectively internalize, (d) polyplexes should disassemble at the right moment, but still protect the cargo against intracellular degradation; (e) they should release the nucleic acid into the corresponding compartment; and (f) they must elicit as low inflammatory or immune response as possible. Although no polyplex or other nonviral gene transfer system exists which would fulfill all these requirements, targeted delivery to the liver, the lungs, or tumors has already been achieved in animal and clinical trials (see Table 18.2).

As one solution toward both a specific and efficient delivery of nucleic acids into target cells, the natural cellular mechanism of receptor-mediated endocytosis has been proposed. Already three decades ago, the utilization of the hepatocytic ASGPR for targeted gene delivery to the liver[44] and the transferrin receptor (TfR) for targeted delivery to tumor cells[45] was reported. Wu and Wu designed DNA/asialoorosomucoid (ASOM)-polylysine complexes for hepatocyte targeting.[44] Using

TABLE 18.2

Examples of Targeted In Vivo Delivery of Polyplexes

Polymer System	Cargo	Delivery Mode	Target Organ	Results	Reference
ASOM-polylysine	pDNA	Intravenous	Liver	First demonstration of ASGPR targeted gene transfer	[44]
GalNAc containing Dynamic polyconjugates	siRNA	Intravenous	Liver	Potent hepatic gene silencing	[26]
PEG-GalNAc trimer conjugates	siRNA	Subcutaneous	Liver	Potent hepatic gene silencing	[27]
TfR-targeted polyplexes	pDNA	Intravenous	Tumor	High gene expression in tumor	[45,50]
TfR-targeted, PEG shielded	pDNA	Intravenous	Tumor	High gene expression in tumor	[51]
TfR-targeted, PEG shielded	siRNA	Intravenous	Tumor	Gene silencing in tumor	[53]
EGFR-targeted, PEG shielded	pDNA	Intravenous	Hepatoma	Gene expression in tumor	[55]
LNPs with sheddable PEG-C$_{14}$ 'indirect targeting' to LDLR	siRNA, mRNA, sgRNA	Intravenous	Liver	Dissociation of PEG in blood circulation triggers binding of apolipoprotein E to LNPs and LDLR mediated uptake	[36]
Alternative LNP formulations identified from barcoded libraries	siRNA, mRNA, sgRNA	Intravenous	Endothelial cells, T cells	Use of DNA barcoded libraries for identification of LNPs that deliver sgRNA and mRNA to cell types different from hepatocytes	[58,59]

Abbreviations: ASGPR—asialoglycoprotein receptor; ASOM—asialoorosomucoid; EGFR—epidermal growth factor receptor; GalNAc—*N*-acetylgalactosamine; LDLR—low density lipoprotein receptor; LNP—lipid nanoparticles; PEG—polyethylene glycol; TfR, transferrin receptor.

this system, hepatocyte-specific in vivo gene transfer of the low density lipoprotein receptor (LDLR) was demonstrated in a rabbit animal model for familial hypercholesterolemia. This resulted in a temporary amelioration of the disease phenotype.[46] In analogous manner, the albumin serum levels of Nagase analbuminemic rats were transiently raised by human albumin gene delivery.[47] More recently, ASGPR targeting was applied also for siRNA delivery. Dynamic polyconjugates of siRNA with reversibly attached targeting ligand GalNAc demonstrated effective knockdown of endogenous genes in mouse liver.[26] The effective knockdown of apolipoprotein B (apoB) and peroxisome proliferator-activated receptor alpha (ppara) resulted in decreased cholesterol levels for up to 10 days after administration. The most recent and at current most successful development for ASGPR targeting[48] presents the design of siRNA-PEG-GalNAc trimer conjugates. Subcutaneous administration resulted in systemic distribution and potent prolonged gene silencing in the liver.[27] Further optimization of the siRNA chemistry for enhanced stability was the basis for the design and approval of Givosiran, the first siRNA drug targeting the ASGPR.[12,13]

Targeting tumors might present a unique opportunity to reach and attack multiple spread metastases. Tumors require iron for proliferation and commonly overexpress the TfR; therefore Wagner and colleague designed Tf-polycations as carrier for pDNA delivery into tumor cells.[45] This strategy was successfully used for transfecting freshly isolated tumor cells from melanoma patients with the interleukin-2 gene, and was translated in form of a cancer vaccine into the first polymer-based ex vivo human gene therapy study.[49] For intravenous in vivo administration, a charge-neutral surface of nanoparticles is essential to minimize non-specific interactions with blood components, allowing greater intravenous circulation time for the nucleic acid to reach its target, and also reducing particle toxicity. Hydrophilic shielding agents include the serum protein transferrin and hydrophilic polymers such as PEG. For example, transferrin-shielded polyplexes[50] or PEG-shielded polyplexes[51,52] demonstrated potential for systemic in vivo targeting of distant subcutaneous tumors of mice with gene expression levels in tumor tissues approximately 100-fold higher than in other organ tissues. Davis and colleagues designed tumor-targeted siRNA polyplexes containing a cationized cyclodextrin polymer/siRNA core and a PEG shield containing transferrin as targeting ligand. Systemic administration enabled specific gene inhibition in TfR-expressing tumors[53,54] that encouraged for entering subsequent clinical development.

In analogous manner, epidermal growth factor (EGF)-PEG-coated pDNA polyplexes were successfully applied for systemic targeting of hepatocellular carcinoma xenografts in mice.[55] Efficient cancer cell targeting has also been achieved for siRNA complexed with sequence-defined lipo-oligomers and coated with folate conjugated PEG for folate receptor (FR) targeting. Tumor target mRNA knockdown was demonstrated upon intravenous administration of these siRNA lipopolyplexes in a FR-positive leukemia mouse model.[56]

A different mechanism of 'indirect receptor targeting' is utilized by LNP systems containing ionizable cationic lipids and a rapidly dissociating PEG C_{14}-lipid for shielding. These nanoparticles take advantage of a 'natural' receptor targeting process. In blood circulation the PEG shield dissociates from LNP with a halftime of <30 min. This triggers incorporation of an apolipoprotein E (ApoE) corona that mediates efficient uptake of the nucleic acid cargo (siRNA, mRNA, sgRNA) into hepatocytes via the low density lipoprotein receptor (LDLR).[36] Recently Dahlman and colleagues introduced DNA barcoding as a strategy for high throughput screening of suitable nanocarriers for in vivo delivery into target tissues that differ from hepatocytes as common LNP targets.[57] Using this concept, LNPs were identified which efficiently deliver nucleic acid cargos to endothelial cells or T cells. This 'chemical target specificity' results from physicochemical properties of LNP and not depending on specific receptor ligands.[58,59]

THERAPIES BY NONVIRAL GENE MEDICINES

Within three decades of therapeutic evaluation including more than 3000 human clinical trials,[60] gene medicines have become a reality, with approximately 20 nucleic acid drugs already approved by medical agencies. This also includes at least eight nonviral oligonucleotide drugs[61] and two siRNA drugs,[11,12] with many more expected to be approved soon.[48] Table 18.3 lists examples of therapeutic concepts evaluated using nonviral gene medicines in pharmacological models and clinical trials.

The polycationic PEI is a potent pDNA transfection agent when locally applied. A series of human clinical trials have been performed using this carrier. In particular, clinical trials were performed in bladder cancer using a pDNA named BC-819. This pDNA expresses the gene for diphtheria toxin-A under regulation of the H19 promoter that is highly expressed in bladder cancer but not in the adult healthy bladder. Intravesical administration into the bladder of BC-819 complexed with PEI completely ablated tumors in a third of the patients and prevented new tumor growth in two-thirds of the patients.[62,63] In a preclinical model of lung cancer, aerosol delivery of PEI/pDNA coding for the p53 gene in combination with 9-nitro-camptothecin resulted in growth inhibition of established B16-F10 lung metastases in mice.[64]

For targeting the liver, based on encouraging results with siRNA-Dynamic poly-conjugates,[26] a GalNAc conjugated melittin-like peptide (GalNAc-MLP) was co-injected with a liver-tropic cholesterol-siRNA conjugate (chol-siRNA) targeting conserved hepatitis B virus (HBV) gene sequences. A multilog reduction of viral gene expression and viral DNA titer was achieved in mouse models of chronic HBV infection.[65] HBV antigens were strongly reduced in some patient groups in a clinical phase 2 trial for the treatment of chronic HBV.[66]

siRNA conjugated with PEG-GalNAc trimer presents an easier but nevertheless highly effective ASGPR-targeted formulation. As the siRNA within these conjugates is not protected by an encapsulating carrier, complete chemical modification of the siRNA backbone for enhanced stability and also enhanced RNAi efficacy was necessary and essential for clinical application. Subcutaneous administration into patients resulted in potent prolonged silencing of therapeutic gene targets in the liver. Givosiran (GIVLAARI), as the first siRNA drug targeting the ASGPR,[12,13] targets aminolevulinic acid synthetase 1 (ALAS1) and received market approval in November 2019 for the treatment of acute hepatic porphyria. This rare genetic disorder causes an induction of hepatic ALA synthase 1 that results in the accumulation of neurotoxic heme intermediates, including ALA and porphobilinogen. Gene silencing of ALAS1 reduces buildup of those toxic metabolites. siRNA PEG-GalNAc conjugates against several other therapeutic gene targets are currently in late stages of clinical development.

The very first siRNA drug formulation reaching the medical market in 2018 is Patisiran (Onpattro), a PEGylated LNP formulation containing siRNA directed against TTR. The drug was approved for therapy of hereditary transthyretin (TTR)-mediated amyloidosis (ATTR amyloidosis).[11] This rapidly progressing, debilitating disease (polyneuropathy, cardiomyopathy) with high mortality is caused by

TABLE 18.3
Examples of Therapeutic Strategies

Polymer System	Cargo	Delivery Mode	Target Gene (Target Organ)	Therapeutic Result	Reference
PEI	pDNA	Intravesical (bladder)	Diphtheria toxin (Bladder cancer)	Diphtheria toxin A expressed under H19 promotor bladder carcinoma, clinical phase 2	[62,63]
PEI	pDNA	Aerosol (Mice)	p53 (Lung metastases)	Growth inhibition of established lung metastases	[64]
Cholesterol-siRNA conjugate plus GalNAc-MLP	siRNA	Intravenous	HBV virus	HBV antigens strongly reduced in some patients in a phase 2 clinical trial for treatment of chronic HBV	[65,66]
siRNA-PEG-GalNAc trimer conjugate	siRNA	Subcutaneous	ALAS1	Givosiran approved for treatment of acute hepatic porphyria, other indications in clinical development	[12,48]
LNPs with sheddable PEG-C14	siRNA	Intravenous	Transthyretin mRNA (Liver)	Patisiran approved for treatment of hereditary transthyretin amyloidosis	[11]
TfR-targeted, shielded	pDNA	Intravenous (Mice)	TNF-alpha (Tumor)	Tumor necrosis, inhibition of tumor growth	[52]
TfR-targeted, PEG shielded	siRNA	Intravenous	Ribonucleotide Reductase Subunit M2 (Tumor)	Tumor patients, clinical phase 1 Reduced expression of RRM2 Protein	[72]
EGFR-targeted, PEG shielded	pDNA	Intravenous (Mice)	TNF-alpha/ chemotherapy NIS/ radiotherapy (Tumor)	Antitumoral gene expression in tumor, antitumor effects of TNF with chemotherapy, or NIS with radioiodide	[68,69]
EGFR-targeted, PEG shielded	Poly (I:C) dsRNA	Local, intravenous (Mice)	(Glioma, others)	Antitumoral by stimulating innate immune response and apoptosis	[70,71]

Abbreviations: ALAS1—aminolevulinic acid synthetase 1; ASGPR—asialoglycoprotein receptor; ASOM—asialoorosomucoid; EGFR—epidermal growth factor receptor; GalNAc—*N*-acetylgalactosamine; HBV—hepatitis B virus; LNP—lipid nanoparticle; MLP—melittin like peptide; NIS—sodium iodide symporter; PEG—polyethylene glycol; PEI—polyethylenimine; Poly (I:C)—polyinosinic-polycytidylic acid; TfR—transferrin receptor; TNF—tumor necrosis factor mutations in the TTR gene (producing a tetrameric protein, also called pre-albumin) resulting in tetramer destabilization, toxic amyloid fibrils, and plaques.

While the established LNPs (indirectly targeting the LDLR) and GalNAc conjugates (targeting the ASGPR) present sound technologies for medical developments when liver hepatocytes are the target cells, further exploration and optimization is required for systemic delivery technologies targeting of other tissues and organs. Preclinical studies demonstrated that TfR- or epidermal growth factor receptor (EGFR)-targeted and shielded polyplexes enable systemic targeting of pDNA or other nucleic acids into distant tumors in mice.[50–53,55] This was further evaluated in pharmacological models using therapeutic genes (see Table 18.3). Repeated systemic application of targeted and shielded polyplexes encoding tumor necrosis factor alpha (TNF-alpha) induced expression in tumor cells close to the feeding blood vessels, which triggers TNF-mediated destruction of the tumor vasculature, tumor necrosis, and inhibition of tumor growth as demonstrated in several tumor models.[52,67,68] EGFR-targeted pDNA polyplexes were applied for imaging and targeted therapy of mice genetically engineered to develop pancreatic ductal adenocarcinoma. For this purpose, pDNA encoding the theranostic sodium iodide symporter (NIS) gene was applied, followed by diagnostic or therapeutic radioiodide.[69] EGFR-targeted, PEG shielded polyplexes were also applied for delivery of double-stranded RNA polyinosinic-polycytidylic (poly (I:C)) as another therapeutic cargo.[70,71] Poly (I:C), when delivered into tumor cells, stimulates an antitumoral innate immune response and apoptosis of tumor cells, as demonstrated by complete cures of mice in an orthotopic glioma model and other tumor models.

A significant step toward human therapy was the first evaluation of tumor-targeted siRNA polyplexes in cancer patients. PEG-shielded and transferrin-containing nanoparticles were formed using a cationic cyclodextrin-based polymer and tumor-suppressive siRNA against the RRM2 (Ribonucleotide Reductase Subunit M2) mRNA. Intravenously administered siRNA demonstrated the intended gene silencing in tumor tissues of treated patients.[72]

CONCLUSION

Targeted delivery to the liver or tumors has been achieved in experimental animals and patients. Therapeutic effects have been demonstrated, and first polymer-containing nucleic acid nanomedicines, Patisiran and Givosiran, have reached the clinical market.[11–13] The limitations of first-generation polymeric carriers (modest activity and significant toxicity) has been addressed by developments of new biodegradable and well defined polycationic polymers, incorporation of targeting and intracellular transport functions, and polyplex formulations that avoid unspecific adverse interactions with the host. The repertoire of therapeutic nucleic acids broadened from classical pDNA to various other active nucleic acid cargos such as mRNA, siRNA, and miRNA or the combination of Cas9 (pDNA, mRNA, or protein) with sgRNA for genome modification. This demands optimization of polymeric delivery systems individually customized for each cargo. A key step presents the further development of polyplexes into 'artificial viruses', i.e. polyplexes possessing virus-like entry functions that are presented by smart polymers and conjugates. These 'smart' polymers in contrast to conventional polymers respond in a more dynamic and controlled manner to alterations in their biological micro-environment such as pH or redox environment, and have to undergo programmed structural changes, to more accurately switch on the individual delivery functions only when required in the individual steps of the delivery process.

REFERENCES

1. Felgner, P.L.; Barenholz, Y.; Behr, J.P.; Cheng, S.H.; Cullis, P.; Huang, L.; Jessee, J.A.; Seymour, L.; Szoka, F.; Thierry, A.R.; Wagner, E.; Wu, G. Nomenclature for synthetic gene delivery systems. Hum. Gene Ther. **1997**, *8*(5), 511–512.
2. Lächelt, U.; Wagner, E. Nucleic acid therapeutics using polyplexes: a journey of 50 years (and beyond). Chem. Rev. **2015**, *115*(19), 11043–11078.
3. Blakney, A.K.; Yilmaz, G.; McKay, P.F.; Becer, C.R.; Shattock, R.J. One size does not fit all: the effect of chain length and charge density of poly(ethylene imine) based copolymers on delivery of pDNA, mRNA, and RepRNA polyplexes. Biomacromolecules **2018**, *19*(7), 2870–2879.
4. Kauffman, A.C.; Piotrowski-Daspit, A.S.; Nakazawa, K.H.; Jiang, Y.; Datye, A.; Saltzman, W.M. Tunability of biodegradable poly(amine-*co*-ester) polymers for customized nucleic acid delivery and other biomedical applications. Biomacromolecules **2018**, *19*(9), 3861–3873.
5. Peng, L.; Wagner, E. Polymeric carriers for nucleic acid delivery: current designs and future directions. Biomacromolecules **2019**, *20*(10), 3613–3626.
6. Freitag, F.; Wagner, E. Optimizing synthetic nucleic acid and protein nanocarriers: the chemical evolution approach. Adv. Drug. Deliv. Rev. **2021**, *168*, 30–54.
7. Hager, S.; Wagner, E. Bioresponsive polyplexes – chemically programmed for nucleic acid delivery. Expert Opin. Drug Deliv. **2018**, *15*(11), 1067–1083.
8. Luo, T.; Liang, H.; Jin, R.; Nie, Y. Virus inspired and mimetic designs in nonviral gene delivery. J. Gene Med. **2019**, *21*(7), e3090.
9. Cheng, Y.; Yumul, R.C.; Pun, S.H. Virus-inspired polymer for efficient in vitro and in vivo gene delivery. Angew. Chem. Int. Ed. Engl. **2016**, *55*(39), 12013–12017.
10. Feldmann, D.P.; Cheng, Y.; Kandil, R.; Xie, Y.; Mohammadi, M.; Harz, H.; Sharma, A.; Peeler, D.J.; Moszczynska, A.; Leonhardt, H.; Pun, S.H.; Merkel, O.M. In vitro and in vivo delivery of siRNA via VIPER polymer system to lung cells. J. Control. Release **2018**, *276*, 50–58.
11. Adams, D.; Gonzalez-Duarte, A.; O'Riordan, W.D.; Yang, C.C.; Ueda, M.; Kristen, A.V.; Tournev, I.; Schmidt, H.H.; Coelho, T.; Berk, J.L.; Lin, K.P.; Vita, G.; Attarian, S.; Plante-Bordeneuve, V.; Mezei, M.M.; Campistol, J.M.; Buades, J.; Brannagan, T.H. 3rd; Kim, B.J.; Oh, J.; Parman, Y.; Sekijima, Y.; Hawkins, P.N.; Solomon, S.D.; Polydefkis, M.; Dyck, P.J.; Gandhi, P.J.; Goyal, S.; Chen, J.; Strahs, A.L.; Nochur, S.V.; Sweetser, M.T.; Garg, P.P.; Vaishnaw, A.K.; Gollob, J.A.; Suhr, O.B. Patisiran, an RNAi therapeutic, for hereditary transthyretin amyloidosis. N. Engl. J. Med. **2018**, *379*(1), 11–21.
12. Balwani, M.; Sardh, E.; Ventura, P.; Peiró, P.A.; Rees, D.C.; Stölzel, U.; Bissell, D.M.; Bonkovsky, H.L.; Windyga, J.; Anderson, K.E.; Parker, C.; Silver, S.M.; Keel, S.B.; Wang, J.D.; Stein, P.E.; Harper, P.; Vassiliou, D.; Wang, B.; Phillips, J.; Ivanova, A.; Langendonk, J.G.; Kauppinen, R.; Minder, E.; Horie, Y.; Penz, C.; Chen, J.; Liu, S.; Ko,

J.J.; Sweetser, M.T.; Garg, P.; Vaishnaw, A.; Kim, J.B.; Simon, A.R.; Gouya, L. Phase 3 trial of RNAi therapeutic givosiran for acute intermittent porphyria. N. Engl. J. Med. **2020**, *382*(24), 2289–2301.

13. Scott, L.J. Givosiran: first approval. Drugs **2020**, *80*(3), 335–339.

14. Wagner, E. Strategies to improve DNA polyplexes for in vivo gene transfer: will "artificial viruses" be the answer? Pharm. Res. **2004**, *21*(1), 8–14.

15. De Smedt, S.C.; Demeester, J.; Hennink, W.E. Cationic polymer based gene delivery systems. Pharm. Res. **2000**, *17*(2), 113–126.

16. Han, S.; Mahato, R.I.; Sung, Y.K.; Kim, S.W. Development of biomaterials for gene therapy. Mol. Ther. **2000**, *2*(4), 302–317.

17. Sonawane, N.D.; Szoka, F.C. Jr.; Verkman, A.S. Chloride accumulation and swelling in endosomes enhances DNA transfer by polyamine-DNA polyplexes. J. Biol. Chem. **2003**, *278*(45), 44826–44831.

18. Uchida, H.; Itaka, K.; Nomoto, T.; Ishii, T.; Suma, T.; Ikegami, M.; Miyata, K.; Oba, M.; Nishiyama, N.; Kataoka, K. Modulated protonation of side chain aminoethylene repeats in *N*-substituted polyaspartamides promotes mRNA transfection. J. Am. Chem. Soc. **2014**, *136*(35), 12396–12405.

19. Schaffert, D.; Troiber, C.; Salcher, E.E.; Fröhlich, T.; Martin, I.; Badgujar, N.; Dohmen, C.; Edinger, D.; Kläger, R.; Maiwald, G.; Farkasova, K.; Seeber, S.; Jahn-Hofmann, K.; Hadwiger, P.; Wagner, E. Solid-phase synthesis of sequence-defined T-, i-, and U-shape polymers for pDNA and siRNA delivery. Angew. Chem. Int. Ed. **2011**, *50*(38), 8986–8989.

20. Lächelt, U.; Kos, P.; Mickler, F.M.; Herrmann, A.; Salcher, E.E.; Rödl, W.; Badgujar, N.; Bräuchle, C.; Wagner, E. Fine-tuning of proton sponges by precise diaminoethanes and histidines in pDNA polyplexes. Nanomedicine **2014**, *10*(1), 35–44.

21. Jarzębińska, A.; Pasewald, T.; Lambrecht, J.; Mykhaylyk, O.; Kümmerling, L.; Beck, P.; Hasenpusch, G.; Rudolph, C.; Plank, C.; Dohmen, C. A single methylene group in oligoalkylamine-based cationic polymers and lipids promotes enhanced mRNA delivery. Angew. Chem. Int. Ed. Engl. **2016**, *55*(33), 9591–9595.

22. McKinlay, C.J.; Vargas, J.R.; Blake, T.R.; Hardy, J.W.; Kanada, M.; Contag, C.H.; Wender, P.A.; Waymouth, R.M. Charge-altering releasable transporters (CARTs) for the delivery and release of mRNA in living animals. Proc. Natl. Acad. Sci. USA **2017**, *114*(4), E448–E456.

23. McKinlay, C.J.; Benner, N.L.; Haabeth, O.A.; Waymouth, R.M.; Wender, P.A. Enhanced mRNA delivery into lymphocytes enabled by lipid-varied libraries of charge-altering releasable transporters. Proc. Natl. Acad. Sci. USA **2018**, *115*(26), E5859–E5866.

24. Miyazaki, T.; Uchida, S.; Nagatoishi, S.; Koji, K.; Hong, T.; Fukushima, S.; Tsumoto, K.; Ishihara, K.; Kataoka, K.; Cabral, H. Polymeric nanocarriers with controlled chain flexibility boost mRNA delivery in vivo through enhanced structural fastening. Adv. Healthc. Mater. **2020**, *9*(16), 2000538.

25. Kowalski, P.S.; Capasso Palmiero, U.; Huang, Y.; Rudra, A.; Langer, R.; Anderson, D.G. Ionizable amino-polyesters synthesized via ring opening polymerization of tertiary amino-alcohols for tissue selective mRNA delivery. Adv. Mater. **2018**, *30*(34), e1801151.

26. Rozema, D.B.; Lewis, D.L.; Wakefield, D.H.; Wong, S.C.; Klein, J.J.; Roesch, P.L.; Bertin, S.L.; Reppen, T.W.; Chu, Q.; Blokhin, A.V.; Hagstrom, J.E.; Wolff, J.A. Dynamic PolyConjugates for targeted in vivo delivery of siRNA to hepatocytes. Proc. Natl. Acad. Sci. USA **2007**, *104*(32), 12982–12987.

27. Nair, J.K.; Willoughby, J.L.; Chan, A.; Charisse, K.; Alam, M.R.; Wang, Q.; Hoekstra, M.; Kandasamy, P.; Kel'in, A.V.; Milstein, S.; Taneja, N.; O'Shea, J.; Shaikh, S.; Zhang, L.; van der Sluis, R.J.; Jung, M.E.; Akinc, A.; Hutabarat, R.; Kuchimanchi,

S.; Fitzgerald, K.; Zimmermann, T.; van Berkel, T.J.; Maier, M.A.; Rajeev, K.G.; Manoharan, M. Multivalent N-acetylgalactosamine-conjugated siRNA localizes in hepatocytes and elicits robust RNAi-mediated gene silencing. J. Am. Chem. Soc. **2014**, *136*(49), 16958–16961.

28. Zintchenko, A.; Philipp, A.; Dehshahri, A.; Wagner, E. Simple modifications of branched PEI lead to highly efficient siRNA carriers with low toxicity. Bioconjug. Chem. **2008**, *19*(7), 1448–1455.

29. Creusat, G.; Zuber, G. Tyrosine-modified PEI: a novel and highly efficient vector for siRNA delivery in mammalian cells. Nucleic Acids Symp. Ser. (Oxf.) **2008**, *52*(1), 91–92.

30. Luo, J.; Wagner, E.; Wang, Y. Artificial peptides for antitumoral siRNA delivery. J. Mater. Chem. B **2020**, *8*(10), 2020–2031.

31. Troiber, C.; Edinger, D.; Kos, P.; Schreiner, L.; Kläger, R.; Herrmann, A.; Wagner, E. Stabilizing effect of tyrosine trimers on pDNA and siRNA polyplexes. Biomaterials **2013**, *34*(5), 1624–1633.

32. Wang, X.-L.; Ramusovic, S.; Nguyen, T.; Lu, Z.-R. Novel polymerizable surfactants with pH-sensitive amphiphilicity and cell membrane disruption for efficient siRNA delivery. Bioconjug. Chem. **2007**, *18*(6), 2169–2177.

33. Luo, J.; Höhn, M.; Reinhard, S.; Loy, D.M.; Klein, P.M.; Wagner, E. IL4-receptor-targeted dual antitumoral apoptotic peptide—siRNA conjugate lipoplexes. Adv. Funct. Mater. **2019**, *29*(25), 1900697.

34. Kuhn, J.; Klein, P.M.; Al Danaf, N.; Nordin, J.Z.; Reinhard, S.; Loy, D.M.; Höhn, M.; El Andaloussi, S.; Lamb, D.C.; Wagner, E.; Aoki, Y.; Lehto, T.; Lächelt, U. Supramolecular assembly of aminoethylene-lipopeptide PMO conjugates into RNA splice-switching nanomicelles. Adv. Func. Mater. **2019**, *29*(48), 1906432.

35. Kuhn, J.; Lin, Y.; Krhac Levacic, A.; Al Danaf, N.; Peng, L.; Höhn, M.; Lamb, D.C.; Wagner, E.; Lächelt, U. Delivery of Cas9/sgRNA ribonucleoprotein complexes via hydroxystearyl oligoamino amides. Bioconjug. Chem. **2020**. *31*(3), 729–742.

36. Cullis, P.R.; Hope, M.J. Lipid nanoparticle systems for enabling gene therapies. Mol. Ther. **2017**, *25*(7), 1467–1475.

37. Akinc, A.; Goldberg, M.; Qin, J.; Dorkin, J.R.; Gamba-Vitalo, C.; Maier, M.; Jayaprakash, K.N.; Jayaraman, M.; Rajeev, K.G.; Manoharan, M.; Koteliansky, V.; Röhl, I.; Leshchiner, E.S.; Langer, R.; Anderson, D.G. Development of lipidoid-siRNA formulations for systemic delivery to the liver. Mol. Ther. **2009**, *17*(5), 872–879.

38. Love, K.T.; Mahon, K.P.; Levins, C.G.; Whitehead, K.A.; Querbes, W.; Dorkin, J.R.; Qin, J.; Cantley, W.; Qin, L.L.; Racie, T.; Frank-Kamenetsky, M.; Yip, K.N.; Alvarez, R.; Sah, D.W.Y.; de Fougerolles, A.; Fitzgerald, K.; Koteliansky, V.; Akinc, A.; Langer, R.; Anderson, D.G. Lipid-like materials for low-dose, in vivo gene silencing. **2010**, *107*(5), 1864–1869.

39. Dong, Y.; Love, K.T.; Dorkin, J.R.; Sirirungruang, S.; Zhang, Y.; Chen, D.; Bogorad, R.L.; Yin, H.; Chen, Y.; Vegas, A.J.; Alabi, C.A.; Sahay, G.; Olejnik, K.T.; Wang, W.; Schroeder, A.; Lytton-Jean, A.K.R.; Siegwart, D.J.; Akinc, A.; Barnes, C.; Barros, S.A.; Carioto, M.; Fitzgerald, K.; Hettinger, J.; Kumar, V.; Novobrantseva, T.I.; Qin, J.; Querbes, W.; Koteliansky, V.; Langer, R.; Anderson, D.G. Lipopeptide nanoparticles for potent and selective siRNA delivery in rodents and nonhuman primates. Proc. Natl. Acad. Sci. **2014**, *111*(11), 3955–3960.

40. Doudna, J.A.; Charpentier, E. Genome editing. The new frontier of genome engineering with CRISPR-Cas9. Science **2014**, *346*(6213), 1258096.

41. Miller, J.B.; Zhang, S.; Kos, P.; Xiong, H.; Zhou, K.; Perelman, S.S.; Zhu, H.; Siegwart, D.J. Non-viral CRISPR/Cas gene editing in vitro and in vivo enabled by synthetic nanoparticle co-delivery of Cas9 mRNA and sgRNA. Angew. Chem. Int. Ed. Engl. **2017**, *56*(4), 1059–1063.

42. Liu, C.; Wan, T.; Wang, H.; Zhang, S.; Ping, Y.; Cheng, Y. A boronic acid-rich den-drimer with robust and unprecedented efficiency for cytosolic protein delivery and CRISPR-Cas9 gene editing. Sci. Adv. **2019**, *5*(6), eaaw8922.
43. Rouet, R.; Thuma, B.A.; Roy, M.D.; Lintner, N.G.; Rubitski, D.M.; Finley, J.E.; Wisniewska, H.M.; Mendonsa, R.; Hirsh, A.; de Oñate, L.; Compte Barrón, J.; McLellan, T.J.; Bellenger, J.; Feng, X.; Varghese, A.; Chrunyk, B.A.; Borzilleri, K.; Hesp, K.D.; Zhou, K.; Ma, N.; Tu, M.; Dullea, R.; McClure, K.F.; Wilson, R.C.; Liras, S.; Mascitti, V.; Doudna, J.A. Receptor-mediated delivery of CRISPR-Cas9 endonuclease for cell-type-specific gene editing. J. Am. Chem. Soc. **2018**, *140*(21), 6596–6603.
44. Wu, G.Y.; Wu, C.H. Receptor-mediated gene delivery and expression in vivo. J. Biol. Chem. **1988**, *263*(29), 14621–14624.
45. Wagner, E.; Zenke, M.; Cotten, M.; Beug, H.; Birnstiel, M.L. Transferrin-polycation conjugates as carriers for DNA uptake into cells. Proc. Natl. Acad. Sci. USA **1990**, *87*(9), 3410–3414.
46. Wilson, J.M.; Grossman, M.; Wu, C.H.; Chowdhury, N.R.; Wu, G.Y.; Chowdhury, J.R. Hepatocyte-directed gene transfer in vivo leads to transient improvement of hypercho-lesterolemia in low density lipoprotein receptor-deficient rabbits. J. Biol. Chem. **1992**, *267*(2), 963–967.
47. Wu, G.Y.; Wilson, J.M.; Shalaby, F.; Grossman, M.; Shafritz, D.A.; Wu, C.H. Receptor-mediated gene delivery in vivo. Partial correction of genetic analbuminemia in Nagase rats. J. Biol. Chem. **1991**, *266*(22), 14338–14342.
48. Debacker, A.J.; Voutila, J.; Catley, M.; Blakey, D.; Habib, N. Delivery of oligonucle-otides to the liver with GalNAc: from research to registered therapeutic drug. Mol. Ther. **2020**, *28*(8), 1759–1771.
49. Schreiber, S.; Kämpgen, E.; Wagner, E.; Pirkhammer, D.; Trcka, J.; Korschan, H.; Lindemann, A.; Dorffner, R.; Kittler, H.; Kasteliz, F.; Küpcü, Z.; Sinski, A.; Zatloukal, K.; Buschle, M.; Schmidt, W.; Birnstiel, M.; Kempe, R.E.; Voigt, T.; Weber, H.A.; Pehamberger, H.; Mertelsmann, R.; Bröcker, E.B.; Wolff, K.; Stingl, G. Immunotherapy of metastatic malignant melanoma by a vaccine consisting of autologous interleukin 2-transfected cancer cells: outcome of a phase I study. Hum. Gene Ther. **1999**, *10*(6), 983–993.
50. Kircheis, R.; Wightman, L.; Schreiber, A.; Robitza, B.; Rössler, V.; Kursa, M.; Wagner, E. Polyethylenimine/DNA complexes shielded by transferrin target gene expression to tumors after systemic application. Gene Ther. **2001**, *8*(1), 28–40.
51. Ogris, M.; Brunner, S.; Schüller, S.; Kircheis, R.; Wagner, E. PEGylated DNA/transferrin-PEI complexes: reduced interaction with blood components, extended circulation in blood and potential for systemic gene delivery. Gene Ther. **1999**, *6*(4), 595–605.
52. Kursa, M.; Walker, G.F.; Roessler, V.; Ogris, M.; Roedl, W.; Kircheis, R.; Wagner, E. Novel shielded transferrin-polyethylene glycol-polyethylenimine/DNA complexes for systemic tumor-targeted gene transfer. Bioconjug. Chem. **2003**, *14*(1), 222–231.
53. Hu-Lieskovan, S.; Heidel, J.D.; Bartlett, D.W.; Davis, M.E.; Triche, T.J. Sequence-specific knockdown of EWS-FLI1 by targeted, nonviral delivery of small interfering RNA inhibits tumor growth in a murine model of metastatic Ewing's sarcoma. Cancer Res. **2005**, *65*(19), 8984–8992.
54. Bartlett, D.W.; Su, H.; Hildebrandt, I.J.; Weber, W.A.; Davis, M.E. Impact of tumor-specific targeting on the biodistribution and efficacy of siRNA nanoparticles mea-sured by multimodality in vivo imaging. Proc. Natl. Acad. Sci. USA **2007**, *104*(39), 15549–15554.

55. Wolschek, M.F.; Thallinger, C.; Kursa, M.; Rössler, V.; Allen, M.; Lichtenberger, C.; Kircheis, R.; Lucas, T.; Willheim, M.; Reinisch, W.; Gangl, A.; Wagner, E.; Jansen, B. Specific systemic nonviral gene delivery to human hepatocellular carcinoma xenografts in SCID mice. Hepatology **2002**, *36*(5), 1106–1114.

56. Klein, P.M.; Kern, S.; Lee, D.J.; Schmaus, J.; Höhn, M.; Gorges, J.; Kazmaier, U.; Wagner, E. Folate receptor-directed orthogonal click-functionalization of siRNA lipopolyplexes for tumor cell killing in vivo. Biomaterials **2018**, *178*, 630–642.

57. Dahlman, J.E.; Kauffman, K.J.; Xing, Y.; Shaw, T.E.; Mir, F.F.; Dlott, C.C.; Langer, R.; Anderson, D.G.; Wang, E.T. Barcoded nanoparticles for high throughput in vivo discovery of targeted therapeutics. Proc. Natl. Acad. Sci. USA **2017**, *114*(8), 2060–2065.

58. Sago, C.D.; Lokugamage, M.P.; Paunovska, K.; Vanover, D.A.; Monaco, C.M.; Shah, N.N.; Gamboa Castro, M.; Anderson, S.E.; Rudoltz, T.G.; Lando, G.N.; Munnilal Tiwari, P.; Kirschman, J.L.; Willett, N.; Jang, Y.C.; Santangelo, P.J.; Bryksin, A.V.; Dahlman, J.E. High-throughput in vivo screen of functional mRNA delivery identifies nanoparticles for endothelial cell gene editing. Proc. Natl. Acad. Sci. USA **2018**, *115*(42), E9944–E9952.

59. Lokugamage, M.P.; Sago, C.D.; Gan, Z.; Krupczak, B.R.; Dahlman, J.E. Constrained nanoparticles deliver siRNA and sgRNA to T cells in vivo without targeting ligands. Adv. Mater. **2019**, *31*(41), 1902251.

60. Ginn, S.L.; Amaya, A.K.; Alexander, I.E.; Edelstein, M.; Abedi, M.R. Gene therapy clinical trials worldwide to 2017 – an update. J. Gene Med. **2018**, *20*(5), e3015.

61. Stein, C.A.; Castanotto, D. FDA-approved oligonucleotide therapies in 2017. Mol. Ther. **2017**, *25*(5), 1069–1075.

62. Sidi, A.A.; Ohana, P.; Benjamin, S.; Shalev, M.; Ransom, J.H.; Lamm, D.; Hochberg, A.; Leibovitch, I. Phase I/II marker lesion study of intravesical BC-819 DNA plasmid in H19 over expressing superficial bladder cancer refractory to bacillus Calmette-Guerin. J. Urol. **2008**, *180*(6), 2379–2383.

63. Gofrit, O.N.; Benjamin, S.; Halachmi, S.; Leibovitch, I.; Dotan, Z.; Lamm, D.L.; Ehrlich, N.; Yutkin, V.; Ben-Am, M.; Hochberg, A. DNA based therapy with diphtheria toxin-A BC-819: a phase 2b marker lesion trial in patients with intermediate risk non-muscle invasive bladder cancer. J. Urol. **2014**, *191*(6), 1697–1702.

64. Gautam, A.; Waldrep, J.C.; Densmore, C.L.; Koshkina, N.; Melton, S.; Roberts, L.; Gilbert, B.; Knight, V. Growth inhibition of established B16-F10 lung metastases by sequential aerosol delivery of p53 gene and 9-nitrocamptothecin. Gene Ther. **2002**, *9*(5), 353–357.

65. Wooddell, C.I.; Rozema, D.B.; Hossbach, M.; John, M.; Hamilton, H.L.; Chu, Q.; Hegge, J.O.; Klein, J.J.; Wakefield, D.H.; Oropeza, C.E.; Deckert, J.; Roehl, I.; Jahn-Hofmann, K.; Hadwiger, P.; Vornlocher, H.P.; McLachlan, A.; Lewis, D.L. Hepatocyte-targeted RNAi therapeutics for the treatment of chronic hepatitis B virus infection. Mol. Ther. **2013**, *21*(5), 973–985.

66. Wooddell, C.I.; Yuen, M.F.; Chan, H.L.; Gish, R.G.; Locarnini, S.A.; Chavez, D.; Ferrari, C.; Given, B.D.; Hamilton, J.; Kanner, S.B.; Lai, C.L.; Lau, J.Y.N.; Schluep, T.; Xu, Z.; Lanford, R.E.; Lewis, D.L. RNAi-based treatment of chronically infected patients and chimpanzees reveals that integrated hepatitis B virus DNA is a source of HBsAg. Sci. Transl. Med. **2017**, *9*(409), eaan0241.

67. Kircheis, R.; Ostermann, E.; Wolschek, M.F.; Lichtenberger, C.; Magin-Lachmann, C.; Wightman, L.; Kursa, M.; Wagner, E. Tumor-targeted gene delivery of tumor necrosis factor-alpha induces tumor necrosis and tumor regression without systemic toxicity. Cancer Gene Ther. **2002**, *9*(8), 673–680.

68. Su, B.; Cengizeroglu, A.; Farkasova, K.; Viola, J.R.; Anton, M.; Ellwart, J.W.; Haase, R.; Wagner, E.; Ogris, M. Systemic TNFalpha gene therapy synergizes with liposomal doxorubicine in the treatment of metastatic cancer. Mol. Ther. **2013**, *21*(2), 300–308.

69. Schmohl, K.A.; Gupta, A.; Grunwald, G.K.; Trajkovic-Arsic, M.; Klutz, K.; Braren, R.; Schwaiger, M.; Nelson, P.J.; Ogris, M.; Wagner, E.; Siveke, J.T.; Spitzweg, C. Imaging and targeted therapy of pancreatic ductal adenocarcinoma using the theranostic sodium iodide symporter (NIS) gene. Oncotarget **2017**, *8*(20), 33393–33404.

70. Shir, A.; Ogris, M.; Wagner, E.; Levitzki, A. EGF receptor-targeted synthetic double-stranded RNA eliminates glioblastoma, breast cancer, and adenocarcinoma tumors in mice. PLoS Med. **2006**, *3*(1), e6.

71. Shir, A.; Ogris, M.; Roedl, W.; Wagner, E.; Levitzki, A. EGFR-homing dsRNA activates cancer-targeted immune response and eliminates disseminated EGFR-overexpressing tumors in mice. Clin. Cancer Res. **2011**, *17*(5), 1033–1043.

72. Davis, M.E.; Zuckerman, J.E.; Choi, C.H.; Seligson, D.; Tolcher, A.; Alabi, C.A.; Yen, Y.; Heidel, J.D.; Ribas, A. Evidence of RNAi in humans from systemically administered siRNA via targeted nanoparticles. Nature **2010**, *464*(7291), 1067–1070.

19 Automated DNA Sequencing

Alexandre Izmailov
Visible Genetics, Inc., Toronto, Ontario, Canada

CONTENTS

INTRODUCTION

DNA sequencing is an important tool for molecular biology, genetic studies, pharmacogenomics, and other areas of fundamental and applied science where information is required about unknown DNA, mutations, DNA fragment size, and for DNA fingerprinting or prediction of drug resistance of viruses. Originally suggested, and widely used, manual sequencers cannot satisfy a growing demand for throughput increase, high reliability, and ease of use. This necessitated the development of automated DNA sequencers. Automation may include data collection and analysis, or as in the case of fully automated sequencers, also sample loading and sieving matrix replacement. The output of automated DNA sequencers is a base calling report, which represents the sequence of the nucleotides in the fragment of a DNA molecule. Different applications may require specific features of a DNA sequencer, but base calling accuracy and read length remain major figures of merit for all sequencers. Clinical applications and research, which involves large volumes of data, require high throughput and a high level of automation. However, automated DNA sequencers targeted to research laboratories must also provide flexibility in their features. Careful assessment of the user requirements allows an optimum selection of DNA sequencer from the wide variety of existing instruments.

DOI: 10.1201/9781003247432-19

SEQUENCING CHEMISTRIES

The goal of DNA sequencing is to obtain information about the sequence of nucleotide bases—adenine (A), guanine (G), cytosine (C), and thymine (T)—that constitute a particular DNA molecule. In order to obtain this information, the sample under investigation is modified in DNA sequencing reactions, which can be based on one of two approaches: enzymatic synthesis of DNA fragments with chain termination by dideoxynucleotides ddNTP (terminators)[1] or by chemical degradation method.[2] The first method is faster, easier to implement, and it is currently the method of choice. Terminators (ddNTP) lack a 3′OH group that is necessary for the extension of DNA's sugar phosphate backbone. Thus, the DNA chain cannot be extended beyond the incorporated ddNTP. A mixture of all four dNTPs and one ddNTP is used in each reaction. For example, in a T-terminated reaction, a ddTTP is added to a mixture of dNTPs. This allows elongation of the DNA chain until ddTTP is occasionally attached to the growing molecule, which stops the reaction. This process is random, and as a result, fragments of different lengths are produced, each terminated at a T-position, for all T's from the primer to the end of the sequence. Four reactions are required for complete description of the DNA sequence, each representing an A-, C-, G-, or T-terminated ladder. In order to enable observation of DNA bands, the fragments are labeled with a tag (radioactive or fluorescent) attached to either the primer (primer chemistry) or ddNTPs (terminator chemistry).

STRUCTURE OF A DNA SEQUENCER

A DNA sequencer is composed of an electrophoresis system that includes a sieving matrix (e.g., polyacrylamide gel), buffer, a high-voltage power supply, and a detection system. When a high voltage (1000 V to 10 kV) is applied to the gel, an electric field forces negatively charged fragments of DNA molecules to move through the mesh created by polymer molecules constituting the gel. The difference in the friction for the DNA fragments of different sizes leads to their spatial separation. Groups of molecules of the same size propagate through the gel as a set of confined bands with the width defined by such factors as diffusion, injection conditions, sample volume, field, temperature gradients, etc.

Manual and automated instruments, in their principle of operation, mainly differ by the method of detection. In manual sequencers, the electrophoretic run is stopped as soon as the first base of interest (typically a primer) reaches the end of the gel. Then, information about DNA sequence is obtained by determining the relative position of the bands in the gel. In automated sequencers, the detection system records variations of the output intensity over time at a fixed location in the gel. The difference in the migration times of various DNA chain lengths results in a set of electrophoretic peaks.

Automated DNA sequencers can be divided into three major categories based on the method used for sieving matrix support: slab gel, capillary, and microarray sequencers.

SIEVING MATRIX

A polymer most frequently used as a sieving matrix in automatic DNA sequencers is polyacrylamide. It can be used in linear or cross-linked form. The porosity of the gel is defined by the concentration of acrylamide and cross-linking agent (typically N,N'-methylene-bis-acrylamide). The concentration of acrylamide may vary from 3% to 10%.[3] The concentration of a cross-linker is ~3%–5% of the total acrylamide concentration. An increase in the acrylamide concentration suggests a higher resolution of the DNA bands and allows calling more bases, but this leads to slower runs. Linear polyacrylamide (LPA), which is more fluid, is typically used in capillary electrophoresis (CE) where the sieving matrix is replaced after each run so the capillary can be reused. The last requirement is a result of the high cost of the capillary arrays and complexity of alignment required after their replacement. Other types of polymers used in CE include hydroxyethyl cellulose (HEC), poly(ethylenoxyde) (PEO), polyvinyl pyrrolidone (PVP), and others.[4]

One of the major figures of merits in the DNA sequencers is the resolution of the DNA bands. A reduction in the width of the DNA band improves resolution, allowing base calling of longer DNA chains. Resolution depends also on the distance that DNA molecules migrate in a sieving matrix. Separation lengths of 10–50 cm are typical, but they may be longer for higher resolution. However, there is a trade-off between increase in resolution and run time: for a given voltage, velocity of DNA molecules in the gel is inversely proportional to the separating length, L, resulting in a migration time proportional to L^2.

For analysis of single-stranded DNA, denaturants such as urea (concentration 4–8 M) or formamide (up to 10%) are added to a gel. Additionally, the electrophoretic runs are conducted at elevated temperature (40–70°C), which helps to prevent self-hybridization.

METHODS OF DETECTION

RADIOLABELING

DNA molecules separated by the sieving matrix need to be detected. In the manual sequencers, an autoradiographic method is used[3] where either a primer or terminators contain a radioisotope (typically ^{32}P or ^{35}S). After completion of the electrophoretic run, the gel is dried and placed in contact with an X-ray film. Exposed film is developed and manually analyzed. Lack of automation in the data analysis and risk associated with using radioactive materials are the most serious drawbacks of this method. Also, the resolution at high base numbers in manual sequencers is lower than that in automatic sequencers because of the much shorter migration distance.

Laser-Induced Fluorescence (LIF)

Laser-induced fluorescence is a standard detection method in a majority of the automated sequencers. This method is extremely sensitive, allows automation of base calling, avoids using radioactive materials, and promotes a throughput increase by using several spectrally distinct labels simultaneously. Laser-induced fluorescence is based on the excitation of a dye label attached to a DNA molecule by laser light. Resulting fluorescence is collected and converted to an electrical signal by a photosensitive element (photomultiplier tube [PMT], photodiode, or charge-coupled device [CCD] element). The output signal (trace) represents a series of peaks each corresponding to a DNA chain of a certain size. One trace provides information about one of the four bases (A, C, G, or T). Therefore, in order to obtain information about a complete sequence, four traces are required. This is achieved either by using four physically separated lanes (or capillaries) or by running four differently labeled samples in a single lane.

If a single laser is used in the sequencer, different parts of the gel or different capillaries are illuminated through a scanning objective lens. Alternatively, a sheath cell[5] can be used with simultaneous illumination of several DNA traces. Some DNA sequencers contain several lasers either with different wavelengths improving excitation[6] or with the same wavelength providing excitation of multiple samples without using moving parts.[7]

Collected fluorescence is converted into digital form and processed by a computer. Software packages used in different sequencers vary in their functionality and efficiency, but all of them involve the following common steps: noise reduction, base line subtraction, color cross-talk compensation, trace alignment, and peak recognition. A typical example of the resulting base calling is shown in Figure 19.1.

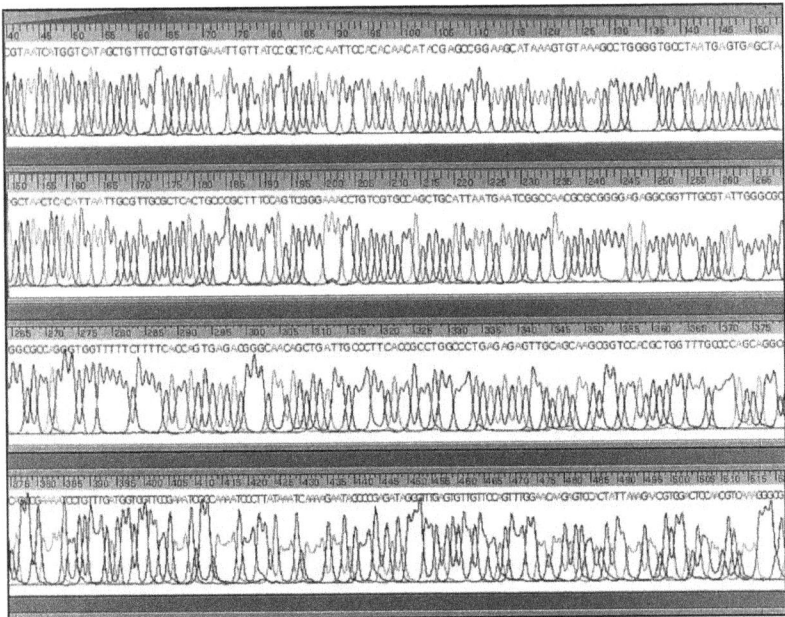

FIGURE 19.1 A typical result of base calling obtained with an automated DNA sequencer. The Ml3 sequencing by ultrathin slab gel electrophoresis. The signal obtained with the Long-Read Tower™ sequencer has been base called with GeneObjects software (From Ref. [7]). The accuracy of base calling is 98.5% up to 500 bases. Similar results are typically produced with other types of automated DNA sequencers, including CE instruments.

Dye Labeling

Dye labeling is typically used for sample identification in the automatic DNA sequencers. The properties of a dye shall satisfy the following major requirements: (1) dye structure shall allow attachment to a primer or ddNTP with a reasonable yield. (2) Absorption spectrum shall match the laser wavelength. (3) Extinction coefficient and quantum yield shall be high enough to provide acceptable sensitivity. (4) Dyes shall be chemically and thermally stable in order to withstand cycle sequencing conditions. (5) Dyes shall be photo stable. (6) Mobility of the DNA fragments labeled with different dyes shall be equalized, which can be achieved by optimization of the labeling process.[8]

Both emission and absorption spectra of the organic dyes employed in DNA labeling are broad (typically 20–50 nm). This relaxes the requirement for spectral content of the excitation light.

Overlapping absorption spectra of several dyes used for labeling allows excitation of all these dyes with a single laser source. However, efficiency of excitation drops off quickly as the laser wavelength deviates from the absorption maximum of the dye and results in a significant drop in sensitivity. To avoid these losses, energy transfer (ET) dyes were introduced.[9]

In the case of multicolor sequencers (when two or more dyes are used for DNA labeling), collected radiation must be spectrally separated by means of a diffractive element or a set of interference filters and is registered by separate PMTs or by a CCD. Alternatively, a single PMT can be used with a set of replaceable filters (a "color wheel"). A more sophisticated detection method used for sensitivity increase is based on single-photon detection.[10]

A complication arises because the emission spectra of the dyes used for labeling different samples overlap and color cross-talk correction of the collected data is required. Improper compensation of the cross-talk results in false peaks in the processed data.

Both visible and near IR dyes can be used for DNA labeling in automated DNA sequencers. The advantage of using red and near IR dyes is lower background fluorescence and decreased laser light scattering at longer wavelengths. This results in better sensitivity. IR dyes can be excited by laser diodes (LD), resulting in a reduction of instrument design complexity.[6,7]

Dye labels can be covalently linked either to a primer or to each of four types of ddNTPs.[11] In the last case, all four sequencing reactions can be carried out in one tube instead of the four used in primer-labeled chemistry. However, dye-labeled terminators may introduce significant variability in the peak intensity, resulting in decreased base calling accuracy and reliability. For clinical applications, dye primer chemistry may be preferred because of the higher reliability of the results. Terminator chemistry provides more flexibility, which may be advantageous for the research laboratories.

In some automated sequencers, less than four colors are used for DNA labeling. For example, a two-dye approach is used in the Long-Read Tower™ and LICOR sequencers. The sample is prepared by running two sequencing reactions in the same tube with differently labeled forward and reverse primers. This approach enables the increase of the read length, accuracy, and reliability of base calling.

TYPES OF DNA SEQUENCERS

SLAB GEL SEQUENCERS

Initially, automated DNA sequencers were based on slab gel electrophoresis with low throughput and automation limited to data collection and analysis. Gel filling and sample loading are done manually. The support glass plates require cleaning after each run. Poor dissipation of heat generated in a slab gel through relatively thick glass plates limits the electric field strength and prevents short run times. Both of these problems are eliminated in the disposable ultrathin slab gels (50 µm thick), which can withstand higher electric field strength, E (100 V/cm), resulting in shorter run times.[7]

Capillary Electrophoresis Sequencers

CE DNA sequencers were introduced to increase analysis throughput. This is a direct result of high electric fields (100–250 V/cm) and multiple capillaries (up to 384 capillaries). Both sieving matrix replacement and sample loading can be automated. The small diameter of capillaries (internal diameter is 50–100 µm) allows for good heat dissipation. Capillaries having smaller diameters (2–10 µm) are not practical as they do not provide enough signal intensity, are difficult to fill with acrylamide, and do not provide reproducible performance.[12] The ability to use replaceable sieving matrices makes CE systems practical as repetitive runs are possible with the same capillary up to 200 times or more.

Electra-kinetic sample loading allows high automation of CE sequencers. But optimization of loading conditions and sample purification is required. These are critical for reducing variability of peak intensity, suppressing decay in gel current and bubble formation, increasing capillary lifetime. Salt concentration, ratio of labeled DNA to unlabeled DNA, and total volume of loaded sample all affect the success of CE analysis.

Typical read length in CE DNA sequencers is 400–600 bases in several hours but can be as high as 1100 bases.[13] Run conditions can be optimized for faster runs, if necessary, by reducing separation lengths, decreasing polymer concentration, increasing voltage or/and gel temperature. Examples of commercial-automated DNA sequencers are shown in Table 19.1.

A variety of automated DNA sequencers is available, providing similar performance characteristics but differing in throughput, flexibility, complexity, and cost. Although generally higher throughput is desirable (especially for clinical labs), for some applications, the major requirement may be short run time, flexibility of characteristics, or low cost of analysis. In this case, smaller DNA sequencers may be advantageous.

Microarray Sequencers

Capillary microarray and microfabricated sequencers are still under development and promise a significant size reduction of automated sequencers providing high throughput and base calling accuracy.[18,19] A 96-channel radial capillary array was demonstrated in Ref. [20]. The microchannel plate provides high resolution due to

TABLE 19.1
Automated DNA Sequencers

Model	Read Length (nt)	Number of Lanes	Run Time (h)	Type
Al31 3731377	850	48/64/96	10	Slab
LI-COR 4300 L	1000–1200	48/64	10	Slab
LI-COR 4300 S/L	800		6	
	500		1.5–3	
Long-Read Tower™	300/500/700	16	0.51/14	Ultrathin slab
ABI 310	500–600	1	3	Capillary
ABI 3700	550	96	5	Capillary
ABI 3730 (xl)	550–1100	48/96	0.5–2	Capillary
CEQ™ 2000	500	8	2	Capillary
CEQ™ 8000	700	8	2	Capillary
MegaBACE™ 1000	550	96	2	Capillary
MegaBACE™ 4000	550	384	2	Capillary

Source: From Refs. [6, 14–17].

the use of tapered turns that provide an effective separation length of 15.9 cm on a compact 150-mm-diameter wafer. An average read length of 430 bases was obtained on this device. In general, miniaturization typically leads to a reduction of separation distance causing poor resolution, insufficient read length, low base calling accuracy, all of which currently limit the application of these devices.

CONCLUSION

Automated DNA sequencers provide important information about DNA structure. Plurality of existing instruments can satisfy a variety of requirements, which may arise in a particular experiment. Although modern automated DNA sequencers are well developed, continuous improvement of their structure, components, and run conditions is still ongoing. The development of new approaches is focused on throughput increase, reduction of sequencer size, and improvement of accuracy, reliability, and read length.

REFERENCES

1. Sanger, F.; Nicklen, S.; Coulson, A.R. DNA sequencing with chain-terminating inhibitors. Proc. Natl. Acad. Sci. U.S.A. **1977**, *74* (12), 5463–5467.
2. Maxam, A.M.; Gilbert, W. A new method for sequencing DNA. Proc. Natl. Acad. Sci. U.S.A. **1977**, *74* (2), 560–564.
3. Andrews, A.T. Electrophoresis. *Theory, Techniques, and Biochemical Clinical Applications*, 2nd Ed.; In Andrews, A.T., Ed.; Oxford: Clarendon Press, **1986**.
4. Quesada, M.A.; Menchen, S. Replaceable polymers for DNA sequencing by capillary electrophoresis. Methods Mol. Biol. **2001**, *162*, 139–166.

5. Swerdlow, H.; Wu, S.L.; Harke, H.; Dovichi, N.J. Capillary gel electrophoresis for DNA sequencing. Laserinduced fluorescence detection with the sheath flow cuvette. J. Chromatogr. **1990**, *516* (1), 61–67.

6. Mcharo, T.M. Associating molecular markers with phenotypes in sweetpotatoes and liriopogons using multivariate statistical modeling. Doctoral dissertation. Louisiana State University. http://bio.licor.com/Pubs/biopub4.htm

7. Yager, T.D.; Baron, L.; Batra, R.; Bouevitch, A.; Chan, D.; Chan, K.; Darasch, S.; Gilchrist, R.; Izmailov, A.; Lacroix, J.-M.; Marchelleta, K.; Renfrew, J.; Rushlow, D.; Steinbach, E.; Ton, C.; Waterhouse, P.; Zaleski, H.; Dunn, M.; Stevens, J. High performance DNA sequencing, and the detection of mutations and polymorphisms, on the Clipper sequencer. Electrophoresis **1999**, *20* (6), 1280–1300.

8. Metzker, M.L.; Lu, J.; Gibbs, R.A. Electrophoretically uniform fluorescent dyes for automated DNA sequencing. Science **1996**, *271* (5254), 1420–1422.

9. Ju, J.; Ruan, C.; Fuller, C.W.; Glazer, A.N.; Mathies, R.A. Fluorescence energy transfer dye-labeled primers for DNA sequencing and analysis. Proc. Natl. Acad. Sci. U.S.A. **1995**, *92* (10), 4347–4351.

10. Gavrilov, D.N.; Gorbovitski, B.; Gouzman, M.; Gudkov, G.; Stepoukhovitch, A.; Ruskovoloshin, V.; Tsuprik, A.; Tyshko, G.; Bilenko, O.; Kosobokova, O.; Luryi, S.; Gortinkel, V. Dynamic range of fluorescence detection and base-calling accuracy in DNA sequencer based on single-photon counting. Electrophoresis **2003**, *24* (7–8), 1184–1192.

11. Wojczewski, C.; Faulstich, K.; Engels, J.W. Synthesis and application of 3′-amino-dye-terminators for DNA sequencing. Nucleotisides Nucleotides **1997**, *16*, 751–754.

12. Lindberg, P.; Stjernstrom, M.; Roeraade, J. Gel electrophoresis of DNA fragments in narrow-bore capillaries. Electrophoresis **1997**, *18* (11), 1973–1979.

13. Carrilho, E.; Ruitz-Martinez, M.C.; Berka, J.; Smimov, I.; Goetzinger, W.; Miller, A.W.; Brady, D.; Karger, B.L. Rapid DNA sequencing of more than 1000 bases per run by capillary electrophoresis using replaceable linear polyacrylamide solutions. Anal. Chem. **1996**, *68*, 3305–3313.

14. Watts, D.; MacBeath, J.R. Automated fluorescent DNA sequencing on the ABI PRISM 310 genetic analyzer. Methods Mol. Biol. **2001**, *167*, 153–170.

15. MacBeath, J.R.; Harvey, S.S.; Oldroyd, N.J. Automated fluorescent DNA sequencing on the ABI PRISM 377. Methods Mol. Biol. **2001**, *167*, 119–152.

16. Izmailov, A.; Yager, T.D.; Zaleski, H.; Darash, S. Improvement of base-calling in multilane automated DNA sequencing by use of electrophoretic calibration standards, data linearization, and trace alignment. Electrophoresis **2001**, *22* (10), 1906–1914.

17. Beckman Coulter Life Sciences. http://www.beckman.com.

18. Schmalzing, D.; Tsao, N.; Koutny, L.; Chisholm, D.; Srivastava, A.; Adourian, A.; Linton, L.; McEwan, P.; Matsudaira, P.; Ehrlich, D. Toward real-world sequencing by microdevice electrophoresis. Genome Res. **1999**, *9*(9), 853–858.

19. Verporte, E. Microfluidic chips for clinical and forensic analysis. Electrophoresis **2002**, *23*, 677–712.

20. Paegel, B.M.; Emrich, C.A.; Wedemayer, J.; Scherer, R.; Mathies, R.A. High throughput DNA sequencing with a microfabricated 96-lane capillary array electrophoresis bioprocessor. Proc. Natl. Acad. Sci. U.S.A. **2002**, *99* (2), 574–579.

20 DNA Sequencing Methods

Scott O. Rogers
Bowling Green State University, Bowling Green, Ohio, U.S.A.

CONTENTS

INTRODUCTION

Sequencing of nucleic acids began in the 1960s, when the sequences of tRNAs were determined. It was accomplished by methods that partially digested different portions of the RNA, and then the individual pieces were joined together on paper manually by the researchers. This was a painstaking process, so much so that the first RNA virus genomes were not fully determined until 1976 (MS2, which is approximately 3.6 kb in length) and 1977 (φX174; approximately 5.4 kb in length). Because most RNAs are single-stranded molecules, obtaining their sequences was limited to degradative methods at that time. In the same year that the RNA sequence of φX174 was published, two different methods for DNA sequencing were published.[1–3] The Maxam and Gilbert method[1] was somewhat analogous to the RNA methods for sequencing, in that it used a set of controlled degradative reactions to create a set of variously sized DNA fragments (Figure 20.1). It employed chemicals that cut at cytosines, another cut at pyrimidines, another cut at guanosines, and the final reaction cut at purines. The resulting products were each loaded into individual lanes of a long polyacrylamide gel, and then subjected to electrophoresis, such that fragments varying by single nucleotides could be distinguished from one another. Furthermore, the starting DNA molecules were labeled with ^{32}P at their 5′ ends, so that only the fragments containing the intact 5′ end could be observed by autoradiography of the gel. This allowed reconstruction of the DNA sequence reading from the 5′ end.

The second method,[2] which became more widely used, was developed by Frederick Sanger and colleagues (Figure 20.2). Instead of being a degradative reaction, this method used DNA polymerase and a set of modified mononucleosides to polymerize DNA fragments of varying lengths. Standard nucleoside triphosphates for DNA synthesis have 2′-deoxyribose sugars with hydroxyls on the 3′-deoxyribose on the growing nucleotide chain, which joins with the 5′ phosphate

DOI: 10.1201/9781003247432-20

191

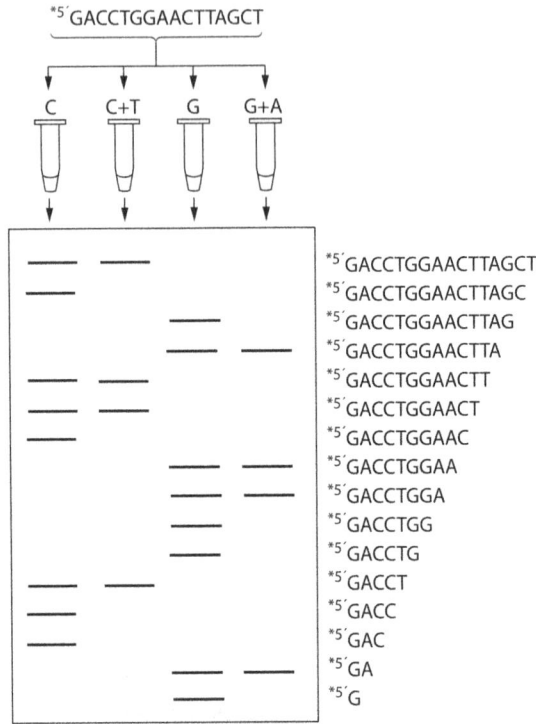

FIGURE 20.1   Diagram of an autoradiogram of a Maxam and Gilbert-type sequencing gel, with an interpretation of the sequence in the right.

group of the incoming nucleoside triphosphate during DNA polymerization. For the Sanger method, a small amount of 2′,3′-dideoxyribose triphosphates were added to each of four reactions (one for each of the four nucleotides). When one of the 2′,3′-dideoxyribose nucleotides was added to a growing DNA chain, no additional nucleotides could be added to the chain, because the end contained no 3′-hydroxyl on the ribose to react with the 5′-phosphate on the incoming nucleoside triphosphate. Thus, the chain was terminated when a dideoxynucleotide triphosphate is added to the DNA.

The original Sanger method used radioactive labeling of one or more of the nucleoside triphosphates, followed by gel electrophoresis through thin polyacrylamide gels, and then autoradiography to visualize the bands. By the late 1980s and early 1990s, several methods using nonradioactive labeling methods were developed, one of which was used extensively in semiautomated and fully automated sequencing methods. The labeled nucleotides had side groups attached to them that would fluoresce when excited by specific wavelengths of light, emitted from lasers. The first methods had only one such chromophore, so that four lanes still had to be run for each sequence, corresponding to the adenine, guanine, cytosine, and thymine

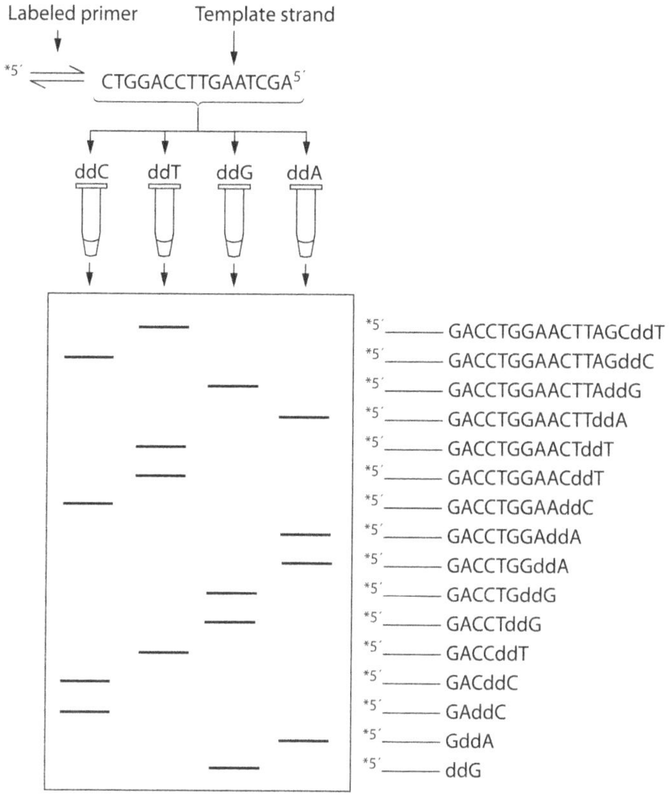

FIGURE 20.2 Diagram of an autoradiogram of a Sanger and Gilbert-type sequencing gel, with an interpretation of the sequence in the right.

reactions. At the bottom of the gel, a laser would move across the gel, along with a light detector, which would send a signal to the attendant computer. Software in the computer would interpret the light signals as nucleotides and their positions in each of the lanes, and a sequence would be generated from the information. Subsequently, a system was developed with four chromophores attached to the dideoxynucleoside triphosphates, one for each type of nucleotide (A, G, C, and T), each of which fluoresced a different color. The advantage of this system was that all four reactions could be mixed into a single reaction and loaded into one lane, as opposed to four lanes for the other systems. Once a fragment reached the bottom of the gel and was struck by the laser light, only one color would fluoresce, corresponding to the specific type of nucleotide. The signal sent to the computer indicated the nucleotide and that position, and the computer built up the sequence results. This system was used to complete the first drafts of the human genome, as well as sequencing for many other applications. Eventually, the gel systems were replaced by capillary tubes filled with a gel-like polymer (Figure 20.3)

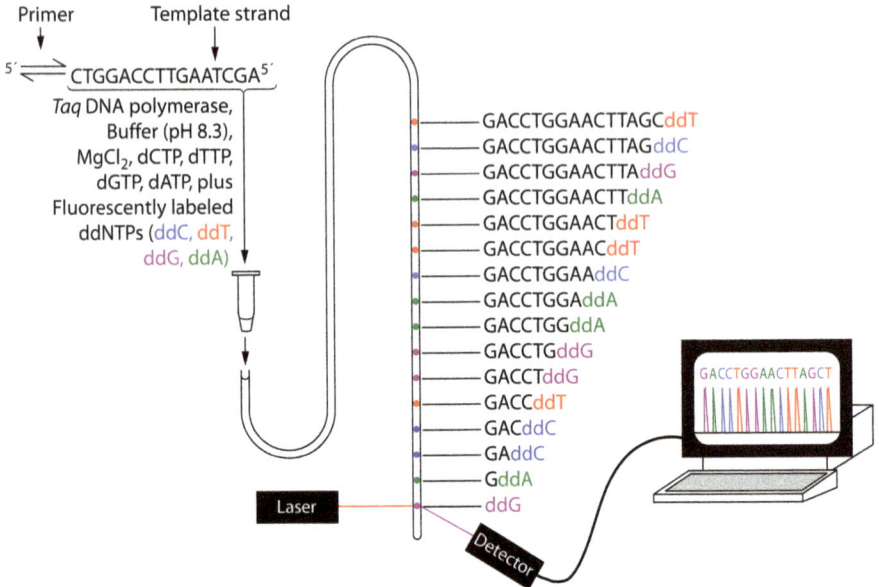

FIGURE 20.3 Sanger sequencing with chromophore labels for each nucleotide type, and electrophoresis through a polymer in a capillary.

NEXT-GENERATION SEQUENCING (NGS)[3]

By the late 1990s and early 2000s, gel and capillary methods were replaced by those that used nanoplates (with small reaction wells), reaction cells, slides, or modified computer chips that allowed many sequencing reactions to proceed in parallel. These were exposed to sensors that recorded the data into computers that interpreted the sequences in each of the reactions taking place synchronously. These methods are collectively termed next-generation sequencing (NGS). Among the first NGS methods to become commercially successful was simply called 454 pyrosequencing (Figure 20.4). As with Sanger sequencing, the DNA was sequenced using DNA polymerase, but polymerization occurred while the template DNAs were immobilized on microbeads. The microbeads had a single-stranded DNA of known sequence covalently bonded to them. The DNA fragments to be sequenced had the complementary sequence attached to them, such that they hydrogen bonded firmly to the single strands on the beads. A second sequence served as the primer for the sequencing reaction. The DNA fragments to be sequenced each had the complementary sequence attached to their opposite ends. To begin the process, the DNA to be sequenced was diluted such that on average only a single-DNA fragment was hybridized to a single bead. These were then subjected to emulsion polymerase chain reaction (emPCR), where the beads and the surrounding aqueous reaction mixture amplified the DNAs, which adhered to the covalently attached DNA on the beads. Each of these small reaction regions was surrounded by mineral oil, which was the other part of the emulsion. The beads were then centrifuged into a microplate that

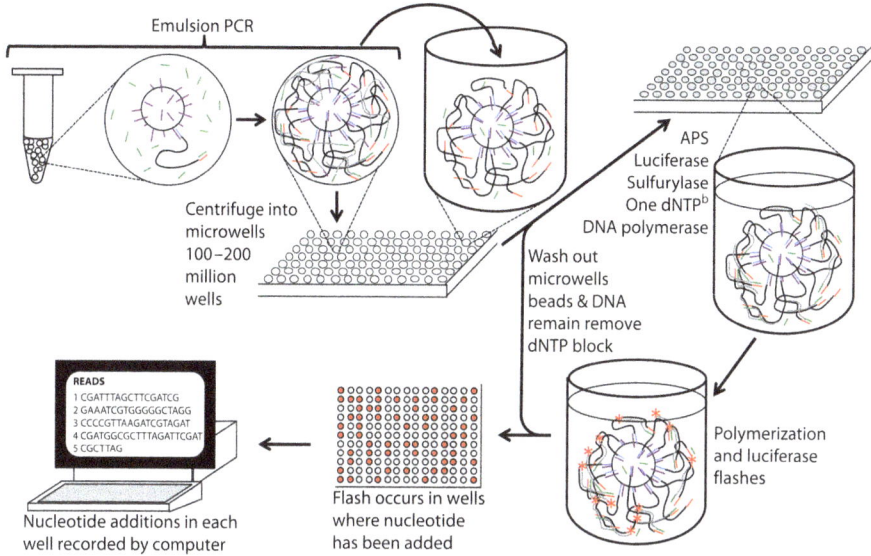

FIGURE 20.4 Summary of 454 pyrosequencing method, including emulsion PCR, sequencing by polymerization, detection by stimulation of luciferase, and recording reads into computer files.

had tiny wells that were only large enough so that one nanobead would fit inside of each well. Polymerization (again, a PCR reaction) was then performed to increase the number of identical DNA fragments on each bead in each well. Finally, Sanger-type sequencing was performed, washing a single type of nucleoside triphosphate over the microplate for each cycle. Also, the enzyme luciferase, the chromogenic compound luciferin, and adenosine-5′-phosphosulfate (APS) also were present in the microplate. When a nucleotide was incorporated into a DNA chain, pyrophosphate was produced. Pyrophosphate reacted with the APS to produce ATP (adenosine triphosphate), which activated luciferase, that acted on luciferase to produce a flash of light. Light sensors and a computer recorded the positions of each of the flashes and correlated this with the nucleoside triphosphate that was added during each cycle. The computer produced a set of DNA sequence fragments, called "reads". Read lengths for each microwell averaged about 200–400 nucleotides, although reads of nearly 1000 nucleotides sometimes resulted. Often more than one million such high-quality reads resulted from a single sample of DNA. However, there was a level of error involved in this method (as well as with all NGS methods), so usually, multiple reads of the same region needed to be produced for each region. During analysis, some nucleotide positions exhibited variation from one sequence to another. A consensus base was chosen by the computer to report in the final sequence. Unfortunately, because of the high cost of the method, in 2014, it was announced that 454 pyrosequencing would no longer be supported and thus has been discontinued in many facilities.

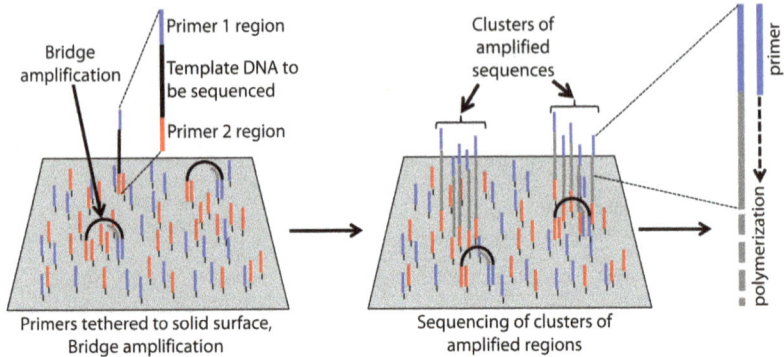

FIGURE 20.5 Summary of Illumina sequencing method.

Other methods, such as Illumina (Solexa), also use specific sequences on each end, but tether the DNA fragments to a slide or chip (similar to a computer chip; Figure 20.5). Again, the DNA fragments are first amplified in number using polymerization reactions, such that each position on the chip contains a high concentration of identical templates. Then the chips are flooded with solutions of each nucleoside triphosphate that have been modified with a fluorophore and blocking group on the nucleoside triphosphate, such that only one nucleotide can be added at a time. After illumination from a laser, the patterns of fluorescence are recorded by a high-resolution digital camera after each cycle, and if there is a light signal for a specific spot on the chip, that nucleotide is added to the data set (the "read") for that spot. Once the data have been recorded, both the fluorophore and the blocking group are chemically removed, and the chip is flooded with the next modified nucleoside triphosphate. Many cycles are repeated, which results in individual reads of 50–300 nucleotides. These can be "assembled" into longer sequences in the computer, which examines the reads for overlaps and then joins them together in longer contigs (contiguous sequences). Other similar NGS methods include those where a single primer is bound to a substrate, which becomes the initiation site for polymerization and detection (Figure 20.6A); a method where the templates are bound to the substrate, and then primed on the opposite end for sequencing (Figure 20.6B); and those that utilize DNA polymerase bound to the substrate (Figure 20.6C). This last method is capable of determining the sequences of DNA fragments over 1 kb in length.

Another sequencing technology is called the Ion Torrent system. This takes advantage of the fact that a hydrogen ion is liberated during the incorporation of a nucleotide into a DNA strand during polymerization. Standard chemistry and nucleoside triphosphates are used in this system, and thus, costs can be lower. The other important part of the system is the use of ion-sensitive field-effect transistors (ISFET), which detect when the hydrogen ions are released during the reaction. The chip containing the ISFET detectors are directly below the microwells into which each DNA template molecule has been deposited. One type of nucleotide at a time is flooded into the plate. When the detector sends a signal to the computer, the position is recorded by the computer, thus determining the nucleotide added for each position, and eventually leading

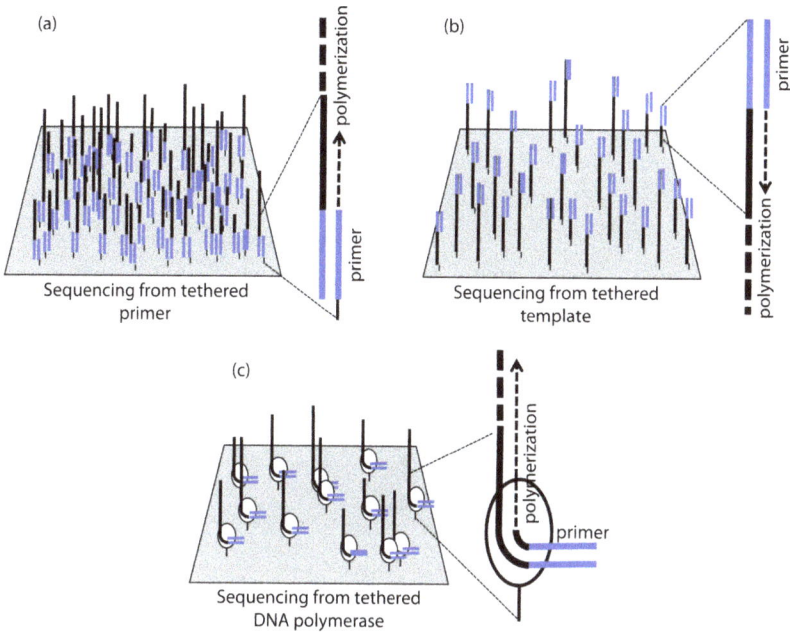

FIGURE 20.6   Summary of other NGS sequencing methods. (a) DNA primers are tethered to a solid substrate. A complementary sequence that has been added to the template DNA to be sequenced is then hybridized to the primer. Part of the remainder of the primer is used to initiate the sequencing reaction. After each labeled nucleotide is added, the plate is illuminated by laser, and each fluorescent spot is recorded to construct reads by the computer (e.g., Helicos BioSciences). (b) DNA templates to be sequenced are tethered to a solid substrate, to which a primer has been added to the opposite end. The primer region is then used to initiate polymerization. Laser illumination and read construction are employed analogous to those described above (e.g., Helicos BioSciences). (c) DNA polymerase molecules are tethered to a solid matrix. Template DNAs, with ligated a primer region, are added. Polymerization, laser illumination, detection and recording are then employed, similar to those above (e.g., Pacific Biosciences).

to a determination of the sequence in each microwell. When more than one nucleotide is added in a particular microwell, the signal sent by the detector is proportional to the number of ions detected and, hence, the number of that particular nucleotide that has been added to the chain. Therefore, it is simpler in some ways than other methods, and because it uses standard DNA polymerase chemistry, it is also less expensive than some of the other methods. When first developed, the reads were short, but several improvements have increased the average read lengths to 200–400 nucleotides, and reads of 800 nucleotides are sometimes produced.

Another step in the technology uses specific nanopore proteins that are immobilized on a membrane (Figure 20.7). The pore is of the size that only one nucleotide at a time of a single-stranded DNA molecule can fit through the pore. Furthermore, because of the size of each of the nucleotides, the time that it takes for each nucleotide to move through the pore differs. Also, ions that are used in the system can pass

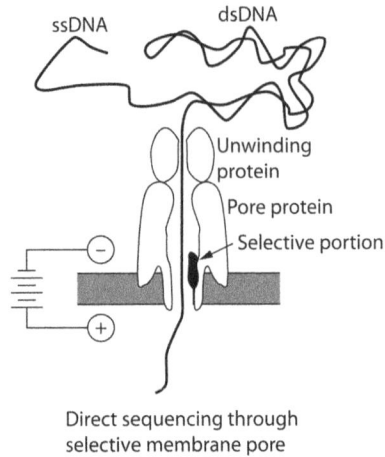

Direct sequencing through
selective membrane pore

FIGURE 20.7   Nanopore sequencing method.

through the pore in addition to the nucleotides. An electrical potential is created by the flow of ions, such that it can be determined when a nucleotide is passing through the pore. Therefore, by recording the electrical potential, and deducing the flow of ions through the pore, as well as the time it takes for the electrical potential to change, the computer can determine the sequence of nucleotides that are passing through the pore.

CONCLUSIONS

Sequencing of nucleic acids developed from degradative and synthesis methods. Detection of the addition of each added nucleotide was first accomplished using radioactively labeled nucleotides and subsequently using fluorescently labeled nucleotides. NGS methods developed from the previous methods, which improved on the rate at which nucleotides could be determined. Currently, DNA (and RNA) sequences are being determined faster than all of them can be analyzed.

REFERENCES

1. Maxam, A.M.; Gilbert, W. A new method for sequencing DNA. Proc. Natl. Acad. Sci. U.S.A. **1977**, *74*, 560–564.
2. Sanger, F.; Coulson, A.R. A rapid method for determining sequences in DNA by primed synthesis with DNA polymerase. J. Mol. Biol. **1977**, *94*, 441–448.
3. Rogers, S.O. Integrated Molecular Evolution. CRC Press, Taylor & Francis Group, Boca Raton, FL, **2017**.

21 Differential Sequencing by Mass Spectrometry

*Christian Jurinke, Christiane Honisch,
Dirk van den Boom*

SEQUENOM Inc., San Diego, California, U.S.A.

CONTENTS

INTRODUCTION

The publication of the human genome sequence in 2001 and the completion of hundreds of thousands of virus, bacterial, archaeal, and eukaryotic genomes by 2020 affirm the success of worldwide genome sequencing projects. The availability of new reference sequence information of various organisms is continuously growing.[1]

Differential sequencing—the systematic comparison of genomic sequences with respect to reference sequences, also called resequencing—represents a central focus of current genome analysis. The exploration of single-nucleotide polymorphisms (SNPs) is one of the most prominent approaches to extract the medical and biological value of genome-sequencing data and to elucidate inter- and intraspecies genetic variations.[2] The analysis of SNPs in the human genome will have a significant impact on the identification of disease susceptibility genes and drug targets and will facilitate the development of new drugs and patient care strategies. Large-scale analysis, detection, and discovery of genetic variability and implementation of dense SNP maps are eminent and push technological development toward high-throughput, cost-efficient applications.

Differential sequencing of infectious agents for fast and reliable identification and typing is an important aspect in the field of molecular diagnostics and epidemiology, including outbreak tracking and classification of pathogens. Fast and highly accurate tools for identification, monitoring, and treatment control are in demand. This chapter describes the use of mass spectrometry (MS) to rapidly identify and localize variable genomic regions based on reference sequences, thus facilitating high-throughput differential sequencing with a gamut of applications.

DOI: 10.1201/9781003247432-21

GENOME ANALYSIS BY MASS SPECTROMETRY:
HISTORY AND STATE-OF-THE-ART

Analysis of nucleic acids by MS has mainly been accomplished by two common soft ionization techniques—electrospray ionization (ESI) and matrix-assisted laser desorption/ionization (MALDI) MS.

The use of MALDI coupled with time-of-flight MS (MALDI-TOF MS) has become one of the leading technologies for high-throughput analysis and high-fidelity measurement of nucleic acid sequence variations. Unambiguous detection of known polymorphic sequences has been demonstrated, even in various formats.[3–5]

Early attempts to apply MALDI-TOF to de novo sequencing as an alternative to separation and detection of Sanger sequencing ladders[6] were hindered by physical analyte fragmentation, limited mass resolution, and mass accuracy in the high mass range. Despite several promising biochemical strategies, which generate truncated DNA sequence fragments of a sequencing primer analogous to dideoxy sequencing, routine-read lengths exceeding 100 bp have never been achieved.[7–9]

In addition to the primer extension-based Sanger sequencing approach, several chemical and enzymatic DNA fragmentation approaches have been proposed to generate short-based specifically cleaved MALDI-TOF analytes.

A chemical cleavage approach utilized P3′-N5′-phosphoamidate-containing DNA replacing dCTP or dTTP by their analog P-N-modified nucleoside triphosphates. Acidic reaction conditions induce base-specific cleavage. However, the required acidic conditions produce unwanted depurination by-products and base loss of adenine and guanine.[10]

A uracil-DNA-glycosylase (UDG)-treatment approach uses strand-separated polymerase chain reaction (PCR) products to generate T-specific abasic sites. Subsequent alkaline and heat treatment induces base-specific DNA cleavage at each T-specific positions.[11] DNA regions of interest require incorporation of dUTP instead of dTTP during PCR. Strand separation is performed by solid-phase separation on streptavidin-coated magnetic beads, which complicates the automatic handling of the assay.

Homogeneous assay formats requiring only subsequent addition of reagents are preferred. Post-PCR in vitro transcription systems combined with base-specific cleavage of the RNA transcripts overcome the issues encountered with classical DNA amplification and primer extension reactions.

The current state-of-the-art differential sequence analysis concept uses a Maxam-Gilbert-like approach: base-specific cleavage of nucleic acid amplification products. PCR amplification of the locus of interest is followed by in vitro transcription and base-specific cleavage.[12] This novel comparative resequencing scheme combines a homogenous in vitro transcription/RNAse system with MALDI-TOF analysis of molecular fragment masses—an intrinsic molecule property. No labeling is required. PCR products of up to 1 kb are subjected to in vitro transcription. Subsequent cleavage of the in vitro transcripts by ribonucleases (e.g., RNase A) generates base-specifically cleaved RNA fragments. Sequence-specific mass signal patterns within a mass range of 1000–9000 Da—equivalent to

3–30 nucleotides—are obtained. The result is a characteristic pattern of RNA fragment masses indicative of the original reference sequence.

RNase T1 (a guanine-specific endonuclease) cleaves in vitro transcripts base—specifically at every G-position.[13] An available alternative is RNase A, which cuts specifically at the 3′-end of the pyrimidine residues C and U.

RNA analytes are more stable and less prone to depurination than DNA during the desorption/ionization process in MALDI-TOF MS because of a balancing effect of the 2′-hydroxy group on the polarization of the N-glycosidic bond of the protonated base.

Figure 21.1 illustrates the robust biochemical scheme for high-throughput differential sequencing. Two PCR reactions of the DNA region of interest introduce a T7-promotor in the forward strand as well in the reverse strand of the amplification product. PCR is followed by shrimp alkaline phosphatase (SAP) treatment to dephosphorylate any unincorporated desoxy-NTPs. RNA polymerase, ribonucleotide, and nuclease-resistant nucleotides are added to the mixture. In vitro transcription generates single-stranded RNA and facilitates further amplification. The RNA is subject to four base-specific cleavage reactions corresponding to each of the four bases. Reactions are driven to completion. This reduces the RNA target to a specific set of RNA compomers in each of the reactions. Analytes are desalted by addition of ion-exchange resin, conditioned, and identified in a single MALDI-TOF MS measurement per reaction. Four sequence-specific mass signal patterns are generated. All cleavage products are consistent with 5′-OH and 3′-phosphate groups, except for the 3′-fragment of the full-length transcript possessing a 3′-OH group.

The experimental set of compomers is used to reconstruct the sequence by cross-comparing the information of the four cleavages to the in silico cleavage pattern of the known reference sequence. Deviations of the pattern indicate sequence changes (Figure 21.1).

Figure 21.2 shows base-specific, cleavage-mediated discovery of a [C/T] sequence change in a 500-bp DNA region of interest. The target region is analyzed by C- and T-specific cleavage of the forward, as well as the reverse strand—equivalent to four base-specific cleavage reactions. Overlays show spectra of the wild-type as well as heterozygous and homozygous mutant samples.

For the wild-type sample [C/C], mass signals of all cleavage reactions can unambiguously be identified based on the reference sequence derived in silico cleavage pattern. Deviations from the in silico pattern, detected as a mass shift, the absence of an existing peak, or the appearance of an additional signal, lead to the identification of the sequence variation.

In the T-specific cleavage reaction of the forward transcript, a sequence change from C to T at position 371 of the target region introduces a new cleavage site and splits the 15-bp wild-type fragment into a 12- and a 3-bp fragment. For the mutant [T/T] sample, this results in the disappearance of a mass signal at 4830.0 Da and the appearance of mutant-specific mass signals at 3850 and 1015.6 Da (signal not shown). Spectra containing all of the signals correspond to the heterozygous sample [C/T].

The C-specific cleavage reaction on the forward strand confirms the observation of the identified [C/T] substitution. A cleavage site is removed from a dimer A[C]

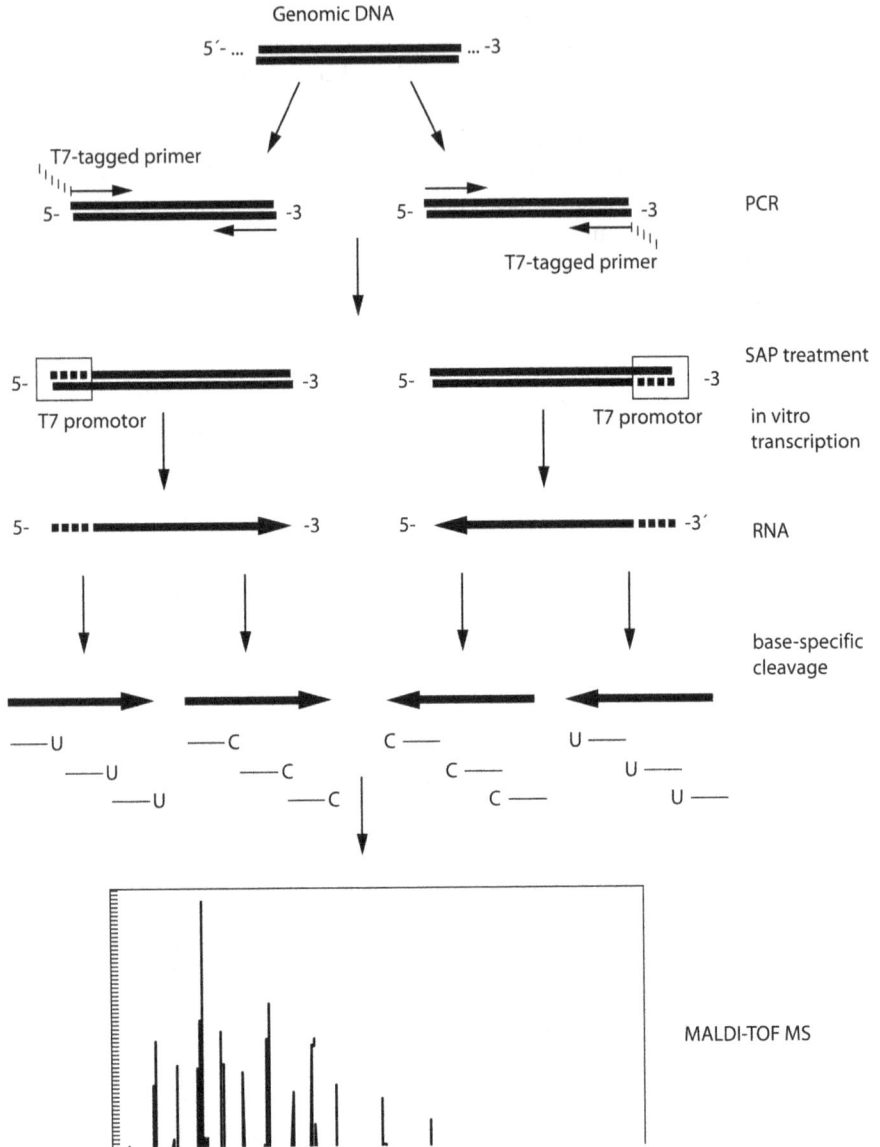

FIGURE 21.1 Base-specific cleavage by MALDI-TOF mass spectrometry. A single-stranded copy of a PCR-amplified target sequence is generated by T7-mediated in vitro transcription and cleaved in four reactions at positions corresponding to each of the four bases. RNA of the forward strand is cleaved at U or C. An A- and G-specific cleavage of the template sequence is facilitated by U or C cleavage of the reverse RNA strand. MALDI-TOF acquisition of spectra of each of the cleavage reactions is followed by comparison to reference sequence derived in silico cleavage pattern.

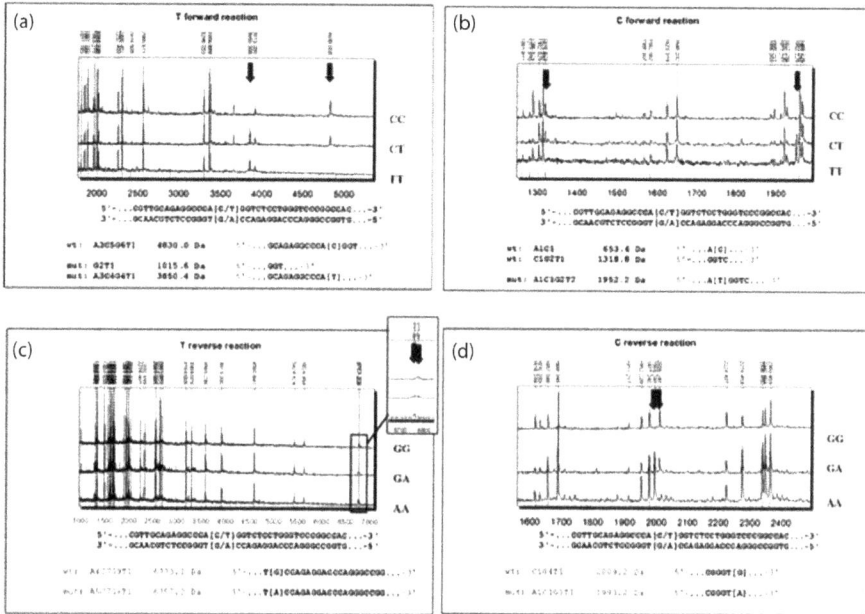

FIGURE 21.2 Identification of a single-nucleotide polymorphism by MALDI-TOF MS analysis. Panels a–d show the four base-specific cleavage reactions. Each spectrum reveals the changes resulting from the substitution of a C wt-allele for a T mutant-allele. Spectra of all three different genotypes are overlaid. Changes are indicated by arrows. (a) T-specific cleavage of the forward transcript. (b) C-specific cleavage of the forward transcript. (c) T-specific cleavage of the reverse transcript. (d) C-specific cleavage of the reverse transcript. Spectra are shown covering all detected signals. All signals affected by the sequence change are indicated.

and generates a 6-bp fragment of 1952.2 Da with the adjacent 4-bp fragment of 1318 Da.

Confirmatory information is generated from the reverse RNA transcript of the target region. Both the T- and the C-specific cleavage reactions generate mass shifts of −16 Da corresponding to an exchange of G vs. A in the affected fragment (Figure 21.2).

In conclusion, a heterozygous sequence change can generate up to five discriminatory observations in a mass spectrum by adding or removing a cleavage site, as well as shift the mass of single products by the mass difference of an exchanged nucleotide. Up to 10 observations might be the result of a homozygous sequence change, because not only additional but also missing signals can be utilized for SNP identification.

Mono-, di-, and trimer nucleotides are usually noninformative. They are excluded from the analysis because of coinciding fragments, and their detection is also diminished because of analyte carrier matrix signals within the low mass range.

The combined observations of all cleavage reactions allow for the unambiguous detection, identification, and localization of almost all sequence changes. The inherent redundancy of information from all cleavage reactions substantiates the reliability of the results. Additional supportive information is obtained when signals are correlated across a multitude of samples.

A simulation of arbitrary 500-bp amplicons in the human genome showed that about 99% of all theoretical sequence changes can be detected and characterized.[12]

A homogeneous 384-assay format enables automated processing with liquid handling devices. Nanoliter dispensing onto matrix-coded chip arrays are utilized for automated, reproducible MALDI-TOF measurements. TOF instruments acquire data with turnaround times of 2 sec per sample at a standard 20-Hz laser repetition rate as opposed to hours of analyte separation in conventional sequencing gel electrophoresis. A single MALDI-TOF mass spectrometer can thus scan 2.5 million bp in every 24 hr.

Real-time quality control of mass spectra results in highly reproducible mass signal patterns facilitating automated peak-pattern interpretation. Signal-to-noise ratio changes can be used as supporting information for sequence changes. A recent software enables automated high-throughput SNP discovery and mutation detection. Time-efficient algorithms calculate the most reasonable explanations for observed mass signal changes, and discover and pinpoint sequence variations based on the information content of the four cleavage reactions.[14]

DIAGNOSTIC APPLICATIONS

MALDI-TOF MS of nucleic acids after base-specific cleavage improves the odds of differential sequence analysis and sets a significant milestone in the field of comparative genomics and genetics. The feasibility of the system exceeds SNP discovery applications. The method facilitates efficient scoring of large numbers of genetic markers in selected populations determining genotypic and phenotypic correlations.

Base-specific cleavage of signature sequences results in species-specific mass signal pattern of the region of interest. These species-specific fingerprints can be utilized to discriminate prokaryotes—bacterial or viral organisms—to the genus-, species-, or strain-specific level and thus facilitate pathogen identification.[15] Figure 21.3 gives an example of the unambiguous identification and differentiation of four mycobacterial species based on their characteristic mass signal fingerprints of the 16S rRNA gene sequences. All expected signals were identified in the spectra and unique identifier signals marked by arrows could be unambiguously assigned to expected discriminatory species-specific fragments (Figure 21.3).

Further expansion of the application portfolio includes the detection of epigenetic modifications, molecular haplotyping, and mutation detection.

Analysis of DNA methylation-mediated gene silencing is seen as a valuable diagnostic tool in cancer research and diagnostics. Qualitative as well as quantitative high-throughput MALDI-TOF DNA methylation analysis relies on bisulfite treatment of genomic DNA, PCR amplification followed by base-specific cleavage. The bisulfite treatment converts nonmethylated cytosine to uracil, while methylated cytosines remain unmodified. As a result, methylated vs. nonmethylated mass signal patterns show significant differences, which allow for differentiation as well as

(a)
r. i.

M. tuberculosis

M. xenopi

M. paraffinicum

M. gordonae

2250   2500   2750   3000   3250   3500   3750   4000   4250   m/z

(b)

	2268.4	2308.4	2348.4	2469.6	2581.6	2597.6	2613.6	2653.6	2870.8	3216.0	3272.0	3561.2	3866.4	3906.4	4115.6	4131.6	4235.6	5456.4
M. tuberculosis																		
M. xenopi																		
M. paraffinicum																		
M. gordonae																		

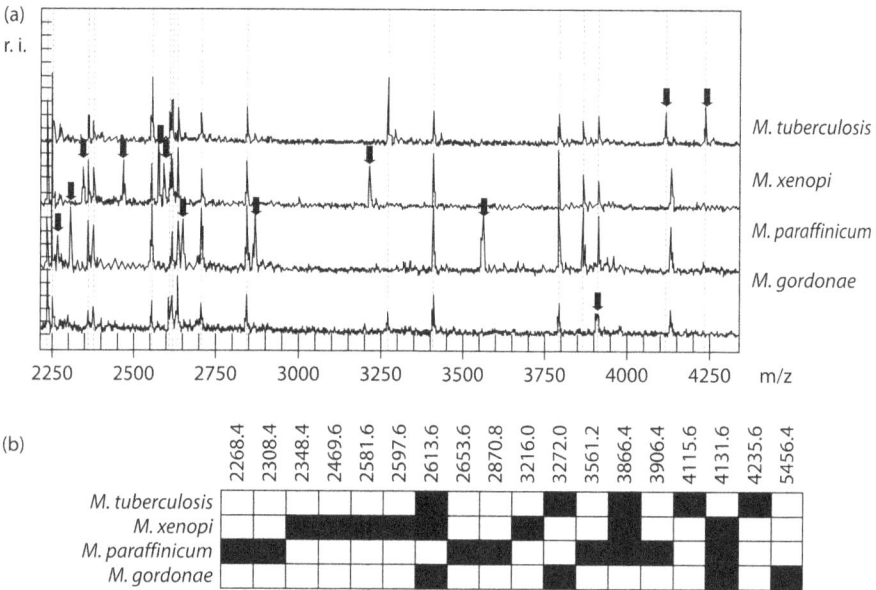

FIGURE 21.3 Pathogen identification by mass spectrometry. (a) Overlay of mass spectra of C-specific forward 16S rDNA fragments of *Mycobacterium tuberculosis* ATCC27294, *Mycobacterium xenopi* DSM43995, *Mycobacterium paraffinicum* DSM44181, and *Mycobacterium gordonae* DSM44160. Identifier peaks are marked with arrow. (b) Barcode of C-specific forward 16S rDNA discriminatory mass fragments.

identification of individual sites of methylation in a target sequence. In addition, comparative quantitation of mass signals can be utilized to determine the relative abundance of methylated vs. nonmethylated target gene regions.[16]

Haplotypes are defined as the collection of genotypes found in a single allele or chromosome. Their unambiguous identification provides additional power in the detection of genes involved in common diseases. This can contribute to a better understanding of the complex etiology of diseases. Long-range, allele-specific PCR using nonextendable exonuclease-resistant competitor oligonucleotides is performed to isolate desired alleles/chromosomes prior to base-specific cleavage and MALDI-TOF detection. The analysis is not exclusive to known markers, and novel associated subhaplotypes can be discovered resulting in comprehensive haplotype information of the loci of interest.[17]

Additional fields of application include large-scale characterization of cDNAs and alternative splice variants—further steps in the attempt to elucidate the genetic code and individual variations.

CONCLUSION

MALDI-TOF MS of base-specifically cleaved nucleic acids is a valuable expansion of the method portfolio in the field of comparative genomics and opens new routes for diagnostic sequencing applications. MALDI-TOF MS combines the determination of an analyte-specific physical property—the molecular mass—with speed of signal acquisition and high accuracy. This results in a high degree of automation and throughput. The inherent redundancy of observations supporting the discovery of sequence changes is up to five times higher compared to sequencing. This feature should propel MALDI-TOF MS as a gold standard in the fields of SNP discovery, mutation detection, pathogen identification, methylation detection, haplotyping, and cDNA characterization.

ACKNOWLEDGMENTS

We would like to thank Patrick Stanssens and Marc Zabeau for their contributions to the development of SNP discovery by base-specific cleavage.

REFERENCES

1. International Human Genome Sequencing Consortium. Initial sequencing and comparative analysis of the mouse genome. Nature **2002**, *409*, 860–921.
2. Melton, L. On the trial of SNPs. Nature **2003**, *422*, 917–923.
3. Graber, J.H.; Smith, C.L.; Cantor, C.R. Differential sequencing with mass spectrometry. Genet. Anal. **1999**, *14* (5–6), 215–219.
4. Tang, K.; Opalsky, D.; Abel, K.; van den Boom, D.; Yip, P.; Del Mistro, G.; Braun, A.; Cantor, C.R. Single nucleotide polymorphism analysis by MALDI-TOF MS. Int. J. Mass Spectrom. **2003**, *226*, 37–54.
5. Jurinke, C.; van den Boom, D.; Cantor, C.R.; Koster, H. The use of MassARRAY technology for high throughput genotyping. Adv. Biochem. Eng. Biotechnol. **2002**, *77*, 57–74.
6. Smith, L.M. The future of DNA sequencing. Science **1993**, *262*, 530–532.
7. Koster, H.; Tang, K.; Fu, D.J.; Braun, A.; van den Boom, D.; Smith, C.L.; Cotter, R.J.; Cantor, C.R. A strategy for rapid and efficient DNA sequencing by mass spectrometry. Nat. Biotechnol. **1996**, *14* (9), 1123–1128.
8. Kirpekar, F.; Nordhoff, E.; Larsen, L.K.; Kristiansen, K.; Roepstorff, P.; Hillenkamp, F. DNA sequence analysis by MALDI mass spectrometry. Nucleic Acids Res. **1998**, *26* (11), 2554–2559.
9. Taranenko, N.I.; Allman, S.L.; Golovlev, V.V.; Taranenko, N.V.; Isola, N.R.; Chen, C.H. Sequencing DNA using mass spectrometry for ladder detection. Nucleic Acids Res. **1998**, *26* (10), 2488–2490.
10. Shchepinov, M.S.; Denissenko, M.F.; Smylie, K.J.; Worl, R.J.; Leppin, A.L.; Cantor, C.R.; Rodi, C.P. Matrix-induced fragmentation of P3′–N5′ phosphoramidate-containing DNA: High-throughput MALDI-TOF analysis of genomic sequence polymorphisms. Nucleic Acids Res. **2001**, *29* (18), 3864–3872.
11. von Wintzingerode, F.; Bocker, S.; Schlotelburg, C.; Chiu, N.H.; Storm, N.; Jurinke, C.; Cantor, C.R.; Gobel, U.B.; van den Boom, D. Base-specific fragmentation of amplified 16S rRNA genes analyzed by mass spectrometry: A tool for rapid bacterial identification. Proc. Natl. Acad. Sci. U.S.A. **2002**, *99* (10), 7039–7044.
12. Stanssens, P.; Zabeau, M.; Meersseman, G.; Remes, G.; Gansemans, Y.; Storm, N.; Hartmer, R.; Honisch, C.; Rodi, C.P.; Böcker, S.; van den Boom, D. High-throughput MALDI-TOF discovery of genomic sequence polymorphisms. **2003**, in press.
13. Hartmer, R.; Storm, N.; Boecker, S.; Rodi, C.P.; Hillenkamp, F.; Jurinke, C.; van den Boom, D. RNase T1 mediated base-specific cleavage and MALDI-TOF MS for high-throughput comparative sequence analysis. Nucleic Acids Res. **2003**, *31* (9), e47.

14. Bocker, S. SNP and mutation discovery using base-specific cleavage and MALDI-TOF mass spectrometry. Bioinformatics **2003**, *19* (Suppl. 1), I44–I53.
15. Lefmann, M.; Honisch, C.; Boecker, S.; Storm, N.; Wintzingerode; von Schlotelburg, F.; Moter, C. A.; van den Boom, D.; Goebel, U.B. A novel mass spectrometry based tool for genotypic identification of mycobacteria. J. Clin. Microbiol. **2003**, in press.
16. Ehrich, M.; Storm, N.; Honisch, C.; van den Boom, D. High-throughput DNA Methylation Analysis by Base-Specific Cleavage of Amplification Products and Analysis by Mass Spectrometry (MALDI-TOF). 94th Annual Meeting Proceedings, Proceedings of the American Association for Cancer Research, Washington, DC, USA, April 5–9, **2003**; #106617.
17. Beaulieu, M.; Kosman, D.; van den Boom, D. Molecular Haplotyping Combining a Novel AS-PCR Technique with Base-Specific Cleavage and Mass Spectrometry. SNPs, Haplotypes and Cancer: Application in Molecular Epidemiology. Conference Proceedings, Key Biscayne, FL, USA, September 13–17, **2003**; A60.

22 Brief Guide to Conducting Biological Database Searches

John Gray
University of Toledo, Toledo, Ohio, U.S.A.

CONTENTS

INTRODUCTION

In 2020, the journal *Nucleic Acids Research* (NAR) released its 27th database issue and updated its Molecular Biology Database Collection to include 1637 actively maintained databases. The proliferation of biological databases underscores the avalanche of data that has accumulated in the past three decades and parallels closely the development of "omics" technologies across many disciplines. For junior researchers entering into the field of modern biological sciences, it is essential for them to learn how to navigate the ocean of information contained in these databases. Even for seasoned researchers, it is necessary to mine such databases to help guide and extend ongoing experimentation.[1] This chapter aims to serve as a brief guide to such researchers and points to general search strategies as well as a survey of a few of the most popular resources [GenBank, protein database (PDB), KEGG, and Ensembl] employed in the biological sciences.

LOCATING DATABASES OF RELEVANCE

Since 2002, there has been a linear increase in the number of new molecular biology databases with about 104 having their debut each year.[2] Therefore, the first task for researchers is to become aware of databases that are of relevance to their

DOI: 10.1201/9781003247432-22

TABLE 22.1
NAR Database Summary Paper Category List[a]

Database Category/ Subcategory	No. of Databases (Totals and Subtotals)
Nucleotide Sequence Database	
International Nucleotide Sequence Database Collaboration	9
Coding and non-coding DNA	49
Gene structure, introns and exons, splice sites	19
Transcriptional regulator site and transcription factors	77
Subtotal	157
RNA Sequence Databases	
Subtotal	104
Protein Sequence Databases	
General sequence databases	13
Protein properties	22
Protein localization and targeting	23
Protein sequence motifs and active Sites	29
Protein domain databases; protein classification	41
Databases of individual protein families	85
Subtotal	214
Structure Databases	
Small molecules	20
Carbohydrates	14
Nucleic acid structure	22
Protein structure	114
	171
Genomics Databases (Non-Vertebrate)	
Genome annotation terms ontologies and nomenclature	23
Taxonomy and identification	14
General genomics databases	29
Viral genome databases	30
Prokaryotes genome databases	71
Unicellular eukaryotes genome databases	21
Fungal genome databases	35
Invertebrate genome databases	54
Model organisms, comparative genomics	2
Subtotal	280

Database Category/ Subcategory	No. of Databases (Totals and Subtotals)
Metabolic and Signaling Pathways	
Enzymes and enzyme nomenclature	
Metabolic pathways	
Protein-protein interactions	
Signaling pathways	
Subtotal	167
Human and Other Vertebrate Genomes	
Invertebrate genome databases	
Model organisms, comparative genomics	
Human genome databases, maps and viewers	
Human ORFs	
Subtotal	176
Human Genes and Diseases	
General human genetics databases	11
General polymorphism databases	44
Cancer gene databases	51
Gene-, system-, or disease-specific databases	58
Subtotal	116
Microarray Data and Other Gene Expression Databases	60
Proteomics Resources	28
Other Molecular Biology Databases	20
Organelle Databases	20
Mitochondrial genes and proteins	
Plant Databases	
General plant databases	
Arabidopsis thaliana	
Rice	
Other plants	
Subtotal	131
Immunological Databases	31
Cell Biology	8
Total databases in NAR collection Nov 2020	1683

[a] https://www.oxfordjournals.org/nar/database/e.

own avenue of research. The freely available Molecular Database Collection maintained by NAR is an excellent starting point to survey the databases that pertain to an individual's area of interest (Table 22.1).

Although the NAR database collection is very comprehensive, it does not include all databases some of which have been announced via newer journals such as *Database: The Journal of Biological Databases and Curation* (https://academic. oup.com/database) and *Bioinformatics* (https://academic.oup.com/bioinformatics). Unfortunately, efforts to create hubs that advertise database resources such as OMICtools[3] have often suffered the fate of individual databases that later become unavailable due to lack of funding. A beginning student should survey the NAR database collection and explore the content of various databases in their area of interest and determine what they have to offer.

POPULAR BIOLOGICAL DATABASES

Of the current diverse set of databases, some are accessed heavily and are highly cited in the biological literature. In Table 22.2, the 30 most heavily cited databases are listed and are likely the ones that most researchers will first access.

One can distinguish between primary (archival) databases (*i.e.,* GenBank) that house experimental data such as sequence information, and secondary databases that house data derived from analyzing primary data (*i.e.,* InterPro). Other databases house a mixture of both types (*i.e.,* SwissProt) and some are more specialized for particular organisms (*i.e.,* FlyBase) or types of data (*i.e.,* RDP). High-value databases are those that start with reliable data and provide robust, accurate, and fast search and retrieval systems. Below we provide a quick guide to searching some of the most popular biological databases.

TABLE 22.2

List of Top 30 Highly Cited Biological Databases Published in NAR 1990–2016[2]

Database Resource Name	Website Access	Debut Year	Total DB Articles	Total Citations
RCSB PDB: Protein Data Bank	http://www.rcsb.org/pdb/	2000	9	17915
KEGG: Kyoto Encyclopedia of Genes and Genomes	http://www.kcgg.jp/	1999	11	16832
Pfam	http://www.sanger.ac.ukJSoftware/Pfam/	1998	11	14408
RDP: Ribosomal Database Project	http://rdp.cme.msu.edu/	1991	14	10183
Uniprot: Universal Protein Resource	http://www.uniprot.org	2004	12	7709
miRBase	http://www.mirbase.org/	2006	4	7672

(Continued)

TABLE 22.2   *(Continued)*
**List of Top 30 Highly Cited Biological Databases Published in NAR
1990–2016[2]**

Database Resource Name	Website Access	Debut Year	Total DB Articles	Total Citations
NCBI GenBank	http://www.ncbi.nlm.nih.gov/	1991	24	7582
NCBI GEO: Gene Expression Omnibus	http://www.ncbi.nlm.nih.gov/ geo/	2002	6	7262
STRING: Search Tool for the Retrieval of Interacting Genes	http ://www.bork.cmbl-heidelbcrg.de/STRING/	2003	7	6139
Ensembl	http://www.ensembl.org/	2002	15	6128
TRANSFAC	http://transfac.gbf.de/	1996	8	6114
PROSITE	http://www.expasy.org/prosite/	1991	13	5812
NCBI COD: Conserved Domain Database	https://www.ncbi.nlm.nih.gov/ structure/cdd/cdd.shtml	2002	8	5537
SWISS-PROT	http://expasy.hcugc.ch	1991	10	5258
SMART: Simple Modular Architecture Research Tool	http://SMART.embl-heidelberg.dc	1999	8	4606
UCSC Genome Browser Database	http://gcnome.ucsc.edu/	2003	13	4500
EMBl.rEBI lnlerPro	http://www.ebi.ac.uk/interpro	2001	7	3743
NCBI Reference Sequences: RefSeq	http://www.ncbi.nlm.nih.gov/ refseq/	2003	9	3655
DrugBank	http://www.drugbank.ca	2006	4	3203
CAZy: Carbohydrate-Active EnZymcs Database	http://www.cazy.org/	2009	2	3103
GO: Gene Ontology	http://www.gencontology.org	2004	7	3063
BioGRID Interaction Database	http://www.thcbiogrid.org	2006	5	2977
Fly Base	http://flybase.org	1994	17	2876
DIP: Database of Interacting Proteins	http://dip.doc-mbi.ucla.edu	2000	4	2825
MEROPS	http://merops.sangcr.ac.uk	1999	10	2774
SCOP: Structural Classification of Proteins	http://scop.mrc lmb.cam.ac.uk/ scop	1997	7	2578
Rfam	http://www.sangcr.ac.uk/ Software/Rfam/	2003	6	2519
NCBI COG: Clusters of Orthologous Groups	http://www.ncbi.nIm.nih. gov/C(Xj/	2000	3	2514
HMDB: Human Metabolome Database	www.hmdb.ca	2007	3	2462
NCBI dbSNP: Database of Single Nucleotide Polymorphisms	http://www.ncbi.nlm.nih.gov/ SNP	2000	2	2456

Source: Data extracted from source file nar_v20_3_plot2.csv. located in Illinois Data Bank https://doi.
org/10.13012/B2IDB-4311325_V1.

SEARCHING NCBI GENBANK USING ENTREZ AND BLAST

The National Center for Biotechnology Information (NCBI) was established in 1988, as a division of the National Library of Medicine (NLM) at the National Institutes of Health (NIH). Although its central mission has been to provide information technologies to aid in the understanding of molecular and genetic processes that control human health and disease, it serves the wider research community in storing genetic information on all living organisms. Furthermore, the NCBI is committed to developing search tools and educational materials to enable use of its resources. For the beginning user, there are a number of helpful tutorials and handbooks (https://www.ncbi.nlm.nih.gov/guide/training-tutorials/) and through their education page (https://www.ncbi.nlm.nih.gov/home/learn/).

Most users first become familiar with NCBI in order to retrieve or compare DNA or protein sequences from GenBank. The GenBank sequence database is an annotated collection of all publicly available nucleotide sequences and their protein translations and is mirrored at the European Molecular Biology Laboratory (EMBL) and the Data Bank of Japan (DDBJ). Since its inception in 1982 has grown to store almost 1.2 trillion base pairs (bp) of traditional GenBank records and more than 15 trillion bp of whole genome and transcriptome sequences (GenBank release 248, February 2022). Sequence records are retrievable by their search engine named Entrez and searchable using the powerful and fast algorithm named Basic Local Alignment Search Tool (BLAST).[4] When searching NCBI databases using Entrez, the search term will automatically be queried against 36 separate databases summarized in Table 22.3.

The diversity of databases (including literature) queried by Entrez means that a wide search net is cast. The resulting links provide a means by which even a beginning researcher can quickly collect information about a given gene or protein. Indeed, it is a worthwhile exercise for any researcher beginning to work on a gene or protein for the first time to conduct such a search using Entrez and NCBI databases. A beginner should peruse the various resource links to develop a sense of how information is scattered across databases and the variety of information that is being collected by various technologies and study types.

A common goal of GenBank users is to find sequences related (homologous) to a search sequence. By finding homologous sequences, one is able to locate related genes both within and across species. In turn, these sequences can be compared to identify conserved functional motifs and learn about gene functionality in organisms outside of the one of study. One of the most powerful and widely used search algorithms to identify homologous sequences is named BLAST. The BLAST algorithm finds regions of local similarity between sequences. It compares nucleotide or protein sequences to sequence databases and also calculates the statistical significance of matches. The BLAST output can be used to infer functional and evolutionary relationships between sequences as well as help identify related members of gene families. The BLAST search tool is accessible via the NCBI homepage (https://blast.ncbi.nlm.nih.gov/Blast.cgi). There are four main BLAST tools with variant versions and nine more specialized BLAST tools (Table 22.4).

TABLE 22.3
Description of Databases Searched via NCBI Entrez Query System

NCBI Database	Web Address	Brief Resource Description
Literature	www.ncbi.nlm.nih.gov...	
Bookshelf	/books	Free online access to books in life science and healthcare
MeSH	/mesh	Medical Subject Headings is a controlled vocabulary thesaurus used
NLM Catalog	/nlmcatalog	NLM bibliographic data for over 1.2 million journals, book, and other materials resident in Locator Plus
PubMed	pubmed.ncbi.nlm.nih.gov	Biomedical literature citation's and abstracts, including Medline
PubMed Central	/pmc	Free, full-text journal articles
Genes	www.ncbi.nlrn.nih.gov...	
Gene	/gene	Gene centered information from a wide range of species
GEO DataSets	/gds	Curated gene expression DataSets from the Gene Expression Omnibus (GEO) repository
GEO Profiles	/geoprofiles	Expression and molecular abundance profiles
HomoloGene	/homogene	Automated system for constructing putative homologous gene groups
PopSet	/popset	DNA sequences derived from population, phylogenetic, mutation, and ecosystem studies
Proteins	www.ncbi.nlrn.nih.gov...	
Conserved domains	/cdd	Resource for the annotation of functional units in proteins
Identical protein groups	/ipg	A single entry for each protein translation found in several sources at NCBI
Protein	/protein	Sequences including translations from annotated coding regions in GenBank, RefSeq TPA, and more
Protein clusters	/proteinclusters	Related protein sequences (clusters) derived from the annotations of whole genomes, organelles, and plasmids
Sparele	/sparele	Functional characterization and labeling of protein sequences grouped by their conserved domain architecture
Structure	/structure	Three dimensional structures providing insight on the biological function of macromolecules
Genomes	www.ncbi.nlrn.nih.gov...	
Assembly	/assembly	Information on assembled genomes, assembly names and links to genomic sequence data
BioCollections	/biocollections	Curated dataset of metadata for culture collections, museums, herbaria, and other natural history collections
BioProject	/bioproject	Collection of biological data related to a single initiative, originating from a single organization or from a consort

(Continued)

TABLE 22.3 *(Continued)*
Description of Databases Searched via NCBI Entrez Query System

NCBI Database	Web Address	Brief Resource Description
BioSample	/biosample	Descriptions of biological source materials used in experimental assays
Genome	/genome	Information on genomes including sequences, maps, chromosomes, assemblies, and annotations
Nucleotide	/nucleotide	Collection of sequences from several sources, including Gen Bank, RefSeq, TPA, and PDB
SRA	/sra	Sequence Read Archive data is the largest publicly available repository of high-throughput sequencing data
Taxonomy	/taxonomy	Curated classification and nomenclature for all of the organisms in the public sequence databases
Clinical	www.ncbi.nlrn.nih.gov...	
ClinicalTrials.gov	/clinicalTrials.gov	Database of privately and publicly funded clinical studies conducted around the world
ClinVar	/clinvar	ClinVar aggregates information about genomic variation and its relations hip to human health
dbGap	/gap	Archives data from studies that have investigated the interaction of genotype and phenotype in humans
dbSNP	/snp	dbSNP contains human single nucleotide variations, microsatellites, and small-scale insertions and deletions
dbVar	/dbvar	Database of human genomic structural variation–large variants >50 bp including insertions and deletions
GTR	/gtr	Genetic Testing Registry is a central location for voluntary submission of genetic test information by providers
MedGen	/medgen	Organizes information related to human medical genetics, such as attributes of conditions with a genetic link
OMIM	/omim	Comprehensive, authoritative compendium of human genes and genetic phenotypes
PubChem	pubchem.ncbi.nlm.nih. gov...	Information on chemical structures, identifiers, chemical and physical properties, biological activities, patents, etc.
BioAssays	/bioassay	Contains small-molecule and RNAi screening data and associated annotation information from contributing organizations
Compounds	/compounds	PubChem compound pages (accession CID) summarize in formation known about a particular chemical
Pathways	/pathways	Integrates pathway information involving the interactions between chemicals, proteins, and genes
Substances	/substances	A substance record might contain a unique identifier, a patent identifier, and a publication reference from which it was

TABLE 22.4
Brief Description of Main, Variant, and Specialized BLAST Search Programs[a]

Nucleotide-Nucleotide

BLASTN	Optimized for very similar sequences (in the same or in closely related species)
megaBLAST	Looks for similar and for more distantly related sequences
BLASTP	Finds homologous proteins. Its algorithm is the basis of other types of BLAST searches (*i.e.*, BLASTX and TI3LASTN)
Quick BLASTP	Adds a preprocessing step to searching the nonredundant (nr) protein database and will find approximately 97% of the database sequences with 70% or more identity to your query
PSI-BLAST	First performs a BLASTP search to collect information that it then uses to produce a Position-Specific-Scoring-Matrix (PSSM). PS1-BLASTcan then search a database of protein sequences with this PSSM
RPS-BLAST	Reverse-Position-Specific BLAST can very quickly search a protein query against a database of PSSMs that were usually produced by PSI-BLAST
DELTA-BLAST	DELTA-BLAST produces a PSSM with a fast RPSBLAST search of the query, followed by searching this PSSM against a database of protein sequences

Nucleotide-Protein

BLASTX	Searches a nucleotide query against a protein database, translating the query on the fly

Protein-Nucleotide

TBLASTX	Searches a protein query against a nucleotide database, translating the database on the fly

Specialized BLAST

Sinari BLAST	Find proteins highly similar to your query
Primer-BLAST	Design primers specific to your PCR template
Global Align	Compare two sequences across their entire span (Needleman-Wunsch)
CDsearch	Find conserved domains in your sequence
IgBLAST	Search immunoglobulins and T-cell receptor sequences
VeeScreen	Search sequences for vector contamination
CDART	Find sequences with similar conserved domain architecture
Multiple Alignment	Align sequences using domain and protein constraints
MOLE-BLAST	Establish taxonomy for uncultured or environmental sequences

[a] Information summarized from NCBI https://blast.ncbi.nlm.nih.gov/Blast.cgi.

Because of the degeneracy of the genetic code, it is always preferable to conduct a homology search using protein sequence or a translated nucleotide sequence. In a typical BLASTP search, a protein sequence is submitted with the goal of finding homologous sequences, and by default, the nr database of nonredundant sequences is searched. Alternatively, one can limit the search to reference sequences only (RefSeq), model organisms, the Swiss PDB (SwissProt), patented sequences (Pataa), protein databank [pdb, metagenomic proteins (env_nr)], or transcriptome assembly proteins (tsa_nr). Additionally, one can choose to include or exclude specific organisms or taxonomic groups. BLAST applies default general, scoring, and masking/filtering parameters that are not discussed here but can be altered to widen or narrow the specificity of the search.

An example of the output from a BLASTP search is provided in Figure 22.1. The output is a list of "hits" in the database that are ranked by the optimized "bit score" (final score in Figure 22.1). A score >80 (>160 for nucleotide searches) generally signifies that the protein query and subject are evolutionarily related. The Expect value (E value) is a statistic that estimates the number of alignments with a bit score >S that you expect to find by chance. This is a useful number, and the general rule of thumb is that an E value below 0.01 signifies that the query and aligned sequence are derived from a common ancestral molecule, and a value close to 0 signifies a perfect or near perfect match. The descriptor and accession links provide access to the alignments and GenBank accession files respectively and often provide the first clues as to the function of a novel sequence. Graphic summary, Alignments, and Taxonomy tabs permit the query-subject alignments to be viewed individually and in relation to the other aligned subjects.

A BLASTP search often serves as the starting point for further investigation of a protein function and its relationship to other proteins. Downloaded sequences can be aligned to make multiple sequence alignments and phylogenetic trees that define protein families within and across organisms. The graphic summary also reveals that a conserved domain search is automatically conducted with the query against the conserved domain database (cdd). Identifying conserved functional domains is particularly valuable in assigning function to novel proteins. Further output details

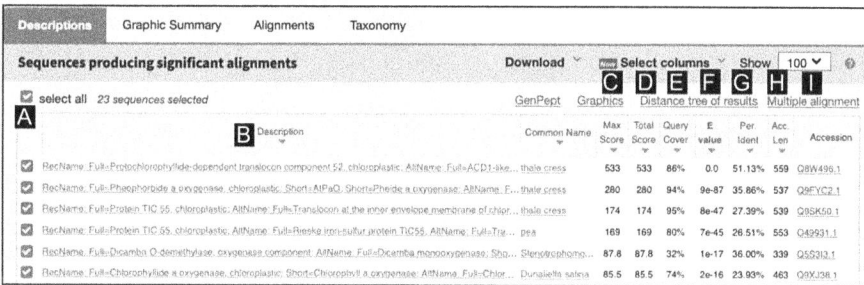

FIGURE 22.1   Sample output from BLASTP search. BLASTP search was conducted using GenBank NP_001147604 as query against the SwissProt database using default parameters. Output information highlighted in letters **A**: selection button to choose some or all sequences for further analysis/download in a variety of formats, **B**: the description/title of matched database sequence, **C**: the highest alignment score (max score) from that database sequence, **D**: the total alignment scores (total score) from all alignment segments, **E**: the percentage of query covered by alignment to the database sequence, **F**: the best (lowest) expect value (E value) of all alignments from that database sequence, **G**: the highest percent identity (Per. Ident) of all query-subject alignments, **H**: the length of the alignment, and **I**: the GenBank accession number of the matched database sequence.

can be obtained by reading the BLAST "how to" document from NCBI https://ftp.ncbi.nlm.nih.gov/pub/factsheets/HowTo_NewBLAST.pdf.

Although not discussed here, other BLAST programs generally follow similar steps and the speed and accuracy of the algorithms are the main reasons that they are preferred over slower but more thorough methods such as FASTA. Together the NCBI resources provide an immense opportunity for discovery and indeed more and more scientific discoveries are being made through database analysis and data mining as data accumulates faster than it can be analyzed in detail.

SEARCHING THE RCSB PROTEIN DATABASE (PDB)

The PDB (http://www.rcsb.org) was originally established in 1971 at Brookhaven National Laboratory and in 1998 came under management by the Research Collaboratory for Structural Bioinformatics (RCSB). As of early 2022, the PDB housed >187,000 macromolecular structures that are freely and publicly available and the number per species, enzyme type and level of resolution are summarized in Figure 22.2. Macromolecular structures provide one of the most powerful means by which molecular functions of proteins and their ligands can be probed and understood and therefore a beginning researcher should become familiar with this resource. It has been proposed that the entire diversity of proteins found in nature can be explained by the fact that they are built from combinations of <20,000 different functional domains.[5,6]

While there is an emphasis on human and mouse structures due to applied biomedical and pharmaceutical interests, model organisms such as *Escherichia coli* and yeast (*Saccharomyces cerevisiae*) have been key to advancing this discipline (Figure 22.2A). Although many enzymes can be categorized into nine types, there are many other protein types in PDB (Figure 22.2B). Understanding the functions of these proteins and their functional domains is best when a high resolution structure

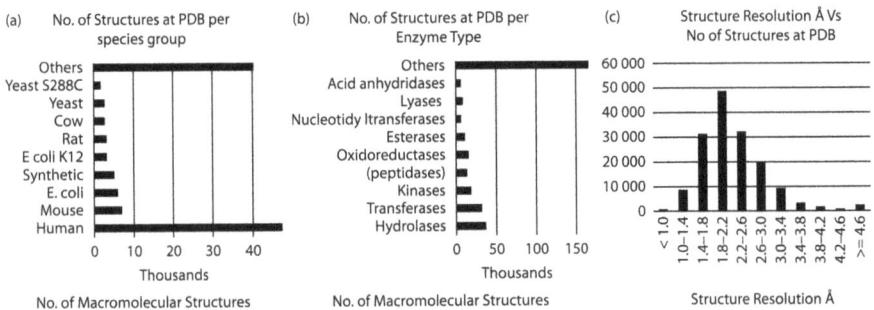

FIGURE 22.2 Summary of number of structures, enzyme types, and structure resolution at PDB. (a) Graph of number of macromolecular structures versus species type. (b) Graph of number of macromolecular structures versus main enzyme type. (c) Graph of structure resolution versus number of macromolecular structures.

Source: Data retrieved from http://www.rcsb.org, November 2020.

is obtained in the range below 2 Å resolution, although this can be difficult to obtain in practice for many less organized proteins (Figure 22.2C).

When searching PDB, a basic search can be conducted using macromolecule names, but a more advanced search permits searches based on structural attributes and return results at the structure, entity, or assembly level, that can be viewed in a variety of formats. Most web browsers now permit the 3D structures to be viewed and rotated, and this is useful for user new to PDB. However, the use of more specialized software is required for detailed structural analysis and comparison. PDB provides an excellent educational portal named PDB-101 (http://pdb101.rcsb.org). This resource is organized into four main browsing categories, Health and Disease, Molecules of Life, Biotech and Nanotech, and Structures and Structure Determination. Within each category are many subcategories, each with learning and teaching resources. The striking visual presentation of many of the resources but especially the SciArt gallery section is both attractive and helpful to the beginning user of this resource.

SEARCHING KEGG: KYOTO ENCYCLOPEDIA OF GENES AND GENOMES

KEGG (www.kegg.jp) is a database resource developed for understanding high-level functions and utilities of the biological system, such as the cell, the organism, and the ecosystem. The database provides largely molecular-level information, especially large-scale molecular datasets generated by genome sequencing and other high-throughput experimental technologies. The updated database contains 19 databases, the content of which is summarized in Table 22.5.[7] Of these, the PATHWAY database is the central resource and consists of manually drawn pathways that describe

---

TABLE 22.5
Summary of 19 Databases Accessible Via KEGG[a]

Category	Entry Point	Content
Systems Information		
	KEGG PATHWAY	A collection of manually drawn pathway maps
	KEGG BRJTE	A collection of hierarchical classification systems capturing functional hierarchies
	KEGGMODULE	Modules which are manually defined functional units of gene sets and reaction sets
Genomic Information		
	KO (KEGG Orthology)	A database of molecular functions represented in terms of functional orthologs
	KEGG GENOME	Is a collection of KEGG organisms, which are the organisms with complete genome sequences
	KEGG GENES	A collection of gene catalogs for all complete genomes generated from public resources, mostly NCBI RefSeq

(Continued)

TABLE 22.5 *(Continued)*
Summary of 19 Databases Accessible Via KEGG^a

Category	Entry Point	Content
	KEGG GLYCAN	Contains the information about amino acid sequence similarities among all protein-coding genes in the complete
Chemical Information		
	KEGG COMPOUND	A collection of small molecules, biopolymers, and other chemical substances that are relevant to biological systems
(KEGG LIGAND)	KEGG GLYCAN	A collection of experimentally determined glycan structures
	KEGG REACTION	Chemical reactions, mostly enzymatic reactions, containing all reactions that appear in the KEGG metabolic pathway maps and only in the EC number system
	KEGGRCLASS	Contains classification of reactions based on the chemical structure transformation patterns of substrate-product pairs
	KEGGENIYME	An implementation of the Enzyme Nomenclature (EC number system) produced by the IUBMB/IUPAC Biochemical Nomenclature Committee
Health Information		
	KEGG MEDICUS	Is an integrated information resource of diseases, drugs, and health-related substances, comprised of the four databases
	KEGG NETWORK	Represents a renewed attempt by KEGG to capture knowledge of diseases and drugs in terms of perturbed molecular
	KEGG VARJANT	A collection of human gene variants that can be associated with human diseases and drugs
	KEGG DISEASE	A collection of disease entries focusing only on the perturbants
	KEGG DRUG	A comprehensive drug information resource for approved Drugs in Japan, the USA, and Europe, unified based on the chemical structure and/or the chemical component of active ingredients
	KEGG DGROUP	Drug groups
	KEGG ENVIRON	A collection of crude drugs, essential oils, and other health-promoting substances, which are mostly natural products
	JAPIC	Japanese drug labels
	DailyMed	FDA drug labels

^a Summarized from KEGG website and from Kaneisha et al. 2021 NAR (gkaa970, https://doi.org/10.1093/nar/gkaa970).

metabolism, genetic information processing, environmental information processing, cellular processes, organismal level systems, and diseases. These pathways are species specific, whereas the MODULES database describes conserved gene sets and biochemical reactions.

The main entry point for searching KEGG is via KEGG2 (www.kegg.jp/kegg/kegg2), which like the ENTREZ query system permits one to query all or selected databases. Alternatively one can data driven or choose organism-oriented or subject-oriented (*i.e.,* cancer, pathogen, virus, etc.) search entry points. The DBGET and LinkDB retrieval systems permit searches not only of KEGG databases but also external databases such as NCBI, ENSEMBL, GO, and organism-specific genome databases (>100 total).

In addition to the extensive databases described above, KEGG also provides tools for computational analysis and genome and metagenome analysis. These include BLAST and FASTA for the analysis of individual genes, but for the advanced user, there are pathway mapping tools such as KEGG Mapper, and tools such as BlastKOALA that permit the mapping and functional characterization of entire genomes.[8] In summary, KEGG continues to be a comprehensive database resource that attempts to be a "computer representation" of the biological systems. It is freely available to academic researchers via its website, but for heavy data users, their FTP download service requires a subscription.

SEARCHING GENOMES USING ENSEMBL AND ENSEMBL-GENOMES

The Ensembl project (www.uswest.ensembl.org) was started in 1999 in anticipation of the completion of the first human genome draft. It is headed by the European Bioinformatics Institute (EBI) at the EMBL (www.ebi.ac.uk). The goal of the project was to automatically annotate the human genome, integrate it with other biological data and make this publicly available via the web. Although the original Ensembl project continues to focus on human and vertebrate genomes (>312 species from alpaca to zig-zag eel), the Ensembl-Genomes project expanded to include >50,000 genomes from bacteria, fungi, metazoa, plants, and protists (http://ensemblgenomes.org).[9] Similar genome browsers are available at NCBI and at the University of California Santa Cruz (UCSC) (genome.ucsc.edu/index). These genomic resources have radically changed the landscape for all students and researchers of biology as gene and genome sequences can serve as the starting point for discovery without having to leave your chair.

A typical search can begin with choosing a genome and a search term (*i.e.,* name of gene or protein). The search result will provide a unique gene identifier and gene location. Selecting the gene identifier will open a graphical interface, which allows users to scroll through a genome and observe the relative location of features such as conceptual annotation (e.g., genes, SNP loci), sequence patterns (e.g., repeats) and experimental data (e.g., sequences and external sequence features mapped onto the genome (Figure 22.3).

In many genomes, a karyotype image is available, and the interface enables the user to zoom in to a region or move along the genome in either direction. Also

(a)

(b)

(c)

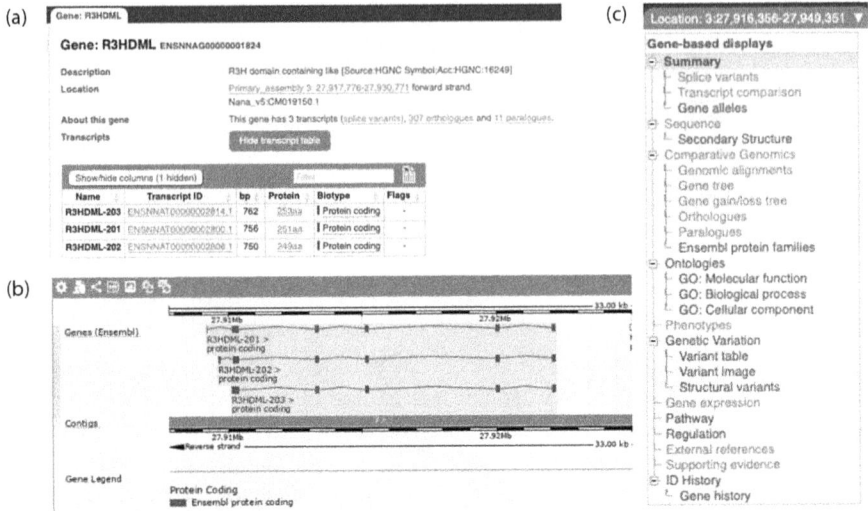

FIGURE 22.3 Sample graphical interface output from Ensembl gene search. In this example, the Indian cobra (*Naja naja*) genome was queried with the term "venom" and the search result listed the gene symbol ENSNNAG00000001824 encoding R3HDML located on chromosome 3 between bp 27,917,776 and 27,930,771. (a) Selecting the gene symbol links to a gene summary with a table of multiple transcripts. (b) An interactive graphical display shows a scaled gene model and its organization and location on chromosome 3. (c) A sidebar provides multiple links to sequences, comparative genomics resources, ontologies, phenotypes, genetic variations, gene expression and pathway and regulatory and literature information where known.

various types of data such as variants, methylation, and expression data are mapped to the genome as data tracks, and individual tracks can be turned on and off, allowing the user to customize the display to suit their research interests. In many cases, data can be exported directly from the interface in a variety of standard file formats for use on a local workstation. Because data is stored on a public MySQL database, the need for downloading large datasets is avoided. Finally, while Ensembl provides an excellent genome browser and interactive interface the website also provides tools such as BLAST, BLAT, BioMart, and the Variant Effect Predictor (VEP) for all supported species. The help page at Ensembl can further help the beginning user to become more familiar with the resources and tools that are available (uswest. ensembl.org/info/index).

CITING DATABASES

The availability of vast amounts of data is unparalleled in human history and is enabling discovery at an astounding rate. When using these data for one's own research, they should be considered legitimate, citable products of research. Data citations should be accorded the same importance in the scholarly record as citations of other research objects, such as publications. The Joint Declaration of Data Citation Principles brought together a group of interested parties to develop a set of principles to encourage data citation (www.force11.org/datacitationprinciples). Therefore, when using such data in research, the user should be sure to cite the appropriate database and use digital object identifiers where applicable.[10]

FINAL RECOMMENDATIONS FOR DATASBASE SEARCHING

This chapter has aimed to provide a brief overview of the value of database searching and how to access a few of the most popular biological databases. In general, the beginning user should adopt an exploratory attitude in order to become familiar with the vast informational resources available to the modern biological student and researcher. They should avail of the help pages to understand the scope of a particular resource and survey sites to find those with particular relevance to their own organismal subject. When conducting searches one should always keep in mind the broader biological context of the gene or protein being investigated. The unifying principle that all living things evolved from a common ancestor is of particular importance when taking advantage of the comparative genomics knowledge that can be gleaned from biological databases.

ACKNOWLEDGMENTS

The author would like to thank the team of faculty at The University of Toledo and Bowling Green State University who teach the graduate program in Bioinformatics, Proteomics, and Genomics for the opportunity to participate in developing materials on the topic covered in this chapter. This work was supported in part by a National Science Foundation IOS-1733633 grant to JG.

REFERENCES

1. Baxevanis, A.D.; Bateman, A. The importance of biological databases in biological discovery. Curr. Protoc. Bioinform. **2015**, *50*, 1.1.1–1.1.8. doi: 10.1002/0471250953. bi0101s50
2. Imker, H.J. 25 Years of molecular biology databases: a study of proliferation, impact, and maintenance. Front. Res. Metrics Anal. **2018**, *3*. doi: 10.3389/frma.2018.00018
3. Henry, V.J.; Bandrowski, A.E.; Pepin, A.S.; Gonzalez, B.J.; Desfeux, A. OMICtools: an informative directory for multi-omic data analysis. Database (Oxford) **2014**. doi: 10.1093/database/bau069
4. Camacho, C.; Coulouris, G.; Avagyan, V.; Ma, N.; Papadopoulos, J.; Bealer, K.; Madden, T.L. BLAST+: architecture and applications. BMC Bioinf. **2009**, *10*, 421. doi: 10.1186/1471-2105-10-421
5. Dohmen, E.; Klasberg, S.; Bornberg-Bauer, E.; Perrey, S.; Kemena, C. The modular nature of protein evolution: domain rearrangement rates across eukaryotic life. BMC Evol. Biol. **2020**, *20*. doi: 10.1186/s12862-020-1591-0
6. Levitt, M. Nature of the protein universe. Proc. Natl. Acad. Sci. U.S.A. **2009**, *106*, 11079–11084. doi: 10.1073/pnas.0905029106

7. Kanehisa, M.; Furumichi, M.; Sato, Y.; Ishiguro-Watanabe, M.; Tanabe, M. KEGG: integrating viruses and cellular organisms. Nucleic Acids Res. **2020**. doi: 10.1093/nar/gkaa970

8. Kanehisa, M.; Sato, Y.; Morishima, K. BlastKOALA and GhostKOALA: KEGG tools for functional characterization of genome and metagenome sequences. J. Mol. Biol. **2016**, *428*, 726–731. doi: 10.1016/j.jmb.2015.11.006

9. Yates, A.D. et al. Ensembl 2020. Nucleic Acids Res. **2020**, *48*, D682–D688. doi: 10.1093/nar/gkz966

10. Neumann, J.; Brase, J. DataCite and DOI names for research data. J. Comput.-Aided Mol. Des. **2014**, *28*, 1035–1041. doi: 10.1007/s10822-014-9776-5

23 Accessing Genomic Databases

Robert J. Trumbly
University of Toledo, Toledo, Ohio, U.S.A

CONTENTS

INTRODUCTION

The revolution in biology has been driven in large part by the exponential increase in available genomic sequences. Knowledge of an organism's genome provides a catalog of all the proteins encoded in the genome and provides a framework for biological investigations. With so many complete genomes available from diverse organisms, we can use comparative genomics for biological insights and evolutionary studies.

A good starting point for learning what databases are available and the latest updates is the annual database issue of Nucleic Acids Research appearing every January. Appended to the first article of this issue is a link to a database of databases: https://www.oxfordjournals.org/nar/database/c/. Some of the groups covered in this listing are as follows: nucleotide sequence databases, RNA sequence databases, protein sequence databases, genomics databases (non-vertebrate), human and other vertebrate genomes, and human genes and diseases.

This review will focus on genomic sequences and the associated gene annotations. Other important data types associated with genomics including genetic variation (SNPs) and gene expression will not be covered.

DOI: 10.1201/9781003247432-23

GENOME SEQUENCES

There are two general types of genomic data, the genomic sequence itself and the annotations of these sequences. Understanding the genetic information for an organism requires its complete genome sequence. Complete genomes are assembled from fragments determined by sequencing projects. The first complete genomes were assembled from sequencing of cloned genomic fragments. Currently genomes are sequenced using whole-genome shotgun sequencing, without the need for the intermediate cloning step. The first complete genome for a cellular organism, as opposed to a virus, was the bacterium *Haemophilus influenzae* in 1995.[1] The first eukaryotic genome to be completely sequenced was the yeast *Saccharomyces cerevisiae* in 1996.[2] The Human Genome Project (HGP) was initiated in 1990 as a large international collaboration to sequence the entire human genome. An essentially complete human genome sequence was announced in 2003.[3] Because certain regions of the human genome that contain highly repetitive domains are difficult to sequence, the definitive genome assembly is still being updated on a regular basis. The most recent human assembly is Human GRCh38/hg38.

There are still some limitations in the most recent human reference genome. There are still gaps in the sequences for all the chromosomes. In addition, the single reference sequence doesn't represent the variation seen in human populations.[4] To address these and other issues, the Human Pangenome Reference Consortium was established (https://humanpangenome.org/). As part of this project, the complete telomere-to-telomere sequence of the human X chromosome was assembled.[5] The current version of the reference human genome, called GRCh38, was derived from only a few individuals and doesn't represent the diversity seen in human populations. One solution is to represent the diversity as a pangenome, with the sequence variation represented graphically, as described above. Another solution is to create a "consensus genome," a single genome that has the most common alleles among thousands of individuals rather than whatever the few individuals used to make the current reference sequence.[6] Alignment of RNA-Seq experiments was significantly improved when aligned to the consensus sequence compared with the standard reference genome.[7]

Genomic databases are repositories of information for specific organisms. The primary data are the nucleotide sequences of the genomes. For many years there have been three main repositories of nucleotide sequences, hosted in different regions of the world. These are the National Center for Biotechnology Information (NCBI),[8] the European Bioinformatics Institute (EBI),[9] and the DNA Data Bank of Japan (DDBJ).[10] All primary sequence information is submitted to the International Nucleotide Sequence Database Collaboration (INSDC),[11] and shared rapidly among the three databases, NCBI, EBI, and the DDBJ (Table 23.1).

The sequencing reads from the many high-throughput sequencing (HTS) projects are housed in Sequence Read Archives (SRAs). The sequences are in either the primary FASTQ files, or as BAM files after alignment to the genome. The availability of the raw sequences allows other researchers to repeat the original analyses or explore alternative analysis methods on the same data. The capillary reads are the raw data from Sanger sequencing experiments. The annotated sequences

TABLE 23.1

Sharing of Information by The International Nucleotide Sequence Database Collaboration (INSDC) Among the Three Major Databases

Data Type[a]	DDBJ	EMBL-EBI	NCBI
Next generation reads	Sequence Read Archive	European Nucleotide	Sequence Read Archive
Capillary reads	Trace Archive	Archive (ENA)	Trace Archive
Annotated sequences	DDBJ		GenBank
Samples	BioSample		BioSample
Studies	BioProject		BioProject

Source: http://www.insdc.org/

[a] The five types of data are accessible in distinct databases in DDBJ and NCBI, but all grouped together in ENA in the EMBL-EBI database.

comprise the sequences and their annotations contributed by research laboratories and projects. These files have a strictly defined structure as flat file plaintext format, described as GenBank format at NCBI and EMBL-Bank format at ENA. The BioSample database contains submitter-supplied information, or metadata, about the biological materials related to the experiments stored in sequence databases. BioSamples could describe a cell line, a primary tissue biopsy, a specific organism, or an environmental isolate. From the NCBI description: "A BioProject is a collection of biological data related to a single initiative, originating from a single organization or from a consortium. A BioProject record provides users a single place to find links to the diverse data types generated for that project." The BioSample and BioProject accessions are linked to their associated experimental data.

GENE ANNOTATIONS

Most databases have annotations for individual genes including the experimental and predicted mRNAs, their protein translation products, and extensive annotations such as gene ontology. The two major programs for annotation are the RefSeq project at NCBI[12] and the Ensembl project at EBI.[13] These will be described in more detail in relation to these two infrastructures. These annotated entities are unique to their respective platforms and are not part of the INSDC shared data. The GENCODE project, a subproject of ENCODE, aims to annotate all evidence-based gene features in the entire human and mouse genomes at a high accuracy (https://www.gencodegenes.org/).[14] GENCODE Basic is a subset of the GENCODE transcript set that contains only 5′ and 3′ complete transcripts. In 2018, NCBI and EMBL-EBI announced a new collaborative project called Matched Annotation from NCBI and EMBL-EBI (MANE). https://www.ncbi.nlm.nih.gov/refseq/MANE/. This project aims to provide a matched set of transcripts for every human protein-coding gene. The transcripts in this set are annotated identically in the RefSeq and the Ensembl-GENCODE gene sets. The MANE Select dataset is a subset of RefSeq Select. For a given human protein-coding gene, the RefSeq Select transcript is designated as the "MANE Select" when it matches the Ensembl "Select" transcript and is included in the public MANE set. In the case of MANE Select transcripts, the keyword "RefSeq Select" is replaced by "MANE Select" in the RefSeq flat file.

RETRIEVING SEQUENCES FROM THE NCBI DATABASES

A common task for biologists is retrieval of gene sequences from databases. NCBI uses Entrez as their database search and retrieval system. From the Entrez start page, queries can search all of the 35 databases covered by Entrez. The 35 databases can be grouped into larger categories: Literature, Health, Genomes, Genes, Proteins, and Chemicals.[8]

As an example, our goal is to obtain the mRNAs and their translation products for the human *ELK1* gene, which has multiple mRNA and protein products, with a role in cancer biology (Patki et al., 2013).[15] If we use ELK1 as the query for the nucleotide database in NCBI Entrez, we will get over a thousand hits. The accessions can be classified as two types, INSDC and RefSeq. INSDC stands for International Nucleotide Sequence Database Collaboration.[11] The INSDC accessions were submitted by researchers and represent their primary sequence information. Many of these sequences are incomplete, redundant, and may differ from the reference genome sequences. NCBI created RefSeq to have standard curated reference sequences derived from the primary sequence data.[12] All RefSeq entries can be recognized by the unique format of their accession numbers: two uppercase letters, followed by an underscore, then a number. The different types of RefSeq entries use different letters to identify the type of sequence. The most commonly used types of RefSeq entries are NM_XXXXXX for mRNAs, NP_XXXXXX for proteins, and XM_XXXXXX for computed mRNA. A new type of RefSeq entity was created recently, NG_XXXXXX, for individual annotated gene sequences.

Starting from the Entrez Gene page may be the best way to obtain the definitive mRNA and protein sequences for that gene. Figure 23.1 shows the middle section of the pages for the ELK1 gene entry, labeled Genomic regions, transcripts, and products. The RefSeq mRNAs are displayed graphically showing the exons present in the individual mRNAs. Selecting a line from the mRNA graph will produce a pop-up with links for downloading the RefSeq mRNA and proteins sequences in FASTA format. On the right side of the page are links to gene-related information (not shown). Choosing RefSeq RNA will produce the list of the RefSeq mRNA accessions related to the gene. The list of protein RefSeq accessions will be produced by choosing RefSeq protein. In the case of ELK1, there are three RefSeq mRNAs and the corresponding three RefSeq protein accessions.

Near the top of the Genomic regions, transcripts, and products section of the Entrez gene pages, the heading "Go to nucleotide:" is followed by three links. The first link, Graphics, will display just the graphics section, already seen in this section. The second link, FASTA, opens a window with a FASTA sequence file with a single genomic sequence containing the exons and introns of the respective gene. The third link, GenBank, will open the GenBank file for this genomic region, with annotations for the mRNAs and the encoded proteins.

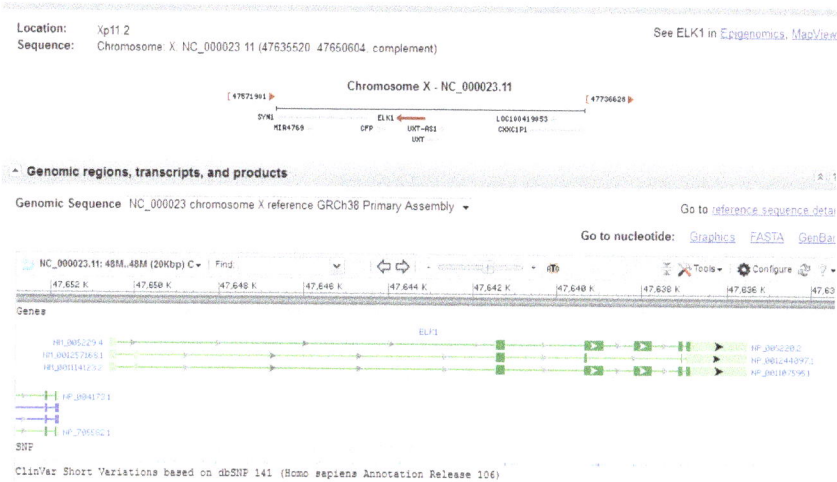

FIGURE 23.1 The middle section of the graphical display for the Entrez Gene page for the human ELK1 gene. The diagram near the top shows the chromosomal context of ELK1. The exon-intron structures for the three ELK1 RefSeq mRNAs are shown, as well as the corresponding encoded proteins.

REFSEQGENE

RefSeqGene, a subset of NCBI's Reference Sequence (RefSeq) project, defines genomic sequences to be used as reference standards for well-characterized genes.[12] From the RefSeqGene description, they provide a "stable foundation for reporting mutations, for establishing conventions for numbering exons and introns, and for defining the coordinates of other variations." The entire gene sequence is displayed, exons and introns, as well as 5-kb upstream of the TSS (Transcription Start Site) and 2-kb downstream of the 3′ exon (Figure 23.2). The RefSeqGene

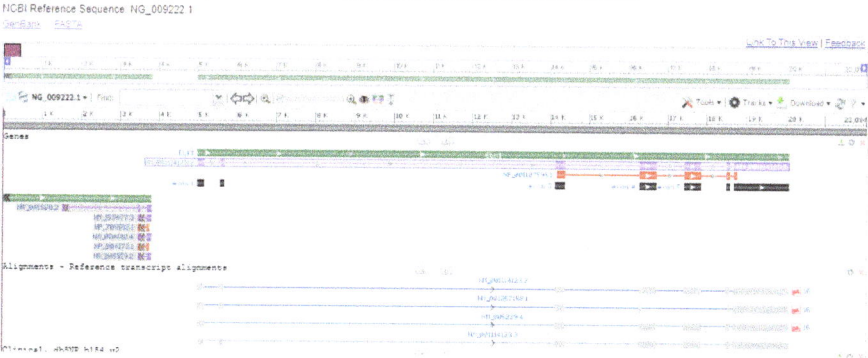

FIGURE 23.2 Shown is the graphical display for the RefSeqGene entry for the human ELK1 gene. The entire gene region is shown including flanking sequences of 5-kb upstream and 2-kb downstream. The multiple mRNA structures are shown at the bottom.

project is associated with the Locus Reference Genomic project (http://www.lrg-sequence.org/). From the LRG site, "A Locus Reference Genomic (LRG) is a manually curated record that contains stable and thus, un-versioned reference sequences designed specifically for reporting sequence variants with clinical implications."

PROGRAMMATIC ACCESS TO NCBI DATA

There is an Application Programming Interface (API) for Entrez functions (the E-utilities) available, with documentation provided at https://eutils.ncbi.nlm.nih.gov/. The API and additional functions are provided in a package Entrez Direct (EDirect) that facilitates access to the NCBI's suite of interconnected databases (publication, sequence, structure, gene, variation, expression, etc.) from a Unix terminal window.

ACCESSING DATA FROM THE UCSC GENOME BROWSER

The UCSC Genome Browser is a preferred method for displaying genomic information. There are now 211 genome assemblies, covering 107 species.[16] From the UCSC Genome Browser home page (https://genome.ucsc.edu/), first choose the desired genome assembly. The most recent human assembly is Human GRCh38/hg38. To select the genome region to display, the gene name of interest can be entered, or the exact sequence location if known. The genome browser now uses the GENCODE genes for specifying gene locations for the human and mouse genomes. For most other genomes, the RefSeq genes are used. Additionally, the genome browser has a very fast sequence search program called BLAT to rapidly identify similar sequences. Many diverse types of information related to specific genomic locations are displayed as tracks. The tracks are organized in groups. The top two groups are Mapping & Sequencing and Genes & Gene Predictions. The extensive results from the ENCODE project are available in the Regulation group.

The tracks shown on the UCSC Genome Browser are displays of data housed in a MySQL database. The underlying data can be accessed via the Table Browser GUI (graphical user interface). The Table Browser User's Guide is available here: https://genome.ucsc.edu/goldenPath/help/hgTablesHelp.html. Because the Table Browser uses the same database as the Genome Browser, the two views are always consistent. By default, the Table Browser displays query results directly in your internet browser window. To redirect the data to a file, type a file name into the output file box before starting the query. As an example of using the Table Browser, say we have the results of a gene expression experiment, with a list of 100 differentially regulated genes. We can use the Table Browser to get the promoter regions for these genes with the following steps: choose the Table Browser from the UCSC Genome Browser home page.

1. Select human hg19 as genome.
2. Group: Genes and Gene Prediction Tracks
3. Track: UCSC genes
4. Table: known Gene
5. Region: genome
6. Identifiers: paste list of gene symbols of the 100 genes.

7. Output format: can first get BED file, to see how many sequences will be returned.
8. To get sequences, change output format to sequence.
9. Get output
10. Sequence Retrieval Region Options: choose Promoter/Upstream 1000 bases
11. Get sequence

The sequences are then available for further downstream analysis such as testing for enrichment of transcription factor binding sites.

Data in the UCSC MySQL database can be accessed directly using MariaDB (MySQL) (https://genome.ucsc.edu/goldenPath/help/mysql.html). The UCSC Genome Browser uses MariaDB as the backend database server. MariaDB is a community-developed, commercially supported fork of the MySQL relational database management system. You must have MariaDB (MySQL) client libraries installed on your computer.

RESOURCES AT THE EUROPEAN BIOINFORMATICS INSTITUE (EMBL-EBI)

The European Bioinformatics Institute (EMBL-EBI), part of the European Molecular Biology Laboratory, provides resources that are comparable to but in some ways different from the NCBI resources.[9] From the EBI site: EBI Search, analogous to Entrez at NCBI, is a search engine that provides access to the biological data resources hosted at the EMBL-EBI.

The European Nucleotide Archive (ENA)[17] houses all of the data that is shared by the INSDC. The ENA houses the sequencing reads from the many HTS projects that are in the SRA sections of NCBI and the DDBJ. The sequences are primarily in either the primary FASTQ files, or as BAM files after alignment to the genome. The ENA also has the trace archive and single sequences shared by INSDC. The single sequences previously comprised a separate database called EMBL-Bank at EBI but has now been incorporated into the ENA. The EMBL-Bank accessions have a flat file plaintext format similar but distinct from the GenBank format. The extensive gene and genome annotations created by EMBL-EBI, analogous to the NCBI RefSeq project, are available at the Ensembl site, described below.

UNIPROT

The EBI hosts the most comprehensive resource for protein sequences and their descriptions at UniProt, also called UniProtKB (https://www.uniprot.org/). UniProt hosts the Swiss-Prot database of manually annotated and reviewed protein accessions.[18] Each accession in the Swiss-Prot database has a comprehensive description of the function, structure, expression, molecular interactions, and links to relevant external resources. A much larger collection of protein sequences in UniProt is the TrEMBL (for translated EMBL) computationally analyzed but not manually curated. The TrEMBL sequences are derived by computationally translating the potential protein-coding sequences in the EMBL nucleotide sequence database.

The UniProt site also hosts three other resources: UniRef, UniParc, and Proteomes. The UniProt Reference Clusters (UniRef) provide clustered sets of sequences from the UniProt Knowledgebase, to simplify searches by reducing redundancy. UniParc is a comprehensive and non-redundant database that contains most of the publicly available protein sequences in the world. Proteins may exist in different source databases and in multiple copies in the same database. UniParc avoids such redundancy by storing each unique sequence only once and giving it a stable and unique identifier (UPI). In the Proteome database each accession has lists and sequences of all the proteins expressed for species with completely sequenced genomes.

ENSEMBL

The Ensembl project as part of EBI uses primary data from sequence resources such as INSDC, dbSNP, and the European Variation Archive (EVA) to create gene and genome annotations. The Ensembl database supplies the annotation of gene structures, regulatory elements and variants for vertebrate species. The most highly represented groups in the latest version are 81 fish, 63 mammal, 50 bird, and 44 rodent species.[13] The Ensembl accession numbers have a standard format starting with ENS. For example, the accession for the human ELK1 gene is ENSG00000126767. The Ensembl genes serve as a hub for gene-related information, analogous to NCBI Entrez Gene. Human transcripts have accessions of the form ENSTXXXXXX that are analogous to the NCBI RefSeq mRNAs. The non-human accession names have identifiers denoting the species. For example, the accession for the mouse ELK1 gene is ENSMUSG00000009406. The Ensembl annotations are displayed on their genome browser that has some similarities to the UCSC Genome Browser. The mRNA structures are aligned to the reference genome. There are extensive options to control which tracks are displayed. Some of the track information is similar to that available on the UCSC Genome Browser but displayed in a different format. The Ensembl browser has a powerful comparative genomics tool that presents phylogenetic trees for genes of interest. There are also extensive resources for gene regulation and genetic variation. Ensembl supports biodiversity genomics and has annotated and integrated 83 new vertebrate genomes across diverse clades in the past year.[13]

The main Ensembl site is limited to vertebrate genomes. There is a separate Ensembl site, (http://www.ensemblgenomes.org), dedicated to non-vertebrate genomes.[19] The databases are organized into five sites, corresponding to major kingdoms of life: bacteria, protists, fungi, plants and metazoan (invertebrate). Information available for all species includes genome sequence and annotations of protein- and noncoding genes. Additional data include transcriptional data, polymorphisms and comparative analysis.

Ensembl data can be downloaded as single files from the main Ensembl site without restriction (https://www.ensembl.org), in bulk from their FTP site (ftp://ftp.ensembl.org), and programmatically via their REST API (https://rest.ensembl.org).[20] A convenient way to access Ensembl data is through their BioMart interface.[21] As shown in Figure 23.3, the user makes a series of choices to get the desired information. First, the database is chosen, in this case Ensembl genes, the

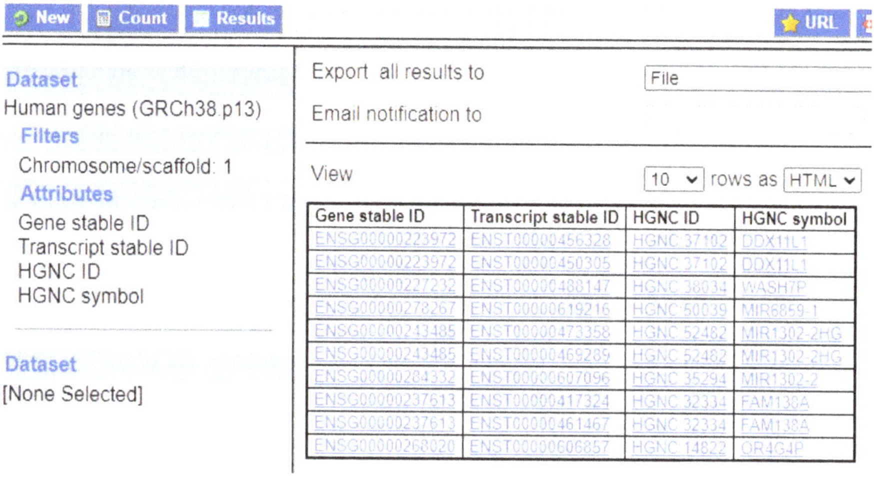

Gene stable ID	Transcript stable ID	HGNC ID	HGNC symbol
ENSG00000223972	ENST00000456328	HGNC:37102	DDX11L1
ENSG00000223972	ENST00000450305	HGNC:37102	DDX11L1
ENSG00000227232	ENST00000488147	HGNC:38034	WASH7P
ENSG00000278267	ENST00000619216	HGNC:50039	MIR6859-1
ENSG00000243485	ENST00000473358	HGNC:52482	MIR1302-2HG
ENSG00000243485	ENST00000469289	HGNC:52482	MIR1302-2HG
ENSG00000284332	ENST00000607096	HGNC:35294	MIR1302-2
ENSG00000237613	ENST00000417324	HGNC:32334	FAM138A
ENSG00000237613	ENST00000461467	HGNC:32334	FAM138A
ENSG00000268020	ENST00000606857	HGNC:14822	OR4G4P

FIGURE 23.3 A screenshot of the results of a simple BioMart search. The user selects the desired data by a sequence of choices: dataset, filters (which genes), and attributes (which features). By default only the first ten rows of the results are shown. A table with all the results can be downloaded by selecting export all results.

human genome version GRCh38. The Filters step specifies which genes are chosen. In this case, all the genes on Chromosome 1 are selected. The Attributes step chooses which features of the genes will be returned. In this example, alternate IDs for the gene and transcripts were selected. There are options for many different features, such as gene ontology and selected parts of the gene sequence. The BioMart functions are available as a Bioconductor package, biomaRt, for implementation in the R software environment.

ENCODE PROJECT

The ENCODE project, standing for Encyclopedia of DNA Elements, aims to catalog as many functional elements in human genome as possible using current technologies. The pilot phase was initiated in 2003, finished in 2007, and covered 1% of the human genome.[22] Using insights from the pilot project, the first production phase covering the entire human genome, also known as ENCODE 2, was carried out from 2007 to 2012. The major results from ENCODE 2 were published together in a special edition of Nature 489 6 September 2012,[23] as well as numerous articles published together in Genome Biology and Genome Research in 2012. A perspective on their findings was published in 2014.[24] The second production phase, ENCODE 3, occupied the years 2012–2017. A special issue of Nature, Volume 583 Issue 7818, 30 July 2020, carried 9 articles summarizing major results from ENCODE 3.[25,26] The list of publications stemming directly from the ENCODE projects can be found at https://www.encodeproject.org/publications/. The current phase, ENCODE 4, started in 2017 and is ongoing.

The ENCODE projects have used an evolving set of experimental tools to characterize chromatin structure, the transcriptome, and more recently, RNA-binding proteins. The experimental data relating to chromatin structure that are currently available include 3D chromatin structure, chromatin accessibility, chromatin interactions, methylome, and chromatin modifications. The data from the pilot phase and ENCODE 2 have been available at the UCSC genome browser (https://genome.ucsc.edu/encode/). All results from ENCODE 2 and subsequent studies are available at the main ENCODE data portal at encodeproject.org. Because of the large and diverse datasets available, the ENCODE portal has developed efficient methods to search and obtain target datasets. The Getting Started section of the ENCODE portal presents guide for searching and displaying data. There they show the example search for a collection of ChIP-seq data: "Experiments targeting H3K4me3 and performed on in vitro differentiated cell samples and originating from human or mouse donors and not tissue samples." The resulting datasets can be downloaded directly. There are also several modes of accessing the data programmatically.

CONCLUSION

This chapter describes the main databases for genomic sequences and their annotations. Individual or multiple sequences can be downloaded for analysis and comparison using convenient web interfaces or various programmatic tools. Gene annotations are available that describe gene features and structures. I have described three main resources, NCBI, the UCSC Genome Browser, and EBI-EMBL, including Ensembl. The UCSC and Ensembl sites feature graphical displays termed genome browsers that allow for visualization of selective gene attributes. The amount of genomic sequences continues to increase rapidly, presenting challenges for its storage and description. The intensive annotation projects such as RefSeq, Ensembl, and GENCODE help provide understanding of the ever-increasing supply of genomic data.

REFERENCES

1. Fleischmann, R.D.; Adams, M.D.; White, O.; Clayton, R.A.; Kirkness, E.F.; Kerlavage, A.R., ... et al. Whole-genome random sequencing and assembly of *Haemophilus influenzae* Rd. Science. **1995**, *269*(5223), 496–512. doi:10.1126/science.7542800
2. Goffeau, A.; Barrell, B.G.; Bussey, H.; Davis, R.W.; Dujon, B.; Feldmann, H.; ... ; Oliver, S.G. Life with 6000 genes. Science **1996**, *274*(5287), 546, 563–547. doi:10.1126/science.274.5287.546
3. Lander, E.S.. Initial impact of the sequencing of the human genome. Nature **2011**, *470*(7333), 187–197. doi:10.1038/nature09792
4. Offord, C. The Pangenome: Are Single Reference Genomes Dead? The Scientist. Dec. **2016**.
5. Miga, K.H.; Koren, S.; Rhie, A.; Vollger, M.R.; Gershman, A.; Bzikadze, A.; ...; Phillippy, A.M. Telomere-to-telomere assembly of a complete human X chromosome. Nature **2020**, *585*(7823), 79–84. doi:10.1038/s41586-020-2547-7
6. Knutsen, A. A new human reference genome represents the most common sequences. The Scientist. Dec. **2020**.

7. Kaminow, B.; Gillis, B.S.; Dobin, J. A. Virtue as the mean: Pan-human consensus genome significantly improves the accuracy of RNA-seq analyses. bioRxiv. **2020**. doi: 10.1101/2020.12.22.423111

8. Sayers, E.W.; Beck, J.; Bolton, E.E.; Bourexis, D.; Brister, J.R.; Canese, K.; ... ; Sherry, S.T. Database resources of the National Center for Biotechnology Information. Nucleic Acids Res. **2021**, *49*(D1), D10–D17. doi:10.1093/nar/gkaa892

9. Cantelli, G.; Cochrane, G.; Brooksbank, C.; McDonagh, E.; Flicek, P.; McEntyre, J.; ... Apweiler, R. The European Bioinformatics Institute: empowering cooperation in response to a global health crisis. Nucleic Acids Res. **2021**, *49*(D1), D29–D37. doi:10.1093/nar/gkaa1077

10. Fukuda, A.; Kodama, Y.; Mashima, J.; Fujisawa, T.; Ogasawara, O. DDBJ update: streamlining submission and access of human data. Nucleic Acids Res. **2021**, *49*(D1), D71–D75. doi:10.1093/nar/gkaa982

11. Arita, M.; Karsch-Mizrachi, I.; Cochrane, G.. The international nucleotide sequence database collaboration. Nucleic Acids Res. **2021**, *49*(D1), D121–D124. doi:10.1093/nar/gkaa967

12. O'Leary, N.A.; Wright, M.W.; Brister, J.R.; Ciufo, S.; Haddad, D.; McVeigh, R.; ... ; Pruitt, K.D. Reference sequence (RefSeq) database at NCBI: current status, taxonomic expansion, and functional annotation. Nucleic Acids Res. **2016**, *44*(D1), D733–D745. doi:10.1093/nar/gkv1189

13. Howe, K.L.; Achuthan, P.; Allen, J.; Allen, J.; Alvarez-Jarreta, J.; Amode, M.R.; ... ; Flicek, P. Ensembl 2021. Nucleic Acids Res. **2021**, *49*(D1), D884–D891. doi:10.1093/nar/gkaa942

14. Frankish, A.; Diekhans, M.; Jungreis, I.; Lagarde, J.; Loveland, J.E.; Mudge, J.M.; ... ; Flicek, P. GENCODE 2021. Nucleic Acids Res. **2021**, *49*(D1), D916–D923. doi:10.1093/nar/gkaa1087

15. Patki, M.; Chari, V.; Sivakumaran, S.; Gonit, M.; Trumbly, R.; Ratnam, M. The ETS domain transcription factor ELK1 directs a critical component of growth signaling by the androgen receptor in prostate cancer cells. J. Biol. Chem. **2013**, *288*(16), 11047–11065. doi:10.1074/jbc.M112.438473

16. Navarro Gonzalez, J.; Zweig, A.S.; Speir, M.L.; Schmelter, D.; Rosenbloom, K.R.; Raney, B.J.; ...; Kent, W.J. The UCSC Genome Browser database: 2021 update. Nucleic Acids Res. **2021**, *49*(D1), D1046–d1057. doi:10.1093/nar/gkaa1070

17. Harrison, P.W.; Ahamed, A.; Aslam, R.; Alako, B.T.F.; Burgin, J.; Buso, N.; ...; Cochrane, G. The European Nucleotide Archive in 2020. Nucleic Acids Res. **2021**, *49*(D1), D82–D85. doi:10.1093/nar/gkaa1028

18. The UniProt Consortium. UniProt: a worldwide hub of protein knowledge. Nucleic Acids Res. **2019**, *47*(D1), D506–D515. doi:10.1093/nar/gky1049

19. Kersey, P.J.; Allen, J.E.; Allot, A.; Barba, M.; Boddu, S.; Bolt, B.J.; ... ; Yates, A. Ensembl Genomes 2018: an integrated omics infrastructure for non-vertebrate species. Nucleic Acids Res. **2018**, *46*(D1), D802–D808. doi:10.1093/nar/gkx1011

20. Madeira, F.; Park, Y.M.; Lee, J.; Buso, N.; Gur, T.; Madhusoodanan, N.; ...; Lopez, R. The EMBL-EBI search and sequence analysis tools APIs in 2019. Nucleic Acids Res. **2019**, *47*(W1), W636–W641. doi:10.1093/nar/gkz268

21. Kinsella, R.J.; Kähäri, A.; Haider, S.; Zamora, J.; Proctor, G.; Spudich, G.; ... Flicek, P.. Ensembl BioMarts: a hub for data retrieval across taxonomic space. Database (Oxford). **2011**, *2011*, bar030. doi:10.1093/database/bar030

22. Birney, E.; Stamatoyannopoulos, J.A.; Dutta, A.; Guigó, R.; Gingeras, T.R.; Margulies, E.H.; de Jong, P.J.. Identification and analysis of functional elements in 1% of the human genome by the ENCODE pilot project. Nature **2007**, *447*(7146), 799–816. doi:10.1038/nature05874

23. ENCODE Project Consortium An integrated encyclopedia of DNA elements in the human genome. Nature **2012**, *489*(7414), 57–74. doi:10.1038/nature11247

24. Kellis, M.; Wold, B.; Snyder, M.P.; Bernstein, B.E.; Kundaje, A.; Marinov, G.K.; … Hardison, R.C. Defining functional DNA elements in the human genome. Proc. Natl. Acad. Sci. U.S.A. **2014**, 111(17), 6131-6138. doi:10.1073/pnas.1318948111

25. Moore, J.E.; Purcaro, M.J.; Pratt, H.E.; Epstein, C.B.; Shoresh, N.; Adrian, J.; … Weng, Z. Expanded encyclopaedias of DNA elements in the human and mouse genomes. Nature **2020**, *583*(7818), 699–710. doi:10.1038/s41586-020-2493-4

26. Snyder, M.P.; Gingeras, T.R.; Moore, J.E.; Weng, Z.; Gerstein, M.B.; Ren, B.; … Myers, R. M.. Perspectives on ENCODE. Nature **2020**, *583*(7818), 693–698. doi:10.1038/s41586-020-2449-8

24 Phylogenetics, Comparative Genomics, and Phylogenomics

Scott O. Rogers
Bowling Green State University, Bowling Green, Ohio, U.S.A.

CONTENTS

INTRODUCTION

Once a DNA (or RNA) sequence is determined for any organism, BLAST similarity searches are often the initial choice for analysis. This provides a relatively rapid determination of taxonomic affinities for the organism. However, the BLAST results are usually just the first step in determining the identity of the organism. Phylogenetic analyses can provide a more precise determination of taxonomic affinities of the organism, cell types, tumors, etc.[1–8] Gene sequences from a number of taxonomically affiliated taxa can be aligned and the number or proportional changes among the sequences can be calculated. Then, a graphical representation of the evolutionary history can be produced (Figure 24.1). This also represents the relative genetic distances among the taxa, which can lead to conclusions regarding the taxonomic affiliations of the sequence of interest.

DOI: 10.1201/9781003247432-24

237

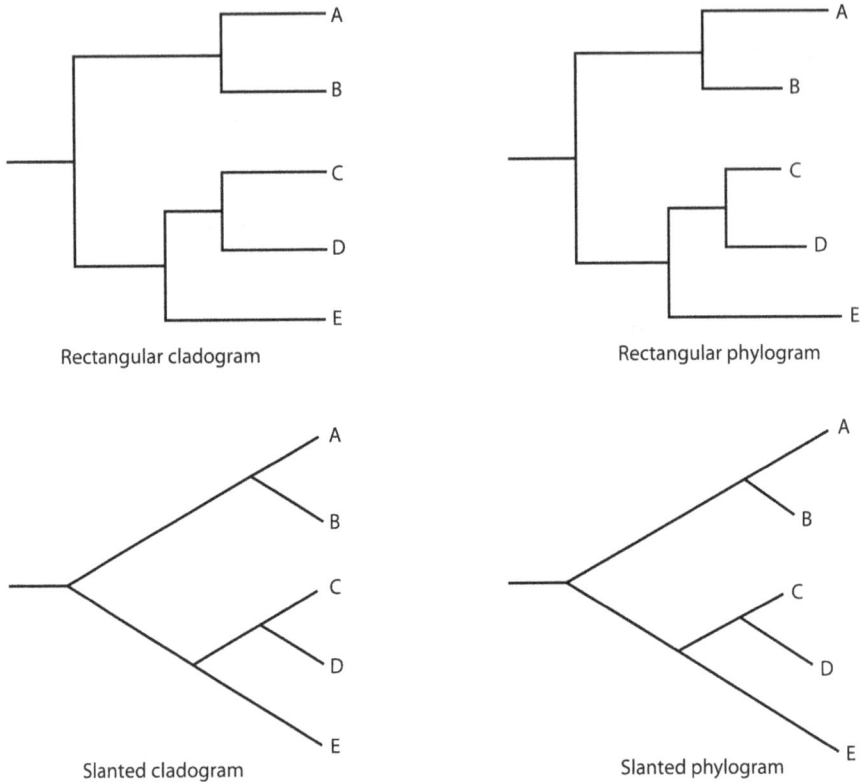

FIGURE 24.1 Types of trees. Rectangular trees are on the top, and slanted trees are on the bottom. Cladograms are on the left. They show branching patterns to indicate evolutionary relationships, but usually do not indicate phylogenetic distances. Phylograms, on the right, show evolutionary relationships, as well as phylogenetic distances. For rectangular phylogenetic trees, only the horizontal branches indicate the phylogenetic distance. The vertical branches are included only to join together the horizontal branches.

Changes in DNA (or amino acid) sequences can be used to determine distinctions between species, as well as among cells within populations and multicellular organisms, including determination of tumor types. The sequences that are obtained from samples can be compared to those from other studies or from national or international sequence data bases using phylogenetic analyses. Depending of the sequences, regions, and species, fine distinctions can be examined, often leading to identification to species, variety, isolate, population, individual, geographical location, or cell (tumor) type (Figure 24.2). Although phylogenetic reconstructions (i.e., trees) were initially developed to examine evolutionary patterns and associations, they can also be used to examine the affinities of species with a sequence from a sample in order to determine the identity and/or characteristics of the organisms/cells associated with the sample.

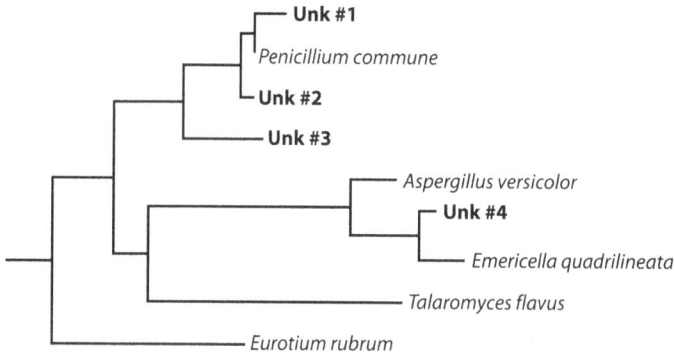

FIGURE 24.2 Phylogenetic tree of isolates (Unk #1–4) compared to sequences from known species based on ribosomal RNA internal transcribed spacers. Unk #1 and Unk #2 are likely *Penicillium commune*, while unk #3 is probably a different species of *Penicillium*. unk #4 is closely related to *Emericella quadrilineata*, but might be a slightly different species. The members of this group produce aflatoxins.

CHOICE OF GENOMIC REGION

Careful selection of the region to be sequenced and the sequences chosen for comparison is paramount to accurate taxonomic determination prior to beginning any phylogenetic analysis. The varied regions within genomes have different rates of nucleotide substitution. Even specific gene loci have regions that may change rapidly and others that are highly conserved. If the question deals with the determination of whether the sample contains a pathogenic bacterium, a particular region of the ribosomal RNA gene locus is generally used. However, if a forensic investigation is attempting to determine whether or not a suspect was at the crime scene, rapidly changing repetitive regions, such as minisatellite regions, are used routinely. If the analysis is to determine whether a particular gene is mutated, then a different region must be sequenced and compared. For some tumors, characteristic regions of the genomes can be informative for identification and treatment options. In genealogical studies, often single-nucleotide polymorphisms (SNPs) are used to determine genetic trees and ancestral geographical origins. In general, broad taxonomic determinations require the use of conserved genomic regions, while identification at the individual or cell level requires the use of rapidly changing regions.

In addition to choosing the appropriate region for comparison, it is equally important to carefully choose the sequences to compare. That is, if there are sets of analogous sequences from a number of taxa, several sequences suspected to be similar to those from the sample, as well as additional sequences from taxa that are more distant taxonomically should be aligned for the phylogenetic analysis. This allows determination of affinities to known organisms/cells (e.g., within genera, species, varieties, individuals) to the sample sequence. It also can indicate ties to more distant organisms/cells. This is useful for determination of pathogens, geographically distinct species/populations, tumor types, and others.

PHYLOGENETIC METHODS

The next consideration is in choosing a method to analyze the distances and simi-larities between the sequences. This depends on the desired level of accuracy, the size of the data set, and the capacity of the computer. It also depends on whether specific tree models will be compared. Some methods use simple calculations to produce distances between every pair of sequences in the alignment, while oth-ers determine all of the possible ways to join all of the sequences together and then test each one to determine the one with the fewest number of nucleotide (or amino acid) changes. Because of the different calculations, some require very little computing time, while others are computationally expensive. Therefore, for small datasets (fewer than 30 sequences), all of the methods can be applied; for moder-ate datasets (30–60 sequences), some of the methods cannot be used; and for large datasets (>60 sequences), only a few of the methods can be used. Of course, these numbers differ if one is using a super computer versus a small laptop computer.

UPGMA (UNWEIGHTED PAIR GROUP METHOD WITH ARITHMETIC MEANS)

This is one of the earliest methods developed, which also relies on simple calcula-tions, such that it can be used on very large datasets. The analysis begins by forming a matrix of the aligned sequences (Figure 24.3). Then, each sequence is compared to every other sequence, and distances are calculated between each pair of sequences, producing a distance matrix. The distances can be based on the total number of differences or proportional distances based on the total differences among all sequences. The calculation can be based on a variety of models, including those

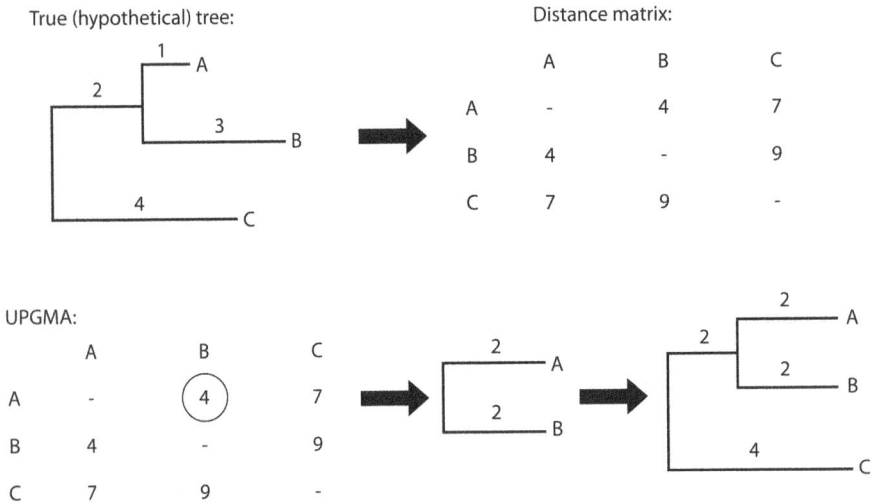

FIGURE 24.3   UPGMA. Upper portion is a hypothetical tree (left) with a distance matrix (right) derived from the tree. The lower portion indicates the first pair (A and B) that are the closest OTUs. Then, the mean of the distances (4/2 = 2) is calculated, and each branch is given that value (2).

that weight transversions differently than transitions. Next, the phylogenetic tree is constructed. Tree construction begins by connecting together the two closest OTUs (operational taxonomic units; i.e., taxon or cell type). Next, distances from each of the other sequences and the mean of the two branches connecting the initial two OTUs are calculated. Then the next closest OTU is connected to the tree. This process is continued until all of the OTUs are added to the tree. The final tree produced by the UPGMA is usually a cladogram rather than a phylogram because the distances between each of the OTUs are usually not accurately depicted by the branch lengths. Because of the averaging of branch lengths, UPGMA can produce inaccurate trees, especially when different rates of sequence change occur on adjacent branches. However, it does produce a tree showing the overall associations among the OTUs.

Neighbor-Joining (NJ)

Another relatively simple, but more powerful method, is Neighbor-Joining (NJ). While this method also calculates distances based on pairs of sequence differences, it keeps track of the nodes (the joining point of two branches of the tree), and therefore does not average the branch lengths, as in UPGMA (Figure 24.4). Because of this,

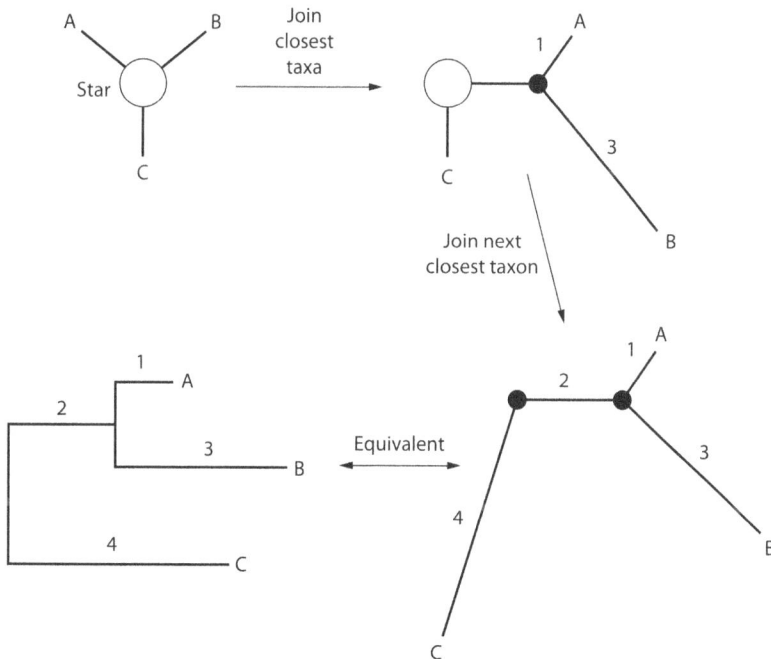

FIGURE 24.4 Neighbor-Joining. Using the same hypothetical tree and distance matrix as in Figure 24.3, a star diagram is constructed, joining all OTUs. Then, the closest OTUs are first joined, but the distances to the next closest OTUs also are used to place the node in the most mathematically appropriate position. Subsequent OTUs are then joined according to their proximity to the partial tree.

the trees are more accurate. As with UPGMA, the computation begins by calculating the distances between every possible pair of sequences. Again, various models can be applied, including those that provide weighting for transitions and transversions. After producing a distance matrix from the sequence matrix, NJ then joins the closest two OTUs together. However, when it joins them, it also considers the distances between these two OTUs and the next closest OTUs. In this way, it can place the node more accurately that does in UPGMA. It then joins the next closest OTU with the initial OTUs based on its original calculation of the distances from those two OTUs. Thus, NJ produces a phylogram, rather than a cladogram (although for most software packages, a cladogram option is provided). NJ is used frequently for moderate and large datasets because it is rapid due to its relatively simple calculations.

Maximum Parsimony (MP)

Parsimony simply means the simplest explanation. For phylogenetic analysis, it means finding the tree that requires the fewest sequence changes for the entire dataset. Maximum parsimony (MP) differs significantly from UPGMA and NJ, which are based on the initial distance calculations. MP begins by creating all of the possible ways to join the OTUs together into bifurcating trees (Figure 24.5). It then searches through the trees to determine which is the shortest tree based on the sequence changes derived by evaluating the sequence alignment data. The number of possible trees increases rapidly as the number of sequences (i.e. OTUs) increase (Figure 24.6), and therefore, this method can be used only for datasets that are small to moderate in size. There are three main search strategies that can be used in MP. The first is an exhaustive search, where every possible tree is calculated and the MP tree(s) is reported. This takes a great amount of computing time and therefore is usually limited to evaluation of a few dozen sequences. Another method is called a heuristic search that begins with a randomly selected tree, and then evaluates

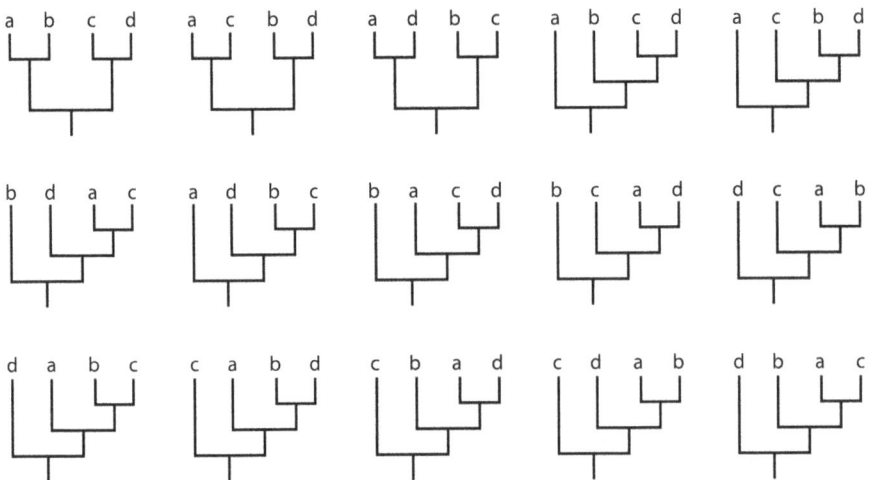

FIGURE 24.5 All possible ways of joining four OTUs for trees rooted at one point for MP.

Joining 4 sequences - unrooted trees

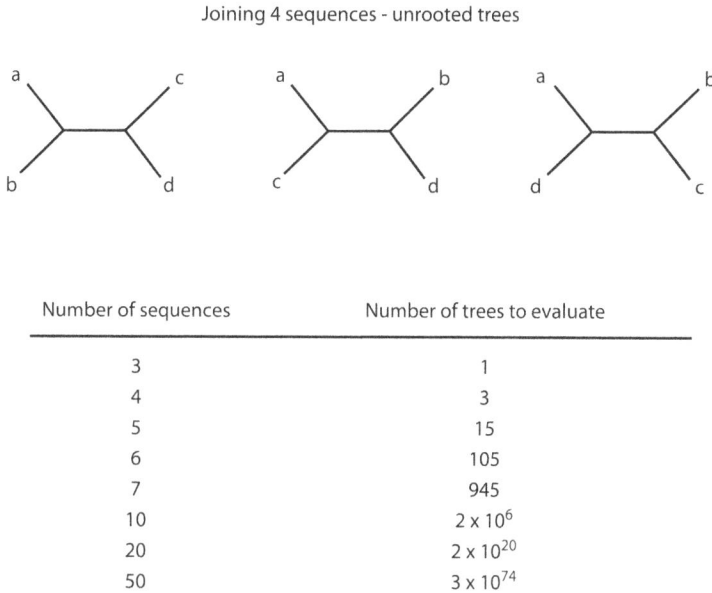

Number of sequences	Number of trees to evaluate
3	1
4	3
5	15
6	105
7	945
10	2×10^6
20	2×10^{20}
50	3×10^{74}

FIGURE 24.6 Numbers of unrooted trees evaluated based on the number of sequences (OTUs) in the dataset.

adjacent trees, choosing the one that is the shortest. While this finds the local MP tree, sometimes it does not find the true MP tree for that dataset. The third method can be termed a semi-exhaustive method, because it finds the MP tree, but does not examine every single tree. It is out of the scope of this chapter to explain here and is described elsewhere in detail. Although this is a good method for determining the MP tree, and is much faster than an exhaustive search, it still requires a great deal of computing time, and therefore, can be used only for small datasets (50–100 sequences).

Maximum Likelihood (ML)

This method is based on the statistical likelihood of sequence differences as determined by the frequencies of changes of each of the types of bases (or amino acids). The algorithm either will determine this based on the frequencies of each base in the dataset provided or the user can provide a model for base pair change and frequencies (Figure 24.7). Then, the natural log (ln) of the likelihood (L) is calculated for the branches joining each of the OTUs together, based on the changes in the sequence along those branches, beginning with the two closest OTUs, and then joining the next closest OTU sequentially. The ln likelihood of the entire tree is the sum of the ln likelihoods of each of the branches. One of the other useful capabilities of maximum likelihood (ML) is in testing hypothetical trees. User supplied trees can be compared to the ML tree to determine whether the hypotheses are supported or rejected, based on likelihood calculations and comparisons. That is, one can test whether the data support the identity of a sample sequence being that of a known species, variety,

e.g. AGCTCCC ——————————————→ AGCTTTT

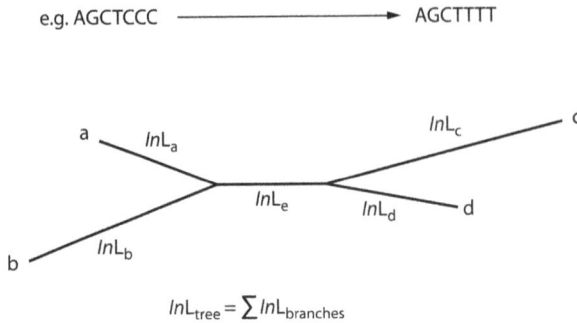

$$lnL_{tree} = \sum lnL_{branches}$$

FIGURE 24.7 Maximum likelihood. A distance matrix is used to join the OTUs beginning with the closest two. Then, the probabilities of each branch are calculated based on the nucleotide changes along that branch. The overall lnL of the tree is the sum of the lnL of the branches.

isolate, individual, tumor type, or cell type (dependent on the genomic region being examined). While this is a powerful statistical method, it requires a great deal of computing power. Because of this, it can only be used for small datasets (usually a maximum of a few dozen sequences).

Bayesian Phylogenetic Methods

Bayesian inference is conceptually analogous to ML, because it relies on the evaluation of probabilities of sequence change along the branches of the trees. However, the search strategies and calculations involved are different than those for ML. In general, this method examines one tree topology based on a previously determined tree that was established previously. The model used to produce the trees can be supplied by the user, or the Bayesian program can produce a model based on the nucleotide frequency and/or a default model. It inches its way through various tree possibilities, testing a current tree with one other tree, determining whether or not it is of higher or lower probability, based on the mutational model chosen. In doing so, it normally evaluates millions of possible trees. As a statistically-based method, it can be used to determine the reliability of the final tree, and it can be used to evaluate and compare more than one tree.

PHYLOGENOMICS AND COMPARATIVE GENOMICS

The rate of generation of genome sequences has been increasing rapidly during the past several decades. Much of this is due to the expansion of next generation sequencing methods. Thousands of complete genome sequences are now available worldwide, improving resolution of taxonomic information for bacteria, archaea, eukaryotes, and viruses. Because of this, it is possible to use a wide variety of genomic regions for comparisons at many different taxonomic levels. Comparative Genomics can be used for identification purposes, as well as many other applications. Computational improvements have also helped greatly to allow comparisons among multiple genomes.

For standard phylogenetic analyses, generally a single gene locus (or concatenated sequences from several gene loci) is being compared among dozens, or sometimes hundreds of taxa (usually tens of thousands or hundreds of thousands of nucleotides per dataset). In phylogenomic[9] and comparative genomics, dozens or hundreds of gene loci are being compared among dozens of taxa (millions of nucleotides in the data set). This represents a change of one to three orders of magnitude in the amount of data to analyze. However, many of the analytical bioinformatics tools used to perform phylogenetic analyses use calculations that increase exponentially as the data set increases in size. Therefore, many of the phylogenetic methods that have been successfully applied to relatively small phylogenetic studies cannot be applied to genome-sized comparisons. For example, MP and ML methods become difficult or impossible to perform on datasets larger than about 10,000 to 100,000 nucleotides in size. While these are accurate phylogenetic methods for smaller datasets, they cannot be easily applied to genomic alignments. However, NJ has been successfully used in many phylogenomic projects. It requires much less computer processing time, because its algorithms are purely computational, and it does not use any searching functions. Bayesian methods have also been used in phylogenomic and comparative genomics research. Although they use search functions that are iterative in nature, the computation time required is less than for MP and ML. Therefore, the advantage of these methods is that it is statistically robust that is similar to ML in this respect.

COMPARISONS

The problems of data set size are limited somewhat because of three factors: repetitive regions, nonhomologous regions, and variations in gene mutation. Repetitive regions are difficult to resolve by NGS, because many of the regions consist of dozens to thousands of identical copies of two to hundreds of base pair units. The reads in NGS results are often too short to encompass the entire region, and while the computer can easily locate many overlaps, often it cannot determine how many repeats constitute the entire region. When the computer attempts to assemble the sequenced fragments into longer contiguous sequences (i.e., contigs), the precise beginnings, middles, and ends of the regions cannot be determined because all of the pieces have regions that are identical with those of other pieces. Also, because of frequent recombination events,[10–12] the number of repeats often changes from cell to cell, such that the DNA that is being sequenced contains many different versions of the repetitive regions. They also may change as a result of DNA polymerase slippage during replication. Therefore, these regions often are left unfinished in the final genomes, and for comparative genomic and phylogenomic studies, they are excluded from the dataset.

Size of the dataset is also limited because some regions cannot be aligned. This often is dependent on relatedness of the sequences/taxa. That is, closely related organisms will have highly similar genomes, and therefore more of the genomes will be amenable to alignment. However, when the organisms are from different domains, or when regions of the genomes have rapid rates of evolution, the sequences have much less in common, and therefore many regions of the genomes must be excluded from analyses, because they cannot be accurately or reliably aligned. A caveat to this is that the most ancient genes are common to most organisms, and therefore, for broad-range phylogenomic studies, primarily ancient genes are analyzed. This necessarily limits the analyses to highly conserved genes, and therefore, the results can only be indicative of evolutionary events that have occurred over billions of years.

The datasets that ultimately can be used for genomic comparisons are reduced somewhat when it contains only bacterial genomes, for many of the reasons outlined above. However, when comparing bacterial and archaeal genomes, the datasets are further reduced, due to their evolutionary distance. The datasets are reduced considerably more for eukaryotic genome comparisons, where only from 1–10% of the genome is composed of mRNA genes, and for comparisons among bacteria, archaea, and eukarya, where the number of genes in common to all is less than the number in each genome. Nonhomologous regions, including introns, also are often disregarded in comparative and phylogenomic studies, unless they are specifically aimed at the study of the introns, splicing, alternative splicing, and/or products (i.e., RNAs) from the introns.

When examining genomic sequences from different organisms, many regions cannot be aligned either because one or more genes are not present in one of the genomes, or the sequences are so different that they cannot be unambiguously aligned. Part of the reason for this is that not all genes, or all regions of each gene, mutate at the same rates, and some organisms may experience large genomic changes compared to other organisms. Punctuated evolutionary events are one of the causes of these differences. Because of this, phylogenomic analyses tend to result in averages of the overall rates of change in each of the sequences, and may fail to indicate some changes that have occurred in specific loci in a particular genome. However, although phylogenomic methods usually cannot utilize complete genomes, and they may miss some changes, due to the sheer volume of nucleotides in each sequence, they provide yet another tool to analyze the evolutionary events that have led to the diversity of life on Earth. Additionally, a phylogenomic analysis analyzes tens or hundreds of genes at a time, and can indicate changes in a number of genes that would require much more time to evaluate were the genes to be analyzed individually using phylogenetic analyses. This has implications when attempting to identify at the species level, where a more limited set of gene regions must be used because they are changing rapidly enough that they differ at the species level, as well as lower taxonomic levels.

SINGLE-NUCLEOTIDE POLYMORPHISMS (SNPS)

Within any populations, and within any multicellular organism, mutations constantly occur. In most instances, they are single nucleotide changes and they cause no deleterious or beneficial effects. Throughout a population, these changes can be mapped as polymorphisms in the population and can be compared to sets of polymorphisms in other populations (or among individuals or cells within an individual). Maps of SNPs have been constructed for the human population (and this work goes on) using DNAs from individuals around the world, such that millions of such SNPs have been recorded and stored in databases. One of the popular uses of SNPs is that of determining ancestry of individuals. One can submit a saliva sample to a company or research lab and the SNP profile can be determined. From the profile it can be determined which SNPs are in the samples that correlate with those from specific populations around the world. For example, a typical sample might indicate that the DNA sample is 40% similar to profiles from Northern Europeans, 30% is similar to profiles from Scandinavians, 20% is similar to profiles from Central Africa, and the remaining 10% is a mixture from various locations in Europe and Asia. SNP profiles are also used in the diagnoses of specific diseases, including malignancies that has led to more precise diagnoses and targeted therapies.

MICROSATELLITES AND MINISATELLITES

Like SNPs, microsatellites and minisatellites are widespread among eukaryotes and have been used to categorize individuals in populations, as well as to characterize populations. Microsatellites consist of short (2–6 bp) tandem repeats, flanked by unique sequences; while minisatellites are tandem copies of 10 bp to over 100 bp, which also are flanked by unique sequences. The number of tandem repeats varies from a few to hundreds in a single locus. Because of their repetitive nature, they are subject to frequent recombination events and DNA polymerase slippage, both of which have the potential to produce variants of the original locus. While they first were classified based on hybridization methods, more recently, molecular methods including PCR, cloning and sequencing have been used for various types of studies. In this sense, they can be used in a phylogenomic studies to identify the origin of individuals within populations, and to infer genetic and geographical origins of individuals. They also have been used to correlate blood samples with suspects in criminal cases. Some regions change so rapidly that the probability of two humans having identical repetitive regions is less than one in several billion. Analogous studies can identify individual trees (and other plants) from microsatellite and minisatellite sequence in their leaves and other plant parts. These have also been used in judicial forensics cases to determine whether the suspects were at a crime scene by examination of the DNA from plant specimens (e.g., leaves, needles, pollen) on a suspect's clothing or automobile.

HOW TO COMPARE

Some of the basic models used for comparisons of single genes can be used for comparing genomes as those used for comparing single genes. For example, the Jukes-Cantor model can be used where the bases all have the same probability of mutating into any other base. More accurate models use different modes of change, weighting transversions more heavily than transitions, because transitions occur more often than do transversions, and C to T changes occur with the highest frequencies. The same considerations that need to be addressed in phylogenetic analyses apply to phylogenomic analyses. That is, homologous regions must be properly aligned, orthologs must be differentiated from paralogs, and regions of sequence ambiguity must be removed from consideration. The differences primarily relate to the sizes of the databases. As mentioned above, this creates some issues regarding the types of analyses that can be performed. Although the models and algorithms utilized in ML and MP are rigorous, they are difficult to perform on very large datasets, due to the number of calculations involved. The primary method of analysis with very large datasets is NJ, as well as Bayesian analyses.

COMPARATIVE GENOMICS

Studies of closely related species can help to clarify both their taxonomic and phylogenetic relationships. For example, rather than being a single species, the taxonomic designation, *Escherichia coli*, is a species complex, and is closely related to species of *Shigella* species. Comparative genomic studies of members of these taxa exhibit a great deal of similarity in their gene contents, but also show a large number of genes unique to each strain or species. The strain *E. coli* K-12 (which is nonpathogenic), is similar but very different from *E. coli* O157:H7 (a virulent pathogen), but it is more similar to *Shigella flexneri* 2a (a pathogen). Additionally, the *E. coli* K-12 genome is 4.6 million base pairs in length that includes nearly 4,300 genes. However, when all *E. coli* strains they can vary by as much as 50% length, and the pan-genome includes more than 16,000 genes. Therefore, in some cases, the molecular comparisons are more informative than determinations from morphological and physiological investigations. Plant genomes have also been used in comparative genomics and phylogenomic studies. Plants have genomes that generally are from about one-eighth the size of the human genome to more than one thousand times larger than the human genome. They also have more genes, but less alternative splicing, and because of this, the total number of proteins that their genomes encode is less than for the human genome, but generally more than for other species. In a comparative genome study of three grass (monocot) species and a eudicot species (*Arabidopsis thaliana*), 55% of the gene families were in common to all four genomes. Another 13% of the gene families were in common only among the grass genomes, indicating their closer phylogenetic relationships relative to *A. thaliana*. Comparative genomics and phylogenomics can help to fine tune our understanding of these organisms that are closely related.

Another mode of comparison is to investigate the amount of synteny (identical gene order along the chromosomes) among the sequences being compared. Chromosomes change relatively rapidly both in size, gene order, and gene content. In many cases, this does not appear to affect the expression of the genes, although in some extreme cases, such as in many cancers, chromosomal translocations are known to cause aberrant gene expression. The mechanisms for chromosomal translocations are primarily unequal crossovers within a single chromosome, or between two chromosomes, either homologs or nonhomologous. The crossovers generate differences in gene order, reversals of gene order (caused by inversions), interruption of vital genes, exchange of parts of genes, and transfers of genes from one chromosome to another. In addition to the transfers of genetic material from one chromosomal location to another, several other changes can occur. Duplications and deletions are frequently found, the positions of the centromeres can change, and chromosomes can split or fuse. These change the chromosome number, but not necessarily the number of genes or genome size. While many of these changes do not change the expression of the genes in the genome, occasionally small to large changes occur. Often this is caused by movement of genes into regions where the chromosome architecture or the local promotors cause changes in the expression of genes in the immediate area of the chromosomal translocation. For example, some cancers are caused by translocations that move constitutive promotors close to normally tightly regulated genes.

CONCLUSIONS

Sequences can be aligned and compared in order to determine fine-scale identification of samples from clinical cases, forensic investigations, or for other purposes. Purely mathematical methods require less computing resources, but can be used for large datasets. Rigorous statistical methods yield highly accurate results, but can be computationally expensive. Phylogenetic, comparative genomics, and phylogenomic methods can produce informative results that can lead to accurate and timely determination of samples that are useful in clinical and forensic analyses that can lead to accurate and timely diagnoses and treatments, as well as information leading to resolution of criminal and other investigations.

REFERENCES

1. Claverie, J.-M.; Notredame, C. *Bioinformatics for Dummies*, 2nd ed. New York, NY: Wiley Publishing, Inc., **2006**.
2. Graur, D.; Li, W.-H. *Fundamentals of Molecular Evolution*, 2nd ed. Sunderland, MA: Sinaur Associates, Inc., Publishers, **2000**.
3. Higgs, P.H.; Attwood, T.K. *Bioinformatics and Molecular Evolution*. Carleton, Victoria, Australia: Blackwell Publishing, **2005**.
4. Li, W.-H. *Molecular Evolution*. Sunderland, MA: Sinaur Associates, Inc., **1997**.
5. Li, W.-H.; Graur, D. *Fundamentals of Molecular Evolution*. Sunderland, MA: Sinaur Associates, Inc., **1991**.
6. Mount, D.W. *Bioinformatics, Sequence and Genome Analysis*, 2nd ed. Cold Spring Harbor, NY: Cold Spring Harbor Laboratory Press, **2004**.
7. Rogers, S.O. *Integrated Molecular Evolution*. Boca Raton, FL: CRC Press, Taylor & Francis Group, **2017**.
8. Prevsner, J. *Bioinformatics and Functional Genomics*, 2nd ed. Hobocken, NJ: Wiley-Blackwell, **2009**.
9. DeSalle, R.; Rosenfeld, J.A. *Phylogenomics, A Primer*. New York, NY: Garland Science (Taylor and Francis), **2013**.
10. Rogers, S.O.; Bendich, A.J. Recombination in *E. coli* between cloned ribosomal RNA intergenic spacers from *Vicia faba*: a model for the generation of ribosomal RNA gene heterogeneity in plants. Plant Sci. **1988**, *55*, 27–31.
11. Rogers, S.O.; Bendich, A.J. Heritability and variability in ribosomal RNA genes of *Vicia faba*. Genetics **1987**, *117*, 285–295.
12. Rogers, S.O.; Bendich, A.J. Ribosomal RNA genes in plants: variability in copy number and in the intergenic spacer. Plant Mol. Biol. **1987**, *9*, 509–520.

25 DNannotator

Annotation Software Tool Kit for Regional Genomic Sequences

Chunyu Liu
University of Chicago, Chicago, Illinois, U.S.A.

CONTENTS

INTRODUCTION

Genome annotation is a process of identifying the location and function of genes and many other elements in a genome. Genome annotation documents the true meaning of the genome sequence and serves ongoing biomedical research that is based on genomics information. Public annotation efforts and data sources have some limitations. High-throughput technologies used in many labs produce more and more local data at an increasing rate. Translating discoveries in a local laboratory into genome annotation in a manner that integrates the local data with the public data is another recurrent challenge that requires general software support. DNannotator provides a series of software tools for local batch annotation. The annotation target can be either user's own genomic DNA (gDNA) sequence or public-assembled human chromosomal sequences. The annotation source can be genes, oligos/primers, sequence tag sites (STSs), and single nucleotide polymorphisms (SNPs). Annotation quality can be relatively easily managed as several log data files are provided by DNannotator.

DOI: 10.1201/9781003247432-25

251

BACKGROUND

Genome annotation is a process that involves procedures in both the wet and the dry lab. The wet lab performs experiments, identifies genes, polymorphisms, and other genomic elements, and studies genes' functions. The dry lab maps those identified elements onto genomic sequences. In this article, we restrict the term of "annotation" to dry lab sequence mapping. The terms annotation and mapping can be used alternatively here. To perform an annotation, two basic inputs are required: gDNA sequence data and interesting genomic elements (feature source data) such as genes and markers. Based on the ATGC sequence data, annotations of genes and their regulatory elements can reveal the true functional units of the genome. Annotations of other elements such as genetic markers, oligos, and primers are essential for genetic studies as they are the basis of many genetic experiments.

Ever since the sequence data were obtained from the human genome sequencing project, scientists started to perform *large-scale systematic genomewide annotation* on those sequence data. Today, the human genome continues to be analyzed. The human genomic sequences were assembled by National Center for Biotechnology Information (NCBI) and Celera. The sequence data have been annotated by several public or private institutes using their own tools and data sources. The data can be accessed at web-based databases, including NCBI's Map Viewer (http://www. ncbi.nlm.nih.gov/mapview/map_search.cgi?taxid=9606), University of California Santa Cruz's (UCSC's) Genome Browser (http://genome.ucsc.edu/) and Sanger Center's Ensembl (http://www.ensembl.org/), and Celera (http://www.celera.com/). For the public annotation, both gDNA sequences and the genomic feature source data are available. For example, the human chromosomal sequences can be found at NCBI's human genome database; human gene data can be found at either Genbank or Unigene database; SNP data can be accessed at dbSNP.

In the course of many research projects, investigators need to perform *regionwide local annotation*. As lab technologies have improved, especially with the introduction of high-throughput experimental methods, many investigators produce large amounts of experimental data and knowledge on genes, regulatory elements, and variants. For example, one ABI 3730 sequencing platform can resequence 1.8-Mb of sequence in 1 day. Hundreds of SNPs can be discovered in 1 day. Besides the data generated in the investigators' own lab, many new data can be found in the latest journal publications. Often enough, these new data cannot be found at the time of publication in the existing public genomic annotation databases because of the significant time gap between the time of data deposit to public databases and the time of releasing an updated genome annotation using the corresponding data, assuming all the data are collected by the public databases. Normally, this kind of delay would be more than 3 months. Unfortunately, this assumption about public data collection is not always true either. SNP data are not required to be deposited into any centralized database when a paper is published. It is frequently observed that some SNPs studied or discovered are absent in dbSNP. But dbSNP is the only SNP database used in all public annotation. An effective research design requires efficient use of existing knowledge. Hence, data produced in an individual investigator's lab and all other data outside of public annotation databases are as valuable as data from public annotation. They need to be integrated with data existing in the current version of public genome annotation. This basic requirement poses an important challenge to current bioinformatics technology because of the following major problems:

1. Heterogeneous annotation target sequence or platform. Because of the human genome sequence assembly, annotation is kept updated at a speed of several months per release; different annotation databases could use different versions of sequence assembly at certain time period. Investigators may want to use different gDNA sequences either for a small regional sequence or a whole chromosome. Sequence assembly is difficult for some genomic regions because of enriched repeat sequences or difficulty of cloning in the public sequencing project. An individual investigator might have better quality sequence assembly of one region than the public assembly. All these different flavors of gDNA sequences created needs of user's local annotation.

2. Enormous amount of data. As described above, an individual lab could generate or collect large amount of data for local annotation.

3. Heterogeneous data formats. Different data source could use very different data format. Preparing input data for annotation tools and formatting output data to meet requirement of data integration and further data mining are not trivial.

4. Annotation quality control. For any intent of feature mapping, the results can be either success or failure. The annotation quality is highly dependent on the sensitivity and specificity of the mapping methods and parameter settings. Any simple bug in the annotation algorithm could cause major defects of the annotation results. The observed problems in public annotation include missing of features which should be mapped, wrong mapping positions, and so on. Because the current public annotation does not provide much log information, it is difficult to detect errors and to find out the causes.

There are a number of tools supporting regionwide local annotation. They can be roughly put into three categories:

1. Use public source data (stored in databases) and some gene prediction algorithm to annotate user's gDNA sequence. Genotator,[1] NIX (Williams et al., http://www.hgmp.mrc.ac.uk/Registered/Webapp/nix/), GeneMachine,[2] GAIA,[3] Alfresco,[4] GESTALT,[5] RUMMAGE,[6] and Oak Ridge National Laboratory Genome Analysis Pipeline (http://compbio.ornl.gov/tools/pipeline) provide integrated annotation, including mapping of known or predicted genes and/or regulatory elements by running multiple gene-prediction programs and searching against static public databases. But none of them incorporate methods for SNP mapping, which is essential for positional cloning projects for complex diseases. None of these systems takes source data supplied by the end user, unless the user can modify the databases in the annotation system.

2. Annotaor's own source data to user's own gDNA sequence. Freeware, such as Artemis,[7] Sequin,[8] and some commercial software such as Vector NTI, provide a good interface to do manual annotation. They do not support batch annotation. Some other programs can be used to assist annotation. For example, BLAST[9] is good at homolog sequence searching. Sim4,[10] est_genome,[11] and Spidey[12] could be used to define the intron-exon structure of a gene. e-PCR[13] can be used to map STSs. However, most of these programs produce data in their own formats which cannot be directly converted into standard format annotation. Therefore, strictly speaking, they are not real annotation tools.

3. DNannotator can use user's own collection of source data (either from public places or generated in the labs) to annotate both user's own gDNA sequence and public chromosomal sequences. DNannotator complements the existing tools mentioned above. It is the first toolkit providing SNP mapping and the only one with the capability to migrate annotations from one sequence platform to another. DNannotator was first described in *Nucleic Acids Research* (2003).[14] It can be accessed at http://sky.bsd.uchicago.edu/DNannotator.htm.

DESIGN AND FUNCTION OF DNANNOTATOR

DNannotator takes annotation source data, such as SNPs, genes, primers, etc., prepared by the user, and/or a specified target of gDNA, and performs de novo annotation. DNannotator can also robustly migrate existing annotations in Genbank format from one sequence to another given that the new sequence covers the same genomic region. The annotation migration function is useful when we are dealing with different versions of sequence assembly or different scope of a region, e.g., one is small regional sequence and another is whole chromosome sequence. The major functions of DNannotator are illustrated in Figure 25.1. The functions of DNannotator are divided into two groups: one for annotation over user's own gDNA sequence and another for annotation over chromosomal sequence (from UCSC Genome Browser's latest version human genome sequence assembly). Both groups can handle similar types of input source data. The input source data can be SNPs, genes, primers/oligos, and STSs. Besides de novo annotation using raw source data, annotated sequence data in Genbank format can be used as input to perform annotation migration, transferring annotated features to another sequence platform—either user's own gDNA or a public chromosomal assembly.

There are four types of optional outputs generated by DNannotator: (1) annotation results in Genbank format; (2) annotation results in tab-delimited text as feature table; (3) reference data reporting features failed at annotation, or features mapped to multiple locations, as well as some sequence analysis raw data such as BLAST's results; (4) besides those, for annotation over a public chromosomal sequence, a GFF format feature data can be provided.

Tab-delimited text output can be imported and managed in a database or spreadsheet and can be combined easily with existing annotation from elsewhere. Genbank format data can be graphically viewed in Artemis (Figure 25.2). GFF format

Major Functions of DNannotator

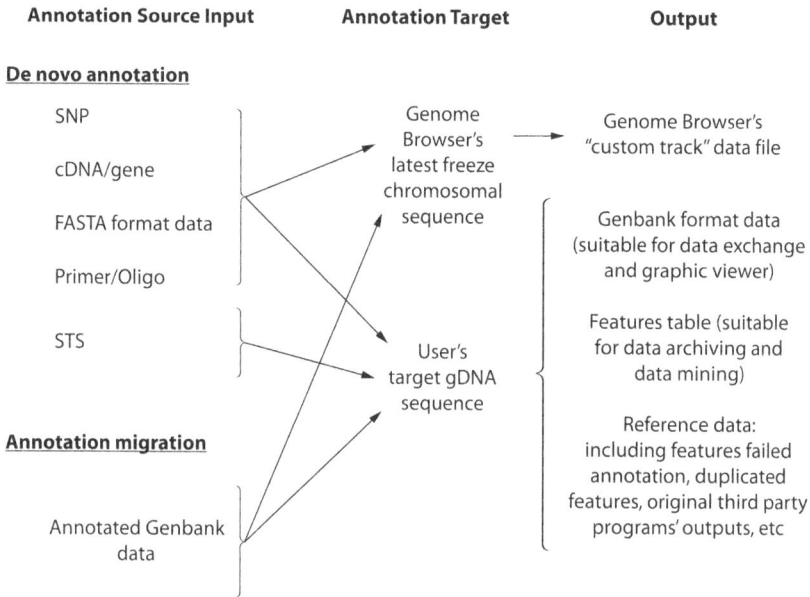

Annotation Source Input	Annotation Target	Output

De novo annotation

SNP

cDNA/gene

FASTA format data

Primer/Oligo

STS

Annotation migration

Annotated Genbank data

Genome Browser's latest freeze chromosomal sequence

User's target gDNA sequence

Genome Browser's "custom track" data file

Genbank format data (suitable for data exchange and graphic viewer)

Features table (suitable for data archiving and data mining)

Reference data: including features failed annotation, duplicated features, original third party programs' outputs, etc

FIGURE 25.1 Major functions of DNannotator. Using input source data of STSs, SNPs, primers, general FASTA, or tab-delimited text data, de novo annotation can be performed on target gDNA sequences through different function modules of DNannotator. A single Genbank format data file can be used as input to perform annotation migration. Similar inputs and outputs are used for the two types of annotation targets.

annotation results can be viewed in Genome Browser side by side with other annotation provided by Genome Browser (Figure 25.3). Reference data can help user to evaluate and improve annotation quality.

MAJOR FEATURES OF DNANNOTATOR

DNannotator has four major features: "local," "batch," "transparent," and "user- and data-friendly."

1. *Local*: All functions of DNannotator were designed to handle the user's own local data collection.
2. *Batch*: All functions were designed to process batch of input source data.
3. *Transparent*: Log data are provided with the annotation results, so that the annotation quality can be managed. Some major analysis parameters can be adjusted through the web interfaces.
4. *User- and data-friendly*: DNannotator not only provides a user-friendly web interface for data input but also accepts data on a large scale needed by a research data pipeline. The input data normally can be prepared or

(a)

(b)

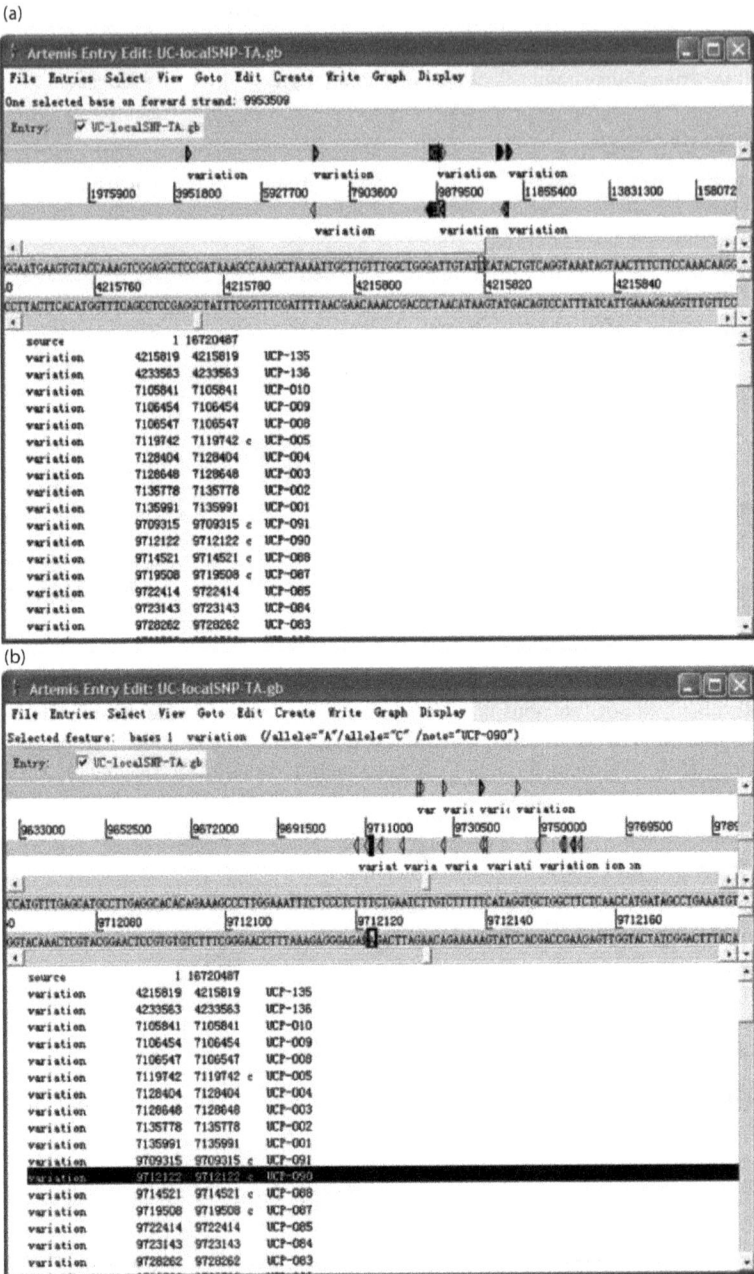

FIGURE 25.2 Graphic view of SNPs annotated onto local sequence assembly. Genbank format output from DNannotator was read in as "Entry" by Artemis. One hundred SNPs discovered locally were displayed. (a) Overview of annotations. (b) Zoom-in view of annotations. The window is composed of three panels. Features listed in lower panel, graphic icons in upper panel, and sequence data in middle panel are internally linked to each other. Clicking one of them will activate highlight of corresponding elements in the other panels.

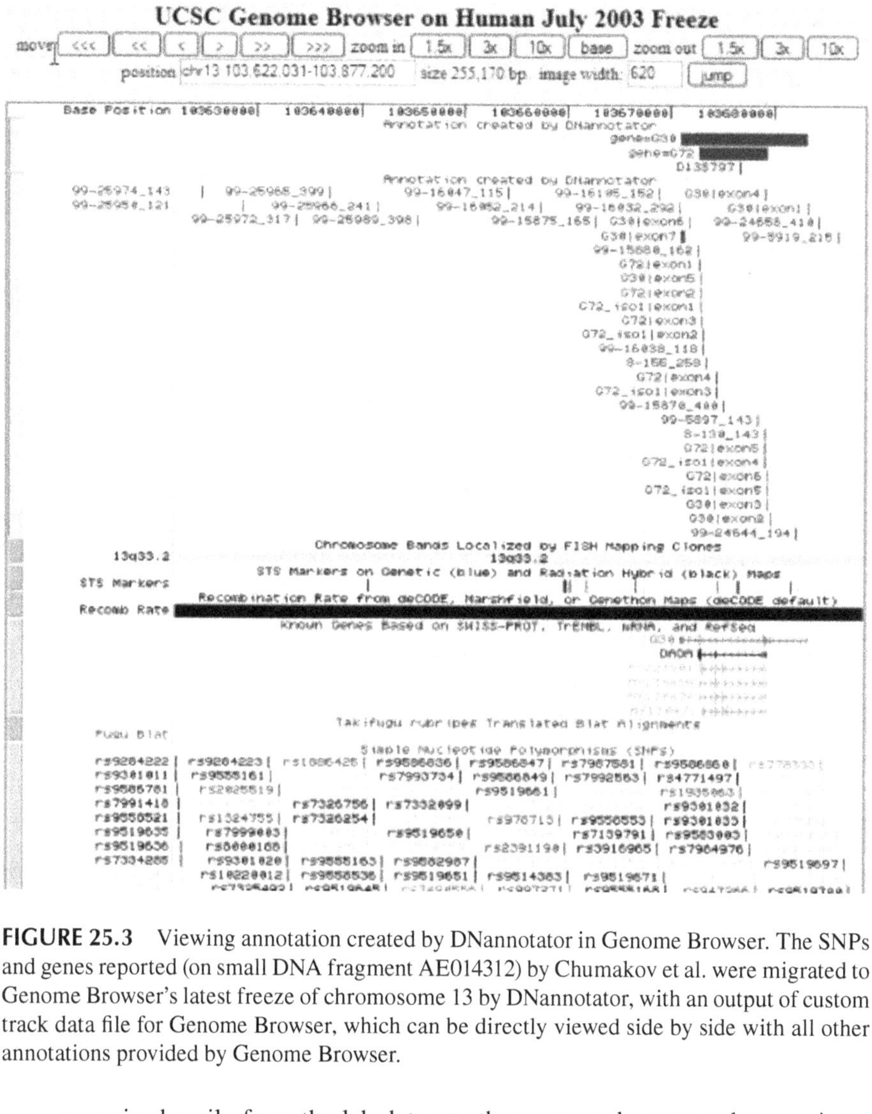

FIGURE 25.3 Viewing annotation created by DNannotator in Genome Browser. The SNPs and genes reported (on small DNA fragment AE014312) by Chumakov et al. were migrated to Genome Browser's latest freeze of chromosome 13 by DNannotator, with an output of custom track data file for Genome Browser, which can be directly viewed side by side with all other annotations provided by Genome Browser.

organized easily from the lab data or other source; the output data are in multiple optional formats either ready for graphic view, or for databasing, or for further data mining or wet-lab assay design.

APPLICATION EXAMPLES OF DNANNOTATOR

DE NOVO ANNOTATION

Example 25.1: From local discovery or newly published data to genome annotation.

In the lab, we sequenced several genes and discovered 100 SNPs. At that time (early 2002), public assembly of human gDNA sequence on 13q32–33 still had a lot of errors. We manually assembled a 17-Mb regional sequence with much better quality, which was later replicated by a subsequent public assembly. DNannotator was then used to annotate all 100 SNPs on this self-made sequence assembly. SNP data were prepared using Microsoft Excel spreadsheet and saved into a tab-delimited text file. DNannotator takes this SNP data file with gDNA sequence in FASTA format and maps all the SNPs to the self-made regional sequence. The Genbank format output is viewed in Artemis as shown in Figure 25.2.

ANNOTATION MIGRATION

Example 25.2: From newly published Genbank data to genome annotation.

Chumakov et al.[15] reported association between the G72 gene and schizophrenia. In that paper, 191 SNPs and two novel genes (G72 and G30) on chromosome 13 were reported. Most of these data could not be found in public annotation. Chumakov et al. deposited a series of Genbank format sequence data for chromosome 13 fragments. DNannotator was used to migrate annotations from these fragments to public assembly of chromosome 13 to find out how these SNPs and genes are mapped on human chromosome 13 and how these SNPs and genes related to the other SNPs and genes. As an example, AE014312 is one of the Chumakov's deposit with 250-kb sequence and contains 20 SNPs, 2 genes, and 1 STS. DNannotator takes this input and produced annotation on human public assembly chromosome 13. All features (SNPs, genes, and STSs) were mapped to chromosome 13 in one step. The GFF format data are viewed in Genome Browser as in Figure 25.3. In this figure, we can view all annotations provided by Chumakov's group with all the other data created by public annotation.

CONCLUSIONS

DNannotator provides services of batch local annotation, which are not available elsewhere. With these tools, user can map SNPs, genes, and STSs to user-specified gDNA sequences including the public-assembled human chromosomal sequences. Thus, integration of genomic data can be performed easily. DNannotator incorporates the ideals of "batch," "local," "transparent," and "user and data-friendly" throughout its tool kits. Lack of these four features is the major shortcoming of many existing bioinformatics tools. We often see new tools handling only a single request, producing outputs without any log information, or ignoring user's own data.

Because of limits of computing power, DNannotator currently only annotates on a single human chromosome per analysis. Therefore, the user could meet an obstacle if the source data's chromosomal source is unknown or it is a mixture from multiple chromosomes. We are working toward having biogrid as our computing platform. Once implemented, DNannotator should be able to perform genome-wide annotation.

ACKNOWLEDGMENTS

This work was supported by Young Investigator Grant to Chunyu Liu from National Alliance for Research in Schizophrenia and Affective Disorders (NARSAD) and NIH R01 MH65560-01 and R01 MH59535 to Elliot S. Gershon. Support from the

Geraldi Norton Memorial Corporation, Mr. and Mrs. Peterson, the Eklund Family, and Anita Kaskel Roe are each gratefully acknowledged.

REFERENCES

1. Harris, N.L. Genotator: A workbench for sequence annotation. Genome Res. **1997**, *7* (7), 754–762.
2. Makalowska, I.; Ryan, J.F.; Baxevanis, A.D. GeneMachine: Gene prediction and sequence annotation. Bioinformatics **2001**, *17* (9), 843–844.
3. Bailey, L.C. Jr.; Fischer, S.; Schug, J.; Crabtree, J.; Gibson, M.; Overton, G.C. GAIA: Framework annotation of genomic sequence. Genome Res. **1998**, *8* (3), 234–250.
4. Jareborg, N.; Durbin, R. Alfresco—A workbench for comparative genomic sequence analysis. Genome Res. **2000**, *10* (8), 1148–1157.
5. Glusman, G.; Lancet, D. GESTALT: A workbench for automatic integration and visualization of large-scale genomic sequence analyses. Bioinformatics **2000**, *16* (5), 482–483.
6. Taudien, S.; Rump, A.; Platzer, M.; Drescher, B.; Schattevoy, R.; Gloeckner, G.; Dette, M.; Baumgart, C.; Weber, J.; Menzel, U.; Rosenthal, A. RUMMAGE—A high-throughput sequence annotation system. Trends Genet. **2000**, *16* (11), 519–520.
7. Rutherford, K.; Parkhill, J.; Crook, J.; Horsnell, T.; Rice, P.; Rajandream, M.A.; Barrell, B. Artemis: Sequence visualization and annotation. Bioinformatics **2000**, *16* (10), 944–945.
8. Benson, A.; Boguski, M.S.; Lipman, D.J.; Ostell, J. GenBank. Nucleic Acids Res. **1997**, *25* (1), 1–6.
9. Altschul, S.F.; Gish, W.; Miller, W.; Myers, E.W.; Lipman, D.J. Basic local alignment search tool. J. Mol. Biol. **1990**, *215* (3), 403–410.
10. Florea, L.; Hartzell, G.; Zhang, Z.; Rubin, G.M.; Miller, W. A computer program for aligning a cDNA sequence with a genomic DNA sequence. Genome Res. **1998**, *8* (9), 967–974.
11. Mott, R. EST_GENOME: A program to align spliced DNA sequences to unspliced genomic DNA. Comput. Appl. Biosci. **1997**, *13* (4), 477–478.
12. Wheelan, S.J.; Church, D.M.; Ostell, J.M.; Spidey: A tool for mRNA-to-genomic alignments. Genome Res. **2001**, *11* (11), 1952–1957.
13. Schuler, G.D. Electronic PCR: Bridging the gap between genome mapping and genome sequencing. Trends Biotechnol. **1998**, *16* (11), 456–459.
14. Liu, C.; Bonner, T.I.; Nguyen, T.; Lyons, J.L.; Christian, S.L.; Gershon, E.S. DNannotator: Annotation software tool kit for regional genomic sequences. Nucleic Acids Res. **2003**, *31* (13), 3729–3735.
15. Chumakov, I.; Blumenfeld, M.; Guerassimenko, O.; Cavarec, L.; Palicio, M.; Abderrahim, H.; Bougueleret, L.; Barry, C.; Tanaka, H.; La Rosa, P.; Puech, A.; Tahri, N.; Cohen-Akenine, A.; Delabrosse, S.; Lissarrague, S.; Picard, F.P.; Maurice, K.; Essioux, L.; Millasseau, P.; Grel, P.; Debailleul, V.; Simon, A.M.; Caterina, D.; Dufaure, I.; Malekzadeh, K.; Belova, M.; Luan, J.J.; Bouillot, M.; Sambucy, J.L.; Primas, G.; Saumier, M.; Boubkiri, N.; Martin-Saumier, S.; Nasroune, M.; Peixoto, H.; Delaye, A.; Pinchot, V.; Bastucci, M.; Guillou, S.; Chevillon, M.; Sainz-Fuertes, R.; Meguenni, S.; Aurich-Costa, J.; Cherif, D.; Gimalac, A.; Van Duijn, C.; Gauvreau, D.; Ouellette, G.; Fortier, I.; Raelson, J.; Sherbatich, T.; Riazanskaia, N.; Rogaev, E.; Raeymaekers, P.; Aerssens, J.; Konings, F.; Luyten, W.; Macciardi, F.; Sham, P.C.; Straub, R.E.; Weinberger, D.R.; Cohen, N.; Cohen, D.; Ouelette, G.; Realson, J. Genetic and physiological data implicating the new human gene G72 and the gene for D-amino acid oxidase in schizophrenia. Proc. Natl. Acad. Sci. U.S.A. **2002**, *99* (21), 13675–13680.

26 ESTAnnotator
A Tool for High-Throughput EST Annotation

Agnes Hotz-Wagenblatt[1] and Thomas Hankeln[2]
[1]Deutsches Krebsforschungszentrum, Heidelberg, Germany
[2]Johannes Gutenberg Universität Mainz, Mainz, Germany

CONTENTS

INTRODUCTION

The protein-coding genetic information of higher eukaryotic genomes represents only a minor percentage of its total DNA, e.g., 1.5% in the human genome, meaning that genes are usually separated by large stretches of noncoding intergenic DNA. Additionally, eukaryotic genes are organized mosaic-style: rather short exons (with an average size of 150 bp in humans), which form the mature mRNA and contain the essential coding information, are separated by introns, which are spliced out from the primary RNA transcripts and do not contribute to protein-coding. Because of this complex architecture of higher organism genomes, it is notoriously difficult to recognize exons in large stretches of genomic DNA sequences, e.g., obtained in the framework of genome projects, and to identify those exons correctly which altogether make up a complete gene. As a "shortcut" alternative to the sequencing and identification of the protein-coding gene repertoire of an organism, Adams et al. introduced a strategy called "EST sequencing" (EST = expressed sequence tag). This approach bypasses all complexities of genome structure by focusing only on the transcribed portions of a genome: the procedure starts with the isolation of an mRNA population from a certain tissue. After converting the mRNA molecules into their complementary DNA (cDNA), all resulting cDNA molecules are cloned into

DOI: 10.1201/9781003247432-26

suitable vector/host systems. Then, usually thousands of clones are being chosen at random for a DNA sequencing of their cDNA integrates, yielding a catalogue of EST sequences which essentially represents a collection of the transcribed portion of the genome (i.e., the genes). Gathering of sequence information from the protein-coding parts of the cDNA can be optimized by producing 5′EST reads instead of 3′EST reads, which usually cover the 3′untranslated gene regions.

The EST approach is extremely cost-effective and fast, and it gives a good overview of those genes that are active in the tissue used as source for the initial RNA preparation. A dedicated freely searchable database (dbEST), a division of GenBank (http://www.ncbi.nlm.nih.gov/Genbank/GenbankOverview.html), has been set up to collect EST data from a huge number of diverse organisms (see http://www.ncbi.nlm.nih.gov/dbEST/dbEST_summary.html), and this database currently contains more than 5.6 million human EST reads alone. Despite this wealth of data, EST sequencing is still an essential tool for the discovery of novel genes that have not been identified by genomic sequencing and gene prediction alone.

The production of a gene catalogue from EST data requires several steps of bioinformatical sequence analysis: (1) Because of the single-pass, "quick but dirty" sequencing strategy, bad sequence data and EST sequences contaminated with vector or repetitive noncoding DNA have to be removed. (2) Overlapping EST reads have to be clustered to obtain a contig sequence of their underlying cDNA. (3) Both EST singletons and EST clusters have to be annotated by searching for similarity to known genes or proteins already existing in nucleotide and protein sequence databases. Here, we describe an EST annotation tool that automates these steps and which was successfully used in categorizing 5000 EST sequence reads during a search for genes involved in differentiation and disease processes in human fetal cartilage tissue.

METHODS AND RESULTS

The basic methodology of generating a gene catalogue by EST sequencing is outlined in Figure 26.1. The reader may refer to Refs. [1–3] for details on the EST production process and for the importance of having EST sequence information in the framework of a genome project.

The EST annotation pipeline of ESTAnnotator[4,5] (Figure 26.2) can be divided into the following subsections: the preparation of quality-checked input EST sequences (shown in the upper part), the initial analysis of EST reads by database searches (shown at the left lower part), and the clustering of overlapping EST sequences as well as subsequent contig analysis (shown in the right lower part).

PREPARATION OF INPUT EST SEQUENCES

ESTAnnotator was designed for processing raw trace files (sequence chromatograms) or sequence text files. As EST sequences are generated by complex cDNA cloning procedures, they are often contaminated with vector and linker sequences. For generating high-quality input EST sequences, vector sequences and naturally occurring repetitive elements are masked by default. If too much sequence information is lost

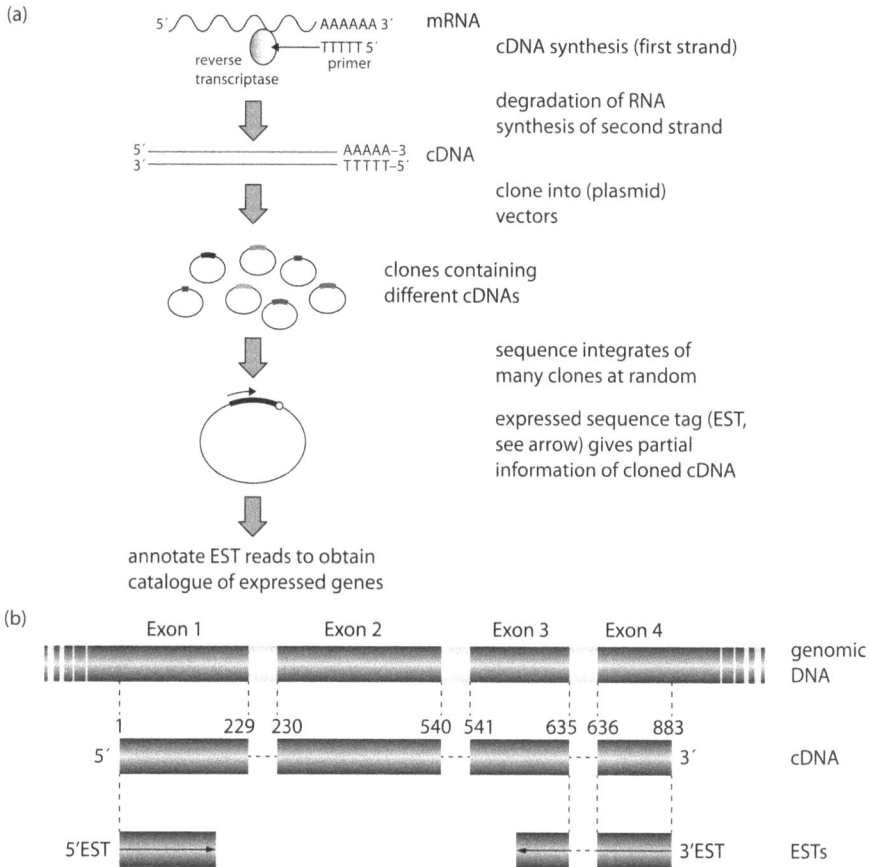

FIGURE 26.1 (a) Outline of an EST sequencing project. Note that the mRNA and cDNA molecules schematically depicted represent the population of all transcribed genes in the analyzed tissue. (b) Positional relationships of exons and introns on the genomic level, the complete cDNA derived from the mature mRNA transcript of the gene, and the partial cDNA sequence information (arrows) present in the 5′ and 3′ EST sequences.

due to quality trimming or masking, no further analysis of the input sequence occurs and the task automatically stops. This was the case for about 2.6% of the EST reads in the cartilage EST project.

BLAST SIMILARITY ANALYSIS

Sequence similarity searching is an important methodology in computational molecular biology. Initial clues to understanding the structure or function of a molecular sequence arise from homologies to other molecules that have been previously studied. Genome database searches reveal biologically significant sequence relationships and suggest future investigation strategies. Database search algorithms are used to compute pairwise comparisons between a candidate query

TASK FLOW CHART

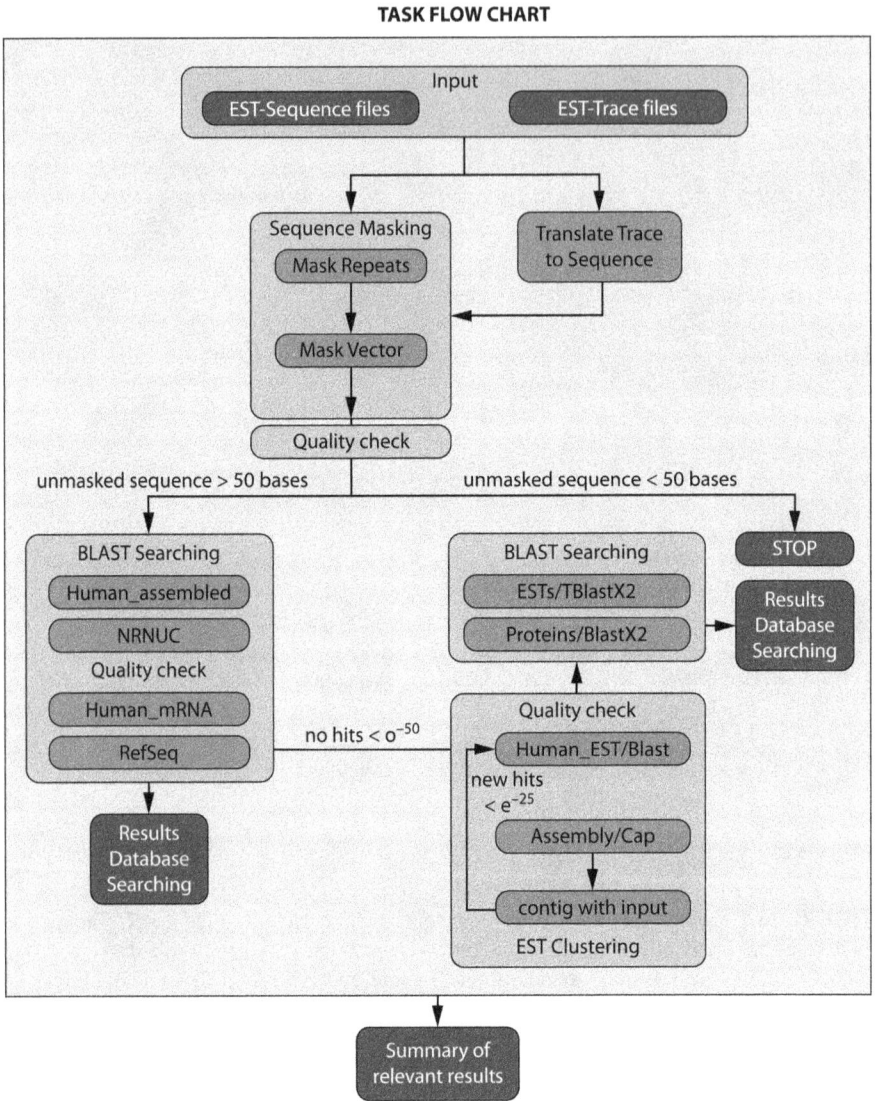

FIGURE 26.2 Flow chart of the ESTAnnotator: programs and rules used for high-throughput annotation of ESTs.

Source: Modified from Ref. [4].

sequence and each of the sequences stored within a database in order to find all the pairs of sequences that have a similarity above a defined threshold. The basic local alignment search tool (BLAST) algorithm is an approximate heuristic algorithm used to compute suboptimal pairwise similarity comparisons. Because of its speed, the BLAST algorithm has become the dominant search engine for biological sequence databases. The EST sequences were run against four different databases using the gapped BLAST program.[6] The search against the NCBI (National Center for Biotechnology Information, Bethesda, USA) nonredundant nucleic acid database (ftp://ftp.ncbi.nih.gov/blast/db) should reveal any similar nucleotide sequence deposited in the public databases. For the chromosomal location of the EST within the human genome, another BLAST search was performed against the NCBI human genomic sequence contig assembly database (ftp://ftp.ncbi.nih.gov/genomes/H_sapiens), which contains the nucleotide assemblies of the human chromosomes. For the identification of known, complete cDNA sequences matching our EST reads, additional searches were performed against the NCBI human mRNA database (ftp://ftp.ncbi.nih.gov/genomes/H_sapiens/RNA), which contains human "model mRNAs" constructed by prediction from genomic sequence data, and the RefSeq database (ftp://ftp.ncbi.nih.gov/refseq/, mRNA part), which contains curated mRNA data from human and other model organisms. If a highly similar mRNA corresponding to our EST read was found in any one of these mRNA databases, no clustering and further database searching were needed. Among the cartilage EST sequences 69.6% showed significant similarity to known genes/mRNAs in the human RefSeq collection, and another 4.8% were found homologous to human model RNAs. Approximately 23% of the cartilage EST sequences could not be identified as known transcripts, but showed significant similarity to genomic regions and/or other anonymous ESTs.[5] A subset of these potentially novel gene sequences is currently under detailed experimental scrutiny for expression in cartilage tissue using RT-PCR, Northern blotting, and mRNA in situ hybridization.

Using the NCBI human assembly database, a corresponding genomic location could be identified for more than 90% of the EST sequences. This information will be valuable in selecting possible candidate genes from regions of the human genome to which diseases related to malformations of the skeleton have been mapped genetically.

CLUSTERING OF OVERLAPPING ESTS AND BLAST ANALYSIS AT THE PROTEIN LEVEL

Expressed sequence tag sequences which could not be reliably assigned to a known mRNA or gene were processed further to finally obtain an annotation. As each gene may have many alternative transcripts that contain exons in different combinations, it is not a trivial task to assign each EST to its progenitor gene. A BLAST search against other human EST sequences from dbEST was therefore used to find homologous, overlapping EST reads in order to extend the original EST sequence by clustering. Assembly of such EST clusters, each containing the input EST sequence, was performed by a contig assembly program (CAP),[7] and the resulting consensus sequence of each cluster was saved. The EST cluster consensus sequences (or the input EST sequence alone, in case no cluster was formed) were used for further similarity searching on the protein sequence level in order to detect even remote similarities to known proteins or other anonymous EST sequences in the databases. These searches were performed by BLASTX against the Swiss-Prot protein database (ftp://ftp.ebi.ac.uk/pub/databases/swissprot/) to check for matching, already annotated proteins, and by TBLASTX against all ESTs of all organisms in dbEST to detect similarities to anonymous coding sequences of other organisms.

THE ESTANNOTATOR REPORT

The final ESTAnnotator report (Figure 26.3) is a web page that displays the database ID and the description line of the top three hits of the BLAST search results if their expectation value is below 0.01. Additionally, a link to the original BLAST output is provided. To illustrate the position of the BLAST hits and the clustered EST sequences, corresponding graphical outputs are displayed in the lower part. The

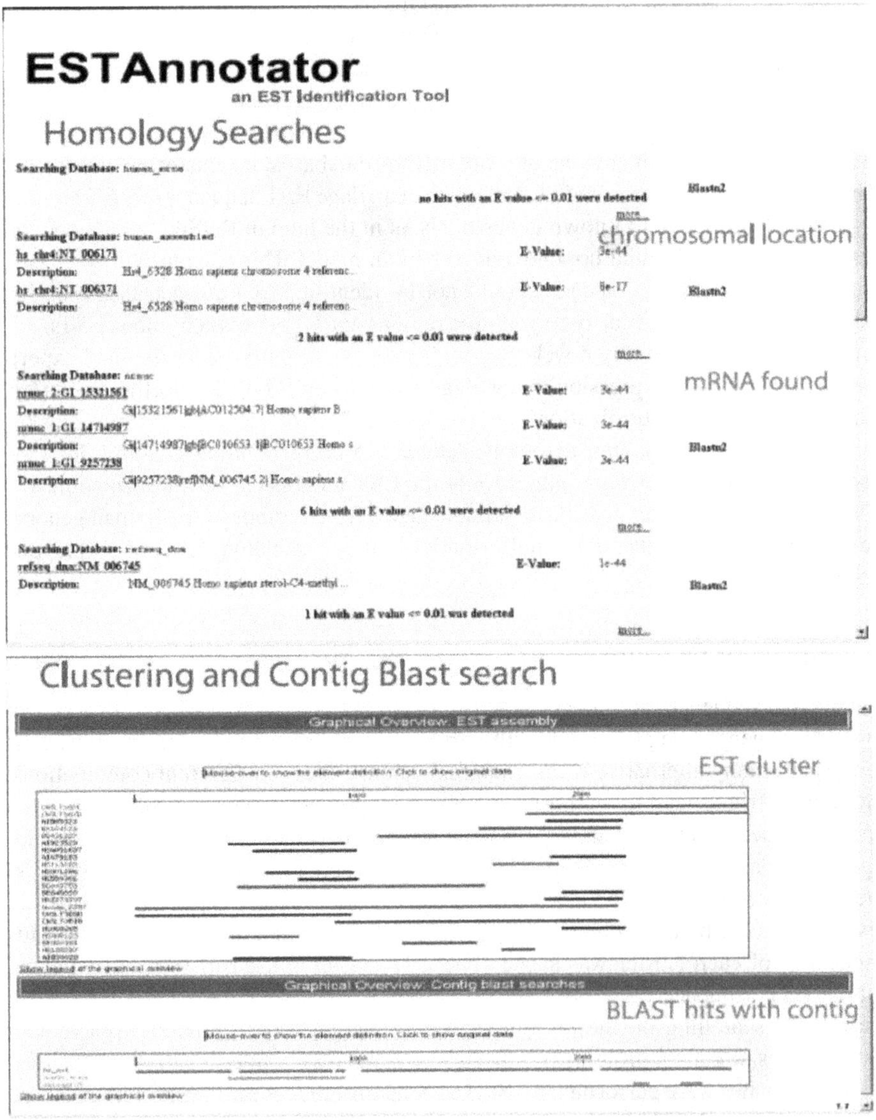

FIGURE 26.3 Parts of the ESTAnnotator report: database search results list and graphical display of the EST contig assembly and protein BLAST hits.

alignment information can be accessed by clicking on the hits within the graphical output. By downloading the XML (Extensible Markup Language) report file from the server, the results from the database searches for each EST sequence can easily be parsed into a database file.

WWW ACCESS BY THE WEB INTERFACE TO HUSAR (W2H)

Using the W3H task system[8] allowed the immediate integration of ESTAnnotator into the W2H web interface.[9] The ESTAnnotator is available at http://genius. embnet.dkfz-heidelberg.de/menu/biounit/open-husar/.

CONCLUSIONS

Expressed sequence tag sequencing and annotation are highly useful for identifying the repertoire of genes transcribed in tissues involved in human diseases, and— even in the well-studied human genome—the approach still represents a valuable tool for the identification of novel genes and alternatively spliced mRNAs. The ESTAnnotator facilitates processing and annotation of medium- to large-scale EST datasets. The successive steps of initial EST read quality control, followed first by the identification of ESTs which correspond to already known genes and mRNAs, and then by the clustering and further annotation by database searching of the remaining EST reads have been automated to avoid manual intervention. ESTAnnotator successfully led to the immediate bioinformatical annotation of about 75% of 5000 EST sequences originating from a human fetal cartilage cDNA library.[5] The tool could be further improved by producing a functional classification of the identified cDNAs (e.g., according to Gene Ontology criteria; http://www.geneontology.org) together with known splice variants and single-nucleotide polymorphisms.

REFERENCES

1. International Human Genome Sequencing Consortium. Initial sequencing and analysis of the human genome. Nature **2001**, *409* (6822), 860–921.
2. Adams, M.D.; Kelley, J.M.; Gocayne, J.D.; Dubnick, M.; Polymeropoulos, M.H.; Xiao, H.; Merril, C.R.; Wu, A.; Olde, B.; Moreno, R.F.; Kerlavage, A.R.; McCombie, W.R.; Venter, J.C. Complementary DNA sequencing: Expressed sequence tags and human genome project. Science **1991**, *252* (5013), 1651–1656.
3. Brandenberger, R.; Wei, H.; Zhang, S.; Lei, S.; Murage, J.; Fisk, G.J.; Li, Y.; Xu, C.; Fang, R.; Guegler, K.; Rao, M.S.; Mandalam, R.; Lebkowski, J.; Stanton, L.W. Transcriptome characterization elucidates signaling networks that control human ES cell growth and differentiation. Nat. Biotechnol. **2004**, *22*, 707–716.
4. Hotz-Wagenblatt, A.; Hankeln, T.; Ernst, P.; Glatting, K.-H.; Schmidt, E.R.; Suhai, S. ESTAnnotator: A tool for high throughput EST annotation. Nucleic Acids Res. **2003**, *31* (13), 3716–3719.
5. Zabel, B.; Schlaubitz, S.; Stelzer, C.; Luft, F.; Schmidt, E.R.; Hankeln, T.; Hermanns, P.; Lee, B.; Jakob, F.; Noeth, U.; Mohrmann, G.; Tagariello, A.; Winterpacht, A. Molecular Identification of Genes and Pathways Involved in Skeletogenesis by EST Sequence Analysis and Microarray Expression Profiling of Human Mesenchymal Stem Cell Differentiation. Abstract 13th Ann. Meeting German Soc. Hum. Genet., Medizinische Genetik: Leipzig, **2002**; Vol. 14, 245.

6. Altschul, S.F.; Madden, T.L.; Schaeffer, A.A.; Zhang, J.; Zhang, Z.; Miller, W.; Lipman, D.J. Gapped BLAST and PSI-BLAST: A new generation of protein database search programs. Nucleic Acids Res. **1997**, *25* (17), 3389–3402.

7. Huang, X. A contig assembly program based on sensitive detection of fragment overlaps. Genomics **1992**, *14* (1), 18–25.

8. Ernst, P.; Glatting, K.H.; Suhai, S. A task framework for the web interface W2H. Bioinformatics **2003**, *19* (3), 278–282.

9. Senger, M.; Flores, T.; Glatting, K.; Ernst, P.; Hotz-Wagenblatt, A.; Suhai, S. W2H: WWW interface to the GCG sequence analysis package. Bioinformatics **1998**, *14* (5), 452–457.

27 The Basics of Omics

Scott O. Rogers
Bowling Green State University, Bowling Green,
Ohio, U.S.A.

CONTENTS

INTRODUCTION

Large amounts of sequence data are becoming available more rapidly with the advent of next generation sequencing (NGS) methods. The first genome sequenced was that of the RNA virus MS2 in 1976. [GENERAL] It consists of 3,569 nucleotides and required several years of work by multiple researchers to determine the complete sequence. In 1977, the first DNA genome sequence was reported, that of φX174, with a genome of 5,386 nuceotides. By the 1990s, bacterial genomes consisting of millions of base pairs and thousands of genes were being determined and published. Again, however, these sequencing projects often required the collaboration of multiple labs working for several years in order to completely sequence and annotate these genomes. By 2000, dozens of genomes had been sequenced.[1–4] Today, many thousands of bacterial genomes have been sequenced, and hundreds of eukaryotic and archaeal genomes have been sequenced. Technological advances have made it possible to determine an entire bacterial genome in a few days.

GENOMICS

Genomics is defined as the determination of the entire genomic sequence of an organism or species. It usually includes annotation (identification and labeling) of genes, or as many as can be determined. For virus genomes, whose genomes range

DOI: 10.1201/9781003247432-27

269

from 3.5 nucleotides to 1.2 million base pairs, this has been completed manually or by using computer algorithms (Figure 27.1). However, computational methods had to be developed for larger genomes, including those of species of Bacteria, Archaea, and Eukarya. Eukaryotic genomes, which range from approximately 400 kb to more than 6×10^{11} bp, required the development of additional strategies, due to their larger sizes, the organizations on multiple chromosomes, the large numbers of introns, the large amounts of non-protein-encoding regions, and to the organellar genomes that accompanied the nuclear genomes. Some of the smallest genomes were sequenced first: *Saccharomyces cerevisiae* (brewer yeast, 12.5 Mb), *Caenorhabditis elegans* (nematode, 100 Mb), *Drosophila melanogaster* (arthropod, 132 Mb), and *Arabidopsis thaliana* (plant, 157 Mb). Conveniently, eukaryotic genomes are subdivided into chromosomes, and while the sizes of chromosomes vary widely, each is roughly the size of some bacterial genomes. Much of the sequencing of eukaryotic genomes was initially accomplished using bacterial artificial chromosomes (BACs). Essentially, large sections of the eukaryotic genomes were cloned as BACs, which were used to transform bacteria so that large amounts of each would be produced by the transformed bacteria. Then, smaller clones were made from those and each was sequenced. Eventually, the entire sequence for each chromosome was determined and then the entire genome was assembled. Each of these projects required enormous efforts from many labs.

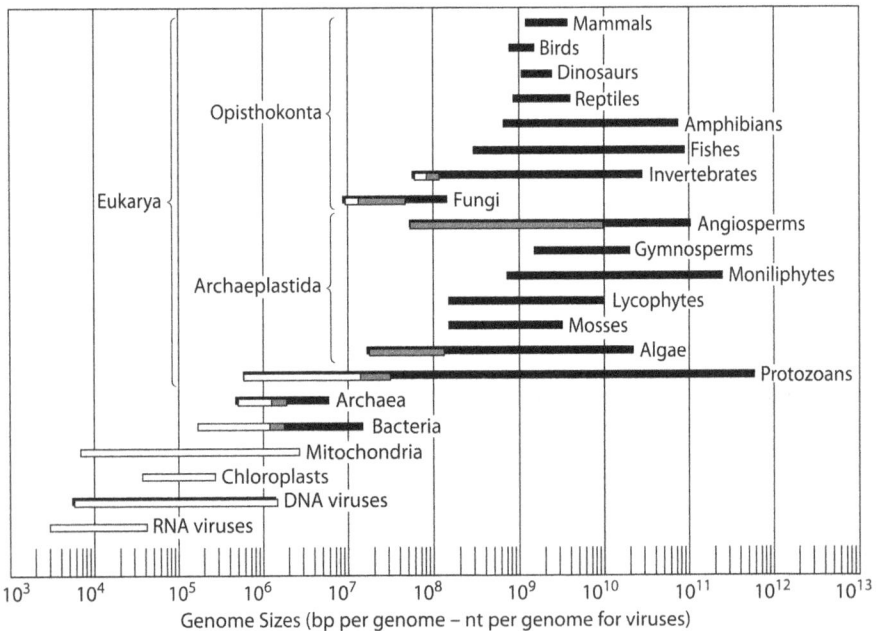

FIGURE 27.1 Genome sizes (log scale) of representative groups of organisms and endosymbiont organelles, arranged phylogenetically. All values are in base pairs, with the exception of viruses, which are in nucleotides (whether the viruses are single or double stranded). For diploid eukaryotes, haploid values are given.

As the human genome project was being planned in the mid-1980s, the scope of the project was beyond the molecular biological and sequencing technologies at the time. However, by the early 1990s, the technology was sufficiently developed to begin the project. The project was huge, and the sequencing facilities that were constructed were essentially large warehouses filled with automated DNA sequencers (all using Sanger-based sequencing methods). The project was more than an order of magnitude larger than the largest genome sequenced by that time. At more than 3 Gb in total length with 23 chromosomes, it represented an almost impossible task when it was begun. For the NIH (National Institutes of Health) sequencing project, led by Francis Collins, standard methods for the time were used, that is, the use of BACs to selectively clone fragments of each chromosome. The locations of each of the large clones was known, which would aid in later assembly of each of the chromosomes, and eventually, the entire genome. Each of the clones was sheared into smaller pieces, cloned, and sequenced. The pieces were assembled *in silico* by overlapping sequences that were identical. One of the main researchers, Craig Venter, thought there was a better way to accomplish this, and so early in the project, he split from the others on the project and began a company to perform the sequencing using his protocol. It consisted of shearing the entire genome into small pieces, cloning and sequencing all of the small pieces, and then reassembling the genome *in silico* by linking up fragments by their overlapping ends. Therefore, he avoided the time-consuming first steps of initial cloning and mapping of large fragments from each of the chromosomes. The computer compared the ends of each of the sequences, and it would link up the sequences if they had identical overlapping sequences. At first, some thought that it was an impossible task, but it worked, and it allowed more rapid sequencing and assembly than did the BAC-based methods. The Venter group began sequencing a year after the NIH group started, and the two groups completed their initial drafts of the human genome at roughly the same time.

The sequencing and partial annotation of the human genome was more-or-less completed by 2003, to coincide with the 50th anniversary of the publication of the 1953 Watson and Crick paper reporting the structure of DNA. Because of the nature of animal genomes (as well as some other eukaryotic genomes), work still is proceeding on the human genome. Only about 1.5–2.0% of the human genome consists of protein-encoding genes, and these were the primary targets to sequence by the NIH group. These genes also often have many introns,[5–7] and some of the introns are extremely long (some are tens of kilobases long, and a few are much longer). Because of this, annotation of the genome has been painstakingly slow for some regions of the genome, even though many researchers are working on this process. Depending on the research group, internet database site, and methods of analyses, estimates for the total number of genes in the human genome range from just over 20,000 to more than 27,000. Researchers are also carefully examining the other 98% of the human genome, which consist of long- and short-repetitive elements, some of which are transposons, retrotransposons, segments of virus genomes, and a number of genes that produce small RNAs, which are important in controlling gene expression.

NGS methods have increased the rates at which genomes can be sequenced by hundreds of times. Analytical algorithms have been developed to increase the annotation and other processes. Currently, the flow of these determinations begins with isolation and growth of the organism or cells from the organism, extraction of the DNA, fragmentation of the DNA, amplification of the DNA fragments, sequencing by NGS methods, assembly of the short sequences into contigs (contiguous sequences), annotation (mapping of genes and other features of the DNA), and distribution of the results into databases and publications.

TRANSCRIPTOMICS

Each species is characterized by its genome, including a set of genes that are transcribed into mRNAs, tRNAs, rRNAs, sRNAs, as well as non-genic regions. However, each organism only expresses portions of its genome at any point in time. Transcriptomics is a field of science that examines the RNAs that are being expressed at different times by the organism, as well as its cells under a variety of circumstances. For example, the human transcriptome in a skin epithelial cell differs significantly from the transcriptome of a T-lymphocyte. The patterns of gene expression change with the stages of development, ages of the cells, tissue type, and organism. Many transcriptomes have been determined in many different types of cells, including human cells. One characteristic of almost all cell types is that there is a set of genes, comprising about 10–30% of the transcriptome (depending on the cell type and organism) that are found expressed constitutively. These are the so-called housekeeping genes, whose products comprise the basic elements of central metabolism (such as glycolysis), replication, transcription, translation, lipid synthesis, etc. Another large set of genes are expressed in similar cell types. For example, they are expressed in most epithelial cells, whether they are in the skin, gut, lungs, etc. Then, there are sets of RNAs that encode proteins that are specific to one or a few cell types. These encode proteins that are characteristic for certain types of cells. For example, B-lymphocytes and plasma cells express specific globulins in high concentrations, which are destined to be released into bodily fluid to inactivate, tag, and eliminate antigens.

Transcriptomes can consist of total RNA, only mRNAs, only rRNAs, or other sets of RNAs (Figure 27.2). In the former case, total RNA is carefully extracted from the sample material, and then reverse transcribed using random DNA primers to produce cDNA copies of the RNA. To select only the mRNAs, the RNA is first sent through a column that has poly-dT (poly-deoxyribose-thymidine) attached to beads in the column matrix. mRNAs will bind to the column, because of the poly-A tails on their 3' ends. Once all of the other RNAs are washed through the column, the mRNA is released from the column using a salt solution. The mRNAs can then be used as a template to create cDNA copies using reverse transcriptase, in conjunction with an oligo dT primer, as well as random primers. The cDNAs can either be cloned as a library, and then sequenced, or they can be directly sequenced using NGS methods. However, in many cases, additional ends must be ligated or attached using PCR methods in order to modify the ends so that they are compatible with the primer or tag sequences for NGS methods. Because rRNA comprises about 80% of the total RNA, for total RNA sequencing projects, usually 80% of the sequences are rRNA sequences, and only about 5% are mRNA sequences. The remainder are tRNA and ncRNA sequences. For sequencing projects where mRNA has been purified, roughly

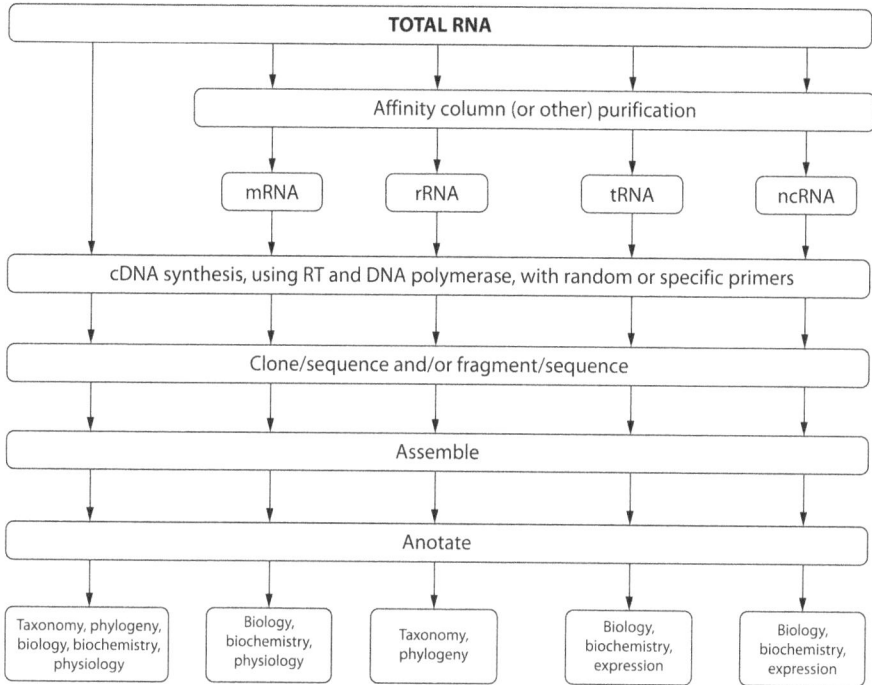

FIGURE 27.2 General scheme for transcriptomic sequencing projects. Initially, total RNA is extracted. This can be used to synthesize cDNA using random hexanucleotide primers, or it can be fractionated using affinity columns, gels, or other methods, into mRNA, rRNA, tRNA, and other ncRNA portions.

90% of the sequences are those from mRNAs. One of the major uses in human medicine is the study of human cancers, whose transcriptome differs greatly from the transcriptomes of normal tissues of the same type.

METAGENOMICS AND METATRANSCTIPOMICS

Genomic and transcriptomic methods have been applied to ecological and environmental scientific questions, aided by NGS methods. These applications are called metagenomics and metatranscriptomics (Figure 27.3). Both involve collection a sample from the environment, extracting the nucleic acids, and then sequencing all of the nucleic acid fragments. These fragments originate from all of the organisms that are present at (or near) the sample site. Metagenomics can determine not only what genes are present in the sample, but it can often lead to a determination of what taxa are present in the sample. This can lead to a scientific reconstruction of the community of organisms and biochemical pathways present at a particular site. Metagenomic studies have been carried out worldwide from hot deserts to frozen deserts, and from oceans, lakes, and streams, to underground caves, mines, and hundreds of other sites.[8-13] Each of the sites examined to date contain unique

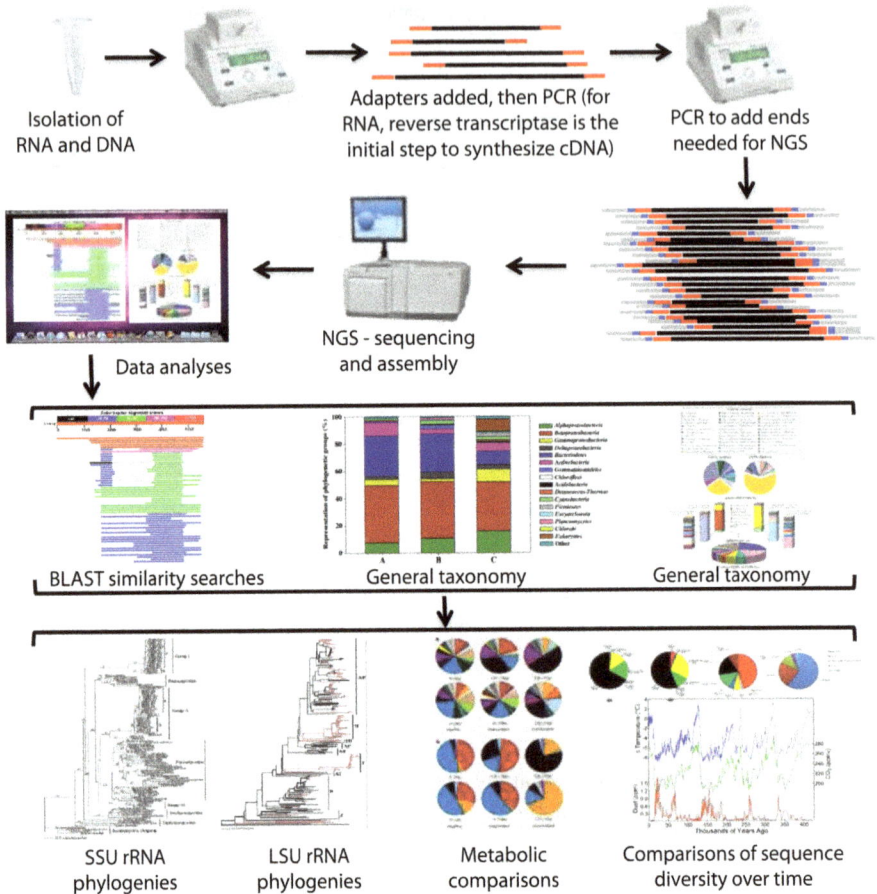

FIGURE 27.3 Overall process for a metagenomic and/or metatranscriptomic project. After careful collection of an environmental (or other) sample, the DNA and/or RNA are carefully and aseptically extracted. For metatranscriptomics, the RNA is used as a template for cDNA synthesis using random hexanucleotide primers.

assemblages of DNA, species, and biochemical networks. Significantly, usually more than half of the sequences are unlike any sequences that have been deposited in the NCBI and EMBL databases. A metagenomic investigation was carried out on the dust in the International Space Station (ISS), which found a large variety of microbes, some of which can be pathogenic on humans, and others that have never previously been described. These studies will be important in the planning of long-range space missions, in order to consider the potential for infection of personnel on those missions.

Metatranscriptomics can answer two basic questions, what are the biochemical pathways that are active at the sample site, and what organisms are living at that site. Because RNA degrades rapidly after cells die, most of the RNAs in a sample come from organisms that were living when the sample was collected. Therefore, the metatranscriptome can indicate the metabolic processes that were ongoing at the time of sampling. These methods have been successfully used to study biological processes in oceans, within and on the surface of humans, in subglacial lakes, in glaciers, in deep mines, in caves, in the atmosphere, in the ISS, and in many other environmental sites. A metagenomic/metatranscriptomic study of ice from subglacial Lake Vostok in Antarctica indicated that a complex ecosystem exists in this lake, which has been buried by kilometers of glacial ice for over 14 million years. An analogous study of the sediment at the bottom of another shallower subglacial lake, Lake Whillans (Antarctica), came to similar conclusions. Therefore, metagenomic and metatranscriptomic studies can provide strong evidence for the complexity of life in many environments, including those that are extreme.

MICROBIOMICS

During the past decade, other types of studies have been undertaken that have yielded interesting results. These are studies of microbiomes. They are essentially metagenomic and/or metatranscriptomic investigations to characterize the consortia of microorganisms that are living together in specific environments. Microbiomes are studied most often in their relation to various niches on (or in) organisms, including humans. For example, many studies have been completed on microbiomes from various parts of the gut from hundreds of healthy people, as well as those who had diseases. From these studies, it has been found that the microbial communities in healthy people of normal weight differ significantly from those who are obese or who have diseases. Also, those, with colitis, colon cancers, and other diseases of the gut, have very different combinations of organisms inhabiting their intestines. An important finding in the studies of gut microbiota is that the microbes not only aid in digesting the food that we eat, but they produce valuable nutrients, including amino acids, important lipids, and cofactors needed for various aspects our metabolism. So, the gut is not just a place where food is broken down and then the parts are absorbed for use, the gut is a bioreactor where parts of the food are used by the microbes to produce vital products that are then absorbed and used.

Microbiomes of other parts of the human body have been published. Microbiomes of many parts of the human skin indicate that each portion of the human body has a different microbiome, and again, these are for healthy individuals. The microbiome of the foot differs from that inhabiting the upper arm, and those differ from the microbiome on the face, etc. Similarly, the mouth, throat, larynx, etc., all have different microbiomes. When individuals are compared, their microbiomes differ with respect to the same body region. However, when studies of gut microbiomes were evaluated as to what metabolic functions each of the microbial species was performing, it was found that the overall biochemistry that was occurring in the sample was nearly identical, regardless of the species that were present. Vaginal microbiomes have been studied, primarily to investigate infection patterns, menstrual cycles, and fertility. From those studies, it was clear that the vaginal microbiome changed during the menstrual cycles of all of the individuals. However, they changed differently among the individuals, such that no firm conclusions could be made to predict what was a normal microbiome. However, when diseases were confirmed, the microbiomes did change.

PROTEOMICS

High-speed computers and methods to rapidly analyze proteins, their cellular locations, and their functions, sometimes a single molecule at a time, have contributed to large-scale studies of proteins, known as the field of proteomics. Proteins can be extracted and purified using a number of methods, although all methods are not amenable to all proteins. This is because some proteins are in the aqueous parts of the cell, called the soluble proteins, while others are bound to membranes. Usually, the extraction buffer contains a detergent, such as SDS (sodium dodecyl sulfate), to solubilize the membranes and the membrane-bound proteins. Following extraction from the cells, the proteins can be further purified by column chromatography, HPLC (high pressure liquid chromatography), gel electrophoresis, ammonium sulfate precipitation, or other methods. For electrophoresis, one-dimension gels are used to initially characterize total proteins in a cell. Two general types of gel conditions are used, native conditions, where the proteins may assume more of their native conformations; or denaturing conditions, using SDS, which stabilizes charges and relaxes the proteins, which causes the proteins to migrate primarily according to their molecular masses. In native gels, the proteins may not necessarily migrate through the gels according to their molecular weight and charges. Secondary structure may cause them to move faster or slower relative to other proteins of similar molecular weights. For SDS gels, the proteins all migrate primarily to their molecular weights, and therefore specific proteins will migrate to the same extent in each experiment.

Isoelectric focus (IEF) gels are useful in determining the charge on a particular protein. Gels are prepared that have a gradient of pH levels. When proteins are loaded onto these gels and subjected to electrophoresis, they will migrate in the gel while they have a charge, but stop migrating when the charge has been neutralized. As they move through the various regions of pH, they will either gain or lose protons. When the protons equalize the total charge on the proteins, the protein will stop moving. This is called its isoelectric point, which helps to characterize proteins. While one-dimension gels are very useful in initial characterizations of protein preparations, two-dimension (2D) gels can yield much more information about the proteins in a sample. For these gels, the entire sample is loaded into on well on one side of the gel. Electrophoresis is then carried out in one direction until the smallest proteins have reached close to the end of the gel. Then, the gel is rotated 90°, and electrophoresis is carried out in that direction. The second dimension often is an IEF procedure. At the end of electrophoresis, the proteins are visualized by staining, or by autoradiography if they were radioactively labeled. The gel can be blotted as well, and western blotting can be carried out with a radioactively labeled antibody that is specific for an epitope of one or more proteins. Often, 2D gel electrophoresis is carried out on more than one sample, each of which originated from different tissues, similar tissues under different conditions, cancer cells versus normal cells, etc. Comparisons are then made to determine the differences in expression between the two samples.

Various types of columns can be used to separate, identify, and/or purify proteins, including those that separate based on molecular mass, solubility, charge, binding to column materials, etc. The columns can also be coupled to equipment that detects the proteins emerging from the end of the column, such as those that measure UV absorption, and those that measure the mass-to-charge ratio (i.e., mass spectrometry). They can also be coupled to a fraction collector, which collects aliquots as they emerge from the column. One of the simplest methods to fractionate a protein sample is to use various concentrations of ammonium sulfate. Various proteins will precipitate from the solution under the different ionic strengths of the solutions of ammonium sulfate, such that a range of proteins can be separated in initial characterizations of the proteome.

The proteome of *Homo sapiens* contains about 90,000 proteins, which are expressed from approximately 22,000–25,000 genes. The difference in number of proteins versus genes is due to extensive alternative splicing (as well as multiple promotors for some genes) in humans (and many animals) that yields several proteins from some genes. For example, the gene that encodes the protein dystrophin, found mainly in muscle cells, also produces at least six other proteins, including utrophin, which is only a bit shorter than dystrophin, and five other short proteins, called dystrobrevins. On the other hand, for most Bacteria and Archaea, one gene encodes a single protein. Therefore, for those organisms, a genome provides a good estimation of the proteome, while the genomes of mammals and other animals can provide only a partial estimation of the proteome. In those cases, much more protein analysis must be performed to accurately determine the proteomes. Categorization of the proteins in the proteome is based on their functions and locations in the cells, such as DNA-binding proteins, cell cycle control proteins, signal transduction proteins, phosphorylases, transporters. Because the proteome determines most of the characteristics of a cell, it is important to understand not only the proteins and their relative concentrations but also to understand their interactions and functions in the cells.

STRUCTURAL GENOMICS

Structural biology attempts to determine the three-dimensional structures of biological molecules. When Watson and Crick reported the structure of DNA, they were engaging in structural biology research. Others, such as Linus Pauling, already had determined that portions of proteins wind up into α-helices, and entire structures of small proteins, such as lysozyme, were determined. Certain structures in proteins, called motifs, appear in many proteins, and most proteins contain motifs found in other proteins. Some motifs are characteristic of binding with DNA, while others are characteristic of embedding in membranes. Part of structural genomics is to search for common motifs in the amino acid sequences of proteins in order to provide estimates of the complete structure of the proteins. This information can be used to plan experiments, using methods such as X-ray crystallography and NMR (nuclear mass resonance), which can determine the structure of the proteins at high resolution. X-ray crystallography has been used to construct high-resolution models of small and large proteins, as well as protein complexes, including the determination of the fine structure of ribosomes. Structural genomics also includes the determination of DNA and RNA structures. By comparing genomic sequence data with structural information from X-ray crystallography studies, structural conclusions can be made regarding the molecular interactions of the sequenced DNAs and RNAs, as well as determination of the interactions among DNAs, RNAs, and proteins.

RNAOMICS

A part of structural genomics is called RNAomics. Although RNAomics might seem superficially similar to transcriptomics, it seeks to use sequence and structural RNA data to determine the important structures and functions of RNA molecules.[14,15] Many RNAs are structural and/or functional, such as rRNAs, tRNAs, sRNAs, and introns. Their structures are dictated by the portions of the RNAs that can fold and hydrogen bond to other parts of the molecules, or to adjacent molecules. Certain structures have been found in many RNAs, such as adenine platforms (consisting of two or more consecutive adenines) and GNRA (each having a G, then any nucleotide, followed by a purine and an A) loops at the ends of double stranded stems. There are many such motifs in RNA molecules, and some of them are catalytic sites, which convert the RNA into a ribozyme (catalytic RNA). One such site is the peptidyl transferase section of the LSU rRNA molecule, which connects amino acids together to form polypeptides during translation. Scientists use structural data from a variety of species in comparative studies to try to determine the structures in the molecules that are conserved, and which have similar functions because of the conservation of the structures. RNAs are more ancient than DNAs and have performed many functions in cells for billions of years. Like proteins, there are motifs found in different RNAs that perform similar functions. RNAomics, and applied research in RNAs, has expanded during the past several decades partly because it has great potential to study disease processes, treat diseases, and rapidly create vaccines to combat infections.

EPIGENOMICS

Epigenetics has become an important field of study in the past several decades, due to the major effects on development and tumorigenesis and due to modifications of histones and nucleotide bases. Epigenetic effects occur as a result of modifications to the structure of the DNA, rather than changes in the genes themselves. These modifications cause changes in the accessibility of the molecules that control transcription of the affected genes, and therefore, depending on the gene being affected, can cause major changes in gene expression and cellular functions. Methylation, demethylation, acetylation, and deacetylation of histones are one of the main mechanisms of epigenetic changes. Methylation of DNA bases (especially cytosine) also affects the structure of DNA and its accessibility to the transcription machinery necessary for expression of the genes in the region. Where high amounts of 5-metylcytosine are present in DNA, Z-form DNA can result, which usually renders the DNA unrecognizable by transcription factors that normally attach to the region to initiate transcription. Epigenomics maps changes in these regions and correlates them with changes in gene expression. The changes and their effects are then studied to determine whether their effects are detrimental, especially in cases where human health is affected, with hopes to be able to develop a control for the malady.

METABOLOMICS

Metabolites are the relatively small molecules that are the result of the cellular reactions and processes that occur in cells. The specific metabolites and their concentrations are characteristic of cells and tissues, and normal versus diseased cells and tissues can produce very different metabolite patterns. Metabolomics is the process of characterizing and categorizing these metabolite patterns and correlating them with specific functions in cells, tissues, organ, species, environments, etc. They have been used extensively in medical fields, due to their sensitive detection of small changes when cells begin to change their patterns of gene expression. In some cases, early signs of disease can be detected using these metabolite signals. In environmental applications, they can be used to detect the initial signals from toxic algal blooms that might affect the organisms in a lake, and in some cases may pose a risk to human health. Because there are many different classes of metabolites that are produced by cells, the methods to detect them also are varied. Some can be detected by spectroscopy, while others require specific chemical assays, and others require more elaborate tests. Because of this, labs must be set up specifically to test a particular subset of metabolites.

Lipidomics is a subset of metabolomics that seeks to identify the lipids, as well as the pathways that produce them. Lipids are extremely important in cells, being the major component of all membranes, as well as begin important storage molecules for carbon and in the production of ATP. Lipids also can be important in the identification of organisms. For example, while Bacteria and Eukarya both use D-glycerol, and fatty acids joined by ester linkages, in their membranes, Archaea use L-glycerol, and isoprene units joined by ether linkages in their membranes. Lipids also are found in lipoproteins embedded in membranes. Often, these are characteristic of certain species and/or cell types. The degree of saturation of the lipids in the membranes can indicate the temperature at which the cells normally grow, or at which they recently were growing. Saturated lipids are often signs of growth in elevated temperatures, while unsaturated lipids often are found in cells growing at lower temperatures, due to their increased fluidity at lower temperatures.

FUNCTIONAL GENOMICS

This branch of genomics utilizes genomic, transcriptomic, proteomic, and metabolomic data to draw conclusions about gene, RNA, protein, and biochemical functions, as well as their interactions. As such, functional genomics often is at the crux of genomics, transcriptomics, proteomics, and metabolomics. Genes and their products (RNAs, and proteins) interact either directly or indirectly, which together determine the characteristics and functions of the encompassing cell. For example, many separate proteins are involved in the control of transcription of each gene, and often include repressors and activators, which can suppress, or speed up (respectively) transcription of the gene. Additionally, disease states, such as carcinogenesis, involve the interactions between proteins on the cell surfaces, which interact with small molecules, hormones, proteins, and other molecules, and transfer the signal to the inside of the cell. The signal is passed on to other proteins and molecules that become activated to produce changes that lead them to pass the signal to other proteins in the process of signal transduction. Many of these signals are passed on to other proteins into the nucleus where they affect gene transcription, most often by activating or repressing transcription of sets of genes. Thus, functional genomics has been extremely important in cancer research, and many therapies have been developed from studying these pathways for specific cancers.

Major branches of functional genomics include studies of DNA-protein, RNA-protein, and protein-protein interactions. Biological molecules almost never act isolated from other molecules, and most form DNA-protein, RNA-protein, and/or protein-protein complexes. These interactions are often crucial to the function of the entire complex. For example, transcription of RNA involves more than a dozen proteins in Bacteria, and two times that number in Eukarya. In addition to histones, there are hundreds of proteins that associate with nucleic acids. DNA-binding proteins usually affect either transcription or replication, while those that associate with RNA include ribosomal proteins, spliceosomal proteins, and those involved in aiding in folding of ncRNAs. These interactions can be studied by polyacrylamide gel methods, assays using antibodies, precipitation experiments, and recombinant DNA methods. Protein-protein interactions have been the focus of many research studies over the past few decades. Most attempts have been to identify the proteins that are involved in specific processes in the cell, and how the proteins interact. Some assays involve cloning the genes for the proteins, and then using a reporter gene (e.g., similar to α-complementation used with bacterial cloning), to determine when and how the proteins interact. Some involve engineering of one or two reporter proteins that interact with another protein of interest. When they interact, the reporter portion reacts, usually by producing a colored or fluorescent signal.

Another important aspect of proteomics and functional genomics is determining how and when proteins are modified. These are post-translational modifications, and include the formation of disulfide bridges (between cysteine residues), phosphorylation (of tyrosines, serines, histidines, and other residues), alkylation (of lysines or arginines), acetylation (on the amino end or on lysines), addition of lipids, addition of sugars, and many others. These modifications greatly increase the number of functions that proteins can carry out. Phosphorylation often activates certain proteins, which are important in signal transduction pathways, and many activate transcription of many genes concurrently during developmental processes. Glycosylated proteins (those with covalently attached sugars) are mostly concentrated in the cell membranes (such as those involved in the ABO blood groups of humans). Many of these are important in cell-to-cell recognition. Lipoproteins also are concentrated in cell membranes and cell walls to provide structural integrity and fluidity to the membranes and walls. Because the proteome is so diverse, with so many interactive components (many of which can be modified in the cell) and so many functions, research groups usually concentrate on one aspect of the proteome in one organism.

CONCLUSIONS

Genomics is the process of sequencing, assembly, and annotation of complete genomes, including coding regions as well as non-coding regions. Transcriptomics is the study of gene expression patterns. Both of these have become important in characterization of disease states, including infections and carcinogenesis. Metagenomics and metatranscriptomics can be used to study all of the organisms and their biochemical processes in a mixed sample of organisms (from the environment or a location within a multicellular organism). Microbiomics is related to these in that it is the study of the microbial community that resides in various locations on (or in) a human or other animal, or in an environment. Proteomics is the characterization of the proteins that are present in a cell, tissue, diseased tissue, etc. Structural Genomics seeks to elucidate the physical structures of biological molecules, including DNA, RNA, proteins, etc. RNAomics is a subset of this, in that it involves determination of RNA structures, and attempts to identify and classify the various motifs in RNA. Epigenomics is the study of changes to DNA (i.e., chromosome remodeling) that cause changes in expression of those regions. Many of these changes occur within individuals, although some can be inherited. Metabolomics uses a variety of methods to characterize the metabolites within cells or tissues. Functional genomics brings many of the above together to build a more complete picture of the cellular and molecular functions ongoing in cells, both healthy and diseased.

REFERENCES

1. Alberts, B.; Bray, D.; Hopkin, K.; Johnson, A.; Lewis, J.; Raff, M.; Roberts, K.; Walter, P. *Essential Cell Biology*, 4th ed. New York, NY: Garland Publishing, Inc. **2013**.
2. Blackburn, G.M.; Gait, M.J., Eds.; *Nucleic Acids in Chemistry and Biology*. New York, NY: Oxford University Press. **1996**.
3. Krebs, J.E.; Goldstein, E.S.; Kirkpatrick, S.T. *Lewin's Genes XI*, 11th ed. Sudbury, MA: Jones and Bartlett Publishers. **2012**.
4. Mount, D.W. Bioinformatics, *Sequence and Genome Analysis*, 2nd ed. Cold Spring Harbor, NY: Cold Spring Harbor Laboratory Press. **2004**.
5. Harris, L.; Rogers, S.O. Splicing by an unusually small group I ribozyme. *Curr. Genet.* **2008**, *54*, 213–222.
6. Harris, L.B.; Rogers, S.O. Finding and characterizing small group I introns. In *Genomics III: Methods, Techniques, and Applications*; Iconcept Press, Ed; Hong Kong, China: iConcept Press, **2014**. pp. 327–341.
7. Harris, L.B.; Rogers, S.O. Evolution of small putative group I introns in the SSU rRNA gene locus of *Phialophora* species. BMC Res. Notes **2011**, *4*, 258–262.
8. D'souza, N.A.; Kawarasaki, Y.; Gantz, J.D.; Lee, R.E. Jr.; Beall, B.F.N.; Shtarkman, Y.M.; Koçer, Z.A.; Rogers, S.O.; Wildschutte, H.; Bullerjahn, G.S.; McKay, R.M.L. Microbes promote ice formation in large lakes. ISME J. **2013**, *7*, 1632–1640
9. Knowlton, C.; Veerapaneni, R.; D'Elia, T.; Rogers, S.O. Microbial analysis of ancient ice core sections from Greenland and Antarctica. Biology **2013**, *2*, 206–232.

10. Gura, C.; Rogers, S.O. Metatranscriptomic and metagenomic analysis of biological diversity in subglacial Lake Vostok (Antarctica). *Biology* **2020**, *9*, 55. doi: 10.3390/biology903055

11. Rogers, S.O.; Shtarkman, Y.M.; Koçer, Z.A.; Edgar, R.; Veerapaneni, R.; D'Elia, T. Ecology of subglacial Lake Vostok (Antarctica), based on metagenomic/metatranscriptomic analyses of accretion ice. Biology **2013**, *2*, 629–650.

12. Shtarkman, Y.M.; Koçer, Z.A.; Edgar, R.; Veerapaneni, R.S.; D'Elia, T.; Morris, P.F.; Rogers, S.O. Subglacial Lake Vostok (Antarctica) accretion ice contains a diverse set of sequences from aquatic, marine and sediment-inhabiting Bacteria and Eukarya. PLoS ONE **2013**, *8*(7), e67221. doi:10.1371/journal.pone.0067221

13. Venkateswaren, K.; Vaishampayan, P.; Cisneros, J.; Pierson, D.L.; Rogers, S.O.; Perry, J. International space station environmental microbiome - microbial inventories of ISS filter debris. Appl. Microbiol. Biotechnol. **2014**, *98*, 6453–6456.

14. Stombaugh, J.; Zirbel, C.L.; Westof, E.; Leontis, N.B. Frequency and isostericity of RNA base pairs. Nucleic Acids Res. **2009**, *37*, 2294–2312.

15. Zirbel, C.L.; Sponer, J.E.; Sponer, J.; Stombaugh, J.; Leontis, N.B. Classification and energetics of the base-phosphate interactions in RNA. Nucleic Acids Res. **2009**, *37*, 4898–4918.

28 Ribotyping

Tim J. Inglis
PathCentre, Perth, Western Australia, Australia

CONTENTS

INTRODUCTION

Ribotyping describes a series of methods used to characterize the ribosomal DNA operon in bacteria and other more complex organisms. In bacteria, rDNA is organized into the 16S, 23S, 5S, and spacer sequences. As ribosomal DNA is present in all bacteria and corresponds to genus and species, it has proved useful for identification and taxonomic and subtyping applications. The importance of rDNA to bacterial survival is reflected in the multiple copies present in most species.

TECHNICAL DESCRIPTION

Early ribotyping methods used digestion of extracted rDNA with restriction endonucleases, separation of rDNA fragments by electrophoresis, and hybridization with a cDNA probe in a Southern blotting process.[1,2] Unlike conventional restriction fragment length polymorphism (RFLP) analysis, which resolves the difference between bacterial isolates by comparison of ethidium bromide-stained DNA bands, ribotyping uses a cDNA probe to resolve differences between isolates by generating a smaller number of bands. The number of bands will depend on how many fragments the cDNA can hybridize to and therefore the number of cutting sites for the restriction endonuclease.[3,4] Restriction enzymes (REs) employed for ribotyping are frequent cutters such as *Eco*R1, *Bam*H1, and *Pvu*2, but some species require less commonly used restriction endonucleases such as *Pvu*2.[5] DNA probes hybridize to either 16S or 23S rDNA or both. Ribotyping probes include an *Escherichia coli* DNA and other sequences.[6,7] The membrane used for blotting

DOI: 10.1201/9781003247432-28

190-19-1 454 Listeria monocytogenes
190-19-2 455 Listeria monocytogenes
190-19-3 456 Listeria monocytogenes
190-19-4 457 Listeria monocytogenes
190-22-1 463 Listeria monocytogenes
190-22-5 524 Listeria monocytogenes
190-24-5 662 Listeria monocytogenes
190-25-2 511 Listeria monocytogenes
190-32-1 739 Listeria monocytogenes
190-32-2 740 Listeria monocytogenes
190-32-3 741 Listeria monocytogenes
190-153-1 1393 Listeria monocytogenes
190-153-2 1394 Listeria monocytogenes
190-153-3 1395 Listeria monocytogenes
190-153-4 1396 Listeria monocytogenes
190-153-5 1397 Listeria monocytogenes
190-153-6 1454 Listeria monocytogenes
190-153-7 1455 Listeria monocytogenes
190-153-8 1456 Listeria monocytogenes
190-170-4 1473 Listeria monocytogenes
190-170-5 1474 Listeria monocytogenes
190-170-6 1475 Listeria monocytogenes
190-170-7 1477 Listeria monocytogenes
190-170-8 1498 Listeria monocytogenes

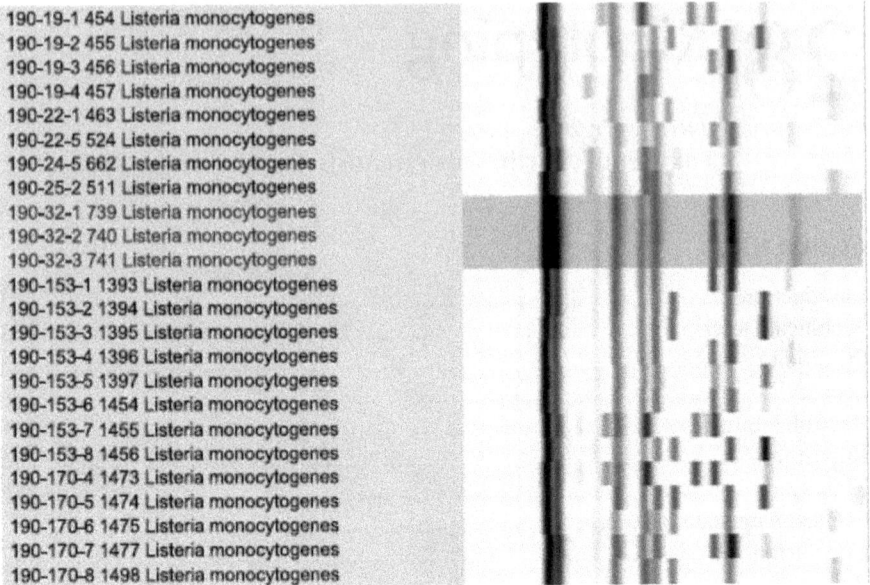

FIGURE 28.1   *Eco*R1 ribotype patterns generated from *Listeria monocytogenes* analyses, showing highlighted cluster of indistinguishable isolates.

Source: From L O'Reilly, PathCentre.

can be either nitrocellulose or nylon, and the labeling system can be a fluorophore, digoxigenin, or a radioisotope.

Several variations have been introduced to adapt ribotyping for wider use. A version of ribotyping based on digestion with *Eco*R1, an *E. coli* cDNA probe, and a fluorescent tracer has been automated to simplify and speed up the procedure (Figure 28.1).[8] In polymerase chain reaction (PCR) ribotyping, PCR primers to highly conserved regions of the 16S–23S intergenic spacer region are used to amplify the product for subsequent length polymorphism analysis.[9]

In outline, the manual ribotyping method is as follows:

1. Bacterial DNA extraction (phenol–chloroform, guanidium, etc.).
2. DNA check (concentration and mini-gel).
3. Digestion with restriction endonuclease.
4. Preparatory gel for blotting.
5. Southern blot.
6. rDNA digoxigenin labeling.
7. DNA marker ladder digoxigenin labeling.
8. Prehybridization.
9. Hybridization.
10. Wash and development.
11. Ribotype pattern recording and analysis.

SPECIFICITY

As all bacteria have rDNA, ribotyping is theoretically practicable for all bacterial species. In the Enterobacteriaceae, the presence of 5–7 rDNA operons results in 10–15 bands, with potential for a good level of discrimination between strains of a given species.[10] On the other hand, Mycobacteria have only 1 or 2 rDNA operons and produce few bands on ribotyping with a correspondingly lower level of discrimination.[11] Southern blotting with alternative probes specific to mycobacteria (e.g., IS 6110 for *Mycobacterium tuberculosis*) has been much more productive for these bacteria.[12] The limited genetic diversity of some important gram-negative pathovars, such as *E. coli* O157, renders ribotyping less suitable for molecular epidemiology of these.[13] On the other hand, bacteria such as *Staphylococcus aureus, Streptococcus pyogenes*, and *Listeria monocytogenes* can be effectively subtyped by ribotyping.[14–16] A large proprietary database of *Eco*R1 ribotypes obtained with the automated system exists for these bacteria. The automated method has been adapted for use with other restriction enzymes (e.g., *Pvu*2 and *Pst*1) but is not easily adapted to the other probes used in manual ribotyping protocols. Clearly, most ribotyping methods have been developed for highly specialized purposes, the details of the method being specific to the application. Each method requires validation against a representative collection of the species in question. The taxonomic value of the information generated by ribotyping therefore depends on the intraspecies variability of the rDNA operon, the number of copies present, the restriction endonuclease used, the cDNA probe, the range of strains already ribotyped, and the epidemiological setting from which the isolates were obtained. Ribosomal DNA has been used as an important arbiter of taxonomic status in determinative bacteriology.[17] Whereas this may be generally true down at species level, there have been reports of species that cannot be distinguished by 16S rRNA studies that have subsequently been resolved by DNA–DNA hybridization.[18] Ribotyping should thus be seen as less discriminating than DNA hybridization studies and therefore taxonomically subordinate. The automated ribotyping procedure is less suited to discriminating closely related fragments from some pathogens than traditional *Eco*R1 ribotyping, thus reducing its capacity to discriminate between strains.[19]

SENSITIVITY

Ribotyping requires only small amounts of high-quality bacterial DNA extract, particularly if the automated process is used. The use of a DNA hybridization probe reduces the sensitivity of band detection compared with conventional RFLP analysis, but, in doing so, improves the readability of the result.

REPRODUCIBILITY

Ribotyping is more reproducible than conventional RFLP analysis. As this is due in part to the specificity of the technique, the ribotype patterns obtained by a given method can give a higher reproducibility when a combination of several REs is used.[20] In the automated procedure, molecular markers and image optimization software are used to increase between-batch reproducibility so that band patterns can be compared with archived patterns using a computerized similarity index.

ROBUSTNESS

The various manual methods of ribotyping depend on a high level of technical skill, take a lot of dedicated laboratory time to complete, and are generally restricted to larger research laboratories.[21] Nevertheless, once established, the method is robust because a positive result depends on hybridization of digested bacterial rDNA with a specific probe. Automation of one of the more commonly used protocols has reduced the technical skill and dedicated time required without affecting the overall robustness of the method.

LIMITS

Ribotyping is limited mainly to analysis of bacterial species and has variable utility depending on the genera in question. Ribotyping can be used to discriminate or identify as far as species level for most bacteria. Some species (e.g., Mycobacteria and some Enterobacteriaceae) cannot be reliably subtyped by this method without resorting to either multiple procedures (each with a different restriction endonuclease) or alternative probes (Southern blotting but not ribotyping[12]). Ribotyping is not a satisfactory alternative to DNA hybridization studies as a final arbiter of taxonomic status[18] but can be used to gain an approximate assessment of taxonomic placing where the hybridization studies have already been performed.[17] At a practical level, the skills, time, and equipment required make manual ribotyping unsuitable for routine bacterial identification or molecular epidemiology service work.[21] Some of these shortcomings are met by the automated EcoR1 protocol.

CLINICAL APPLICATIONS

Ribotyping has seen wide application as a molecular epidemiology tool in clinical bacteriology.[3,8,10,13–16,19–21] It has often been used to complement other molecular typing methods such as DNA macrorestriction analysis when used in epidemiological investigations.[8,10,15,16,21,22] The most common of these applications are in hospital infection control and public health outbreak investigations where ribotyping is used to identify a cluster of isolates to confirm or refute preliminary epidemiological inferences.[23] In this setting, it has been argued that ribotyping is suitable for a wide range of typing applications.[22] More recently, the speed of automated ribotyping has made this version of the method more attractive to public health laboratories with a biosecurity role.[24] Larger reference laboratories have also found a use for ribotyping as one of a series of methods used to complete the identification and typing of species of uncertain status.

CONCLUSIONS

Ribotyping, or the characterization of the ribosomal DNA operon, has become an established analytical tool for bacteria and other more complex organisms. Automation of what was previously a lengthy experimental procedure now ensures that this technique can be used for a wide variety of molecular epidemiology tasks. The use of ribotyping, in connection with other genotyping procedures, is likely to play an increasing part in laboratory-based investigations of hospital- and community-acquired infections.

REFERENCES

1. Snipes, K.P.; Hirsh, D.C.; Kasten, R.W.; Hansen, L.M.; Hird, D.W.; Carpenter, T.E.; McCapes, R.H. Use of an rRNA probe and restriction endonuclease analysis to finger-print *Pasteurella multocida* isolated from turkeys and wildlife. J. Clin. Microbiol. Aug **1989**, *27* (8), 1847–1853.
2. Bialkowska-Hobrzanska, H.; Harry, V.; Jaskot, D.; Hammerberg, O. Typing of coagulase-negative staphylococci by Southern hybridization of chromosomal DNA fingerprints using a ribosomal RNA probe. Eur. J. Clin. Microbiol. Infect. Dis. Aug **1990**, *9* (8), 588–594.
3. Alonso, R.; Aucken, H.M.; Perez-Diaz, J.C.; Cookson, B.D.; Baquero, F.; Pitt, T.L. Comparison of serotype, biotype and bacteriocin type with rDNA RFLP patterns for the type identification of *Serratia marcescens*. Epidemiol. Infect. Aug **1993**, *111* (1), 99–107.
4. Qu, M.; Kan, B.; Qi, G.; Liu, Y.; Gao, S. A study on genetic polymorphism of rRNA gene pattern of *Vibrio cholerae* O139 in China. Zhonghua Liuxingbingxue Zazhi Jun **2002**, *23* (3), 203–205.
5. Bailey, J.S.; Fedorka-Cray, P.J.; Stern, N.J.; Craven, S.E.; Cox, N.A.; Cosby, D.E. Serotyping and ribotyping of Salmonella using restriction enzyme *Pvu*II. J. Food Prot. Jun **2002**, *65* (6), 1005–1007.
6. Garaizar, J.; Kaufmann, M.E.; Pitt, T.L. Comparison of ribotyping with conventional methods for the type identification of *Enterobacter cloacae*. J. Clin. Microbiol. Jul **1991**, *29* (7), 1303–1307.
7. Cookson, B.D.; Stapleton, P.; Ludlam, H. Ribotyping of coagulase-negative staphylococci. J. Med. Microbiol. Jun **1992**, *36* (6), 414–419.
8. Pfaller, M.A.; Wendt, C.; Hollis, R.J.; Wenzel, R.P.; Fritschel, S.J.; Neubauer, J.J.; Herwaldt, L.A. Comparative evaluation of an automated ribotyping system versus pulsed-field gel electrophoresis for epidemiological typing of clinical isolates of *Escherichia coli* and *Pseudomonas aeruginosa* from patients with recurrent gram-negative bacteremia. Diagn. Microbiol. Infect. Dis. May **1996**, *25* (1), 1–8.
9. Kostman, J.R.; Alden, M.B.; Mair, M.; Edlind, T.D.; LiPuma, J.J.; Stull, T.L. A universal approach to bacterial molecular epidemiology by polymerase chain reaction ribotyping. J. Infect. Dis. Jan **1995**, *171* (1), 204–208.
10. De Cesare, A.; Manfreda, G.; Dambaugh, T.R.; Guerzoni, M.E.; Franchini, A. Automated ribotyping and random amplified polymorphic DNA analysis for molecular typing of *Salmonella enteritidis* and *Salmonella typhimurium* strains isolated in Italy. J. Appl. Microbiol. Nov **2001**, *91* (5), 780–785.
11. Poulet, S.; Cole, S.T. Repeated DNA sequences in mycobacteria. Arch. Microbiol. Feb **1995**, *163* (2), 79–86.
12. Cave, M.D.; Eisenach, K.D.; McDermott, P.F.; Bates, J.H.; Crawford, J.T. IS6110: Conservation of sequence in the *Mycobacterium tuberculosis* complex and its utilization in DNA fingerprinting. Mol. Cell. Probes Feb **1991**, *5* (1), 73–80.

13. Martin, I.E.; Tyler, S.D.; Tyler, K.D.; Khakhria, R.; Johnson, W.M. Evaluation of ribotyping as epidemiologic tool for typing *Escherichia coli* serogroup O157 isolates. J. Clin. Microbiol. Mar **1996**, *34* (3), 720–723.
14. Oliveira, A.M.; Ramos, M.C. PCR-based ribotyping of *Staphylococcus aureus*. Braz. J. Med. Biol. Res. Feb **2002**, *35* (2), 175–180.
15. Shundi, L.; Surdeanu, M.; Damian, M. Comparison of serotyping, ribotyping and pulsed-field gel electrophoresis for distinguishing group A Streptococcus strains isolated in Albania. Eur. J. Epidemiol. Mar **2000**, *16* (3), 257–263.
16. Baloga, A.O.; Harlander, S.K. Comparison of methods for discrimination between strains of *Listeria monocytogenes* from epidemiological surveys. Appl. Environ. Microbiol. Aug **1991**, *57* (8), 2324–2331.
17. Ludwig, W.; Klenk, H.P. Overview: A phylogenetic backbone and taxonomic framework for prokaryotic systematics. In *Bergey's Manual of Systematic Bacteriology*; Boone D.R.; Castenholz R.W.; Garrity G.M., Eds.; Springer-Verlag: New York, NY, **2001**; 2, 49–65.
18. Fox, G.E.; Wisotzky, J.D.; Jurtshuk, P. How close is close: 16S rRNA sequence identity may not be sufficient to guarantee species identity. Int. J. Syst. Bacteriol. **1992**, *42*, 166–170.
19. Dalsgaard, A.; Forslund, A.; Fussing, V. Traditional ribotyping shows a higher discrimination than the automated RiboPrinter system in typing *Vibrio cholerae* O1. Lett. Appl. Microbiol. Apr **1999**, *28* (4), 327–333.
20. Blanc, D.S.; Siegrist, H.H.; Sahli, R.; Francioli, P. Ribotyping of *Pseudomonas aeruginosa*: Discriminatory power and usefulness as a tool for epidemiological studies. J. Clin. Microbiol. Jan **1993**, *31* (1), 71–77.
21. Patton, C.M.; Wachsmuth, I.K.; Evins, G.M.; Kiehlbauch, J.A.; Plikaytis, B.D.; Troup, N.; Tompkins, L.; Lior, H. Evaluation of 10 methods to distinguish epidemic-associated *Campylobacter* strains. J. Clin. Microbiol. Apr **1991**, *29* (4), 680–688.
22. Inglis, T.J.; Clair, A.; Sampson, J.; O'Reilly, L.; Vandenberg, S.; Leighton, K.; Watson, A. Real-time application of automated ribotyping and DNA macrorestriction analysis in the setting of a listeriosis outbreak. Epidemiol. Infect. Aug **2003**, *131* (1), 637–645.
23. Bingen, E.H.; Denamur, E.; Elion, J. Use of ribotyping in epidemiological surveillance of nosocomial outbreaks. Clin. Microbiol. Rev. Jul **1994**, *7* (3), 311–327.
24. Inglis, T.J.; O'Reilly, L.; Foster, N.; Clair, A.; Sampson, J. Comparison of rapid, automated ribotyping and DNA macrorestriction analysis of *Burkholderia pseudomallei*. J. Clin. Microbiol. Sep **2002**, *40* (9), 3198–3203.

29 Forensic DNA Typing
Y Chromosome

Lluís Quintana-Murci
Institut Pasteur, Paris, France

CONTENTS

INTRODUCTION

The human Y chromosome is one of the smallest chromosomes, representing 2–3% of a haploid genome. It is the only chromosome that has no homologue at any state or sex, in contrast to the X chromosome that has a homologue in the female sex. The majority of the Y chromosome (95%), termed nonrecombining Y (NRY), does not undergo recombination during male meiosis; however, two regions, pseudoautosomal regions (PAR1 and PAR2), located at the distal portions of the telomeres, recombine with homologous counterparts on the X chromosome. From a functional viewpoint, the Y performs specialized roles that are crucial for males, and therefore for the whole population, such as sex determination and male fertility. The haploid status of the Y and its exclusive paternal transmission make it a very useful tool in different domains as follows: (1) the estimation of historical patterns of population movements and splitting; (2) forensic applications, such as identification of male DNA in cases with male/female stain mixtures; and (3) genealogical studies and paternity testing, especially for deficiency cases, where the alleged father is deceased and his male relatives need to be tested.

DOI: 10.1201/9781003247432-29

Y-CHROMOSOME MARKERS: HAPLOGROUPS
AND HAPLOTYPES

One unique feature of Y-chromosome markers is that they are inherited as a single block in linkage. Therefore Y-linked variations are largely a result of accumulation of de novo mutations over time.[1] This feature is very useful in high-resolution discrimination of individuals. Large-scale sequencing efforts and the development of rapid mutation detection techniques have accelerated the discovery of Y-chromosome variation. Today, more than 200 biallelic polymorphisms and around 30 multiallelic markers are available,[2–5] most of which are suitable for polymerase chain reaction (PCR)-based genotyping techniques. Biallelic markers mainly include single base pair variants (single nucleotide polymorphism [SNPs]) and also a reduced number of small insertions/deletions (indels), whereas multiallelic markers mainly refer to microsatellites and minisatellites. The main difference between the two sets of markers is their mutation rate: biallelic markers present lower mutation rates (about 2×10^{-8} per base per generation) than microsatellite markers ($2-3\times10^{-3}$ per base per generation). Biallelic variants and insertion/deletion polymorphisms almost certainly represent unique molecular events and define stable and deep-rooted branches, known as haplogroups that can be traced back in time over thousands of years. These haplogroups present either a wide geographic distribution or, in some cases, more regionally clustered.[3,4] The multiallelic markers define the internal diversity of these haplogroups, and the combination of the allelic states of different microsatellite loci are known as haplotypes. Microsatellite haplotypes are very useful for microevolutionary studies and to distinguish subtle genetic differences between populations and/or individuals.[5] The high mutation rate of these markers that enables a good discrimination between individuals, makes them the most suitable markers in forensic studies.[6] However, this high mutation rate has a disadvantage: the appearance of recurrent mutations. Thus, it is sometimes difficult to distinguish if two individuals are "identical by state" (share the same allele but not the same ancestry) or "identical by descent" (share the same allele and the same ancestor). This has important repercussions in population genetic studies because, at the population level, the use of these markers alone may lead to an underestimation of genetic distances among populations, thereby distorting genuine population relationships.

THE Y AND THE PAST: HUMAN ORIGINS
AND POPULATION DISPERSALS

Y-chromosome studies have shown that Y-linked variation is nonrandomly distributed among human populations, showing lineage profiles that can be strikingly different among different worldwide populations.[2–4] These observations have obviously important consequences in ascertaining the origin of an individual during forensic investigations. From a human evolution viewpoint, Y-chromosome studies have given a major contribution to a better understanding of human origins and population dynamics. Y-chromosome phylogenies support the African origin of our species around 150,000 years ago. The most divergent branches of the trees are restricted to African populations, and non-African populations find their roots in branches of African origin.[2–4] These observations support the scenario in which modern humans can be traced back to a single African ancestral population that lived in Africa (i.e., the so-called Out of Africa model), from where it dispersed throughout the other continents replacing preexisting human species. Substantial continental and local population structure has been revealed, and hypotheses of early human migrations at different times formulated, such as the mode and tempo of the peopling of Asia and subsequent colonization of the Americas and the colonization of the Pacific.[3]

THE Y AND THE PRESENT: FORENSIC AND GENEALOGICAL STUDIES

SEX ASSIGNMENT

The most obvious use of the Y chromosome in forensics is the sex assignment. The most commonly used test, the amelogenin sex test (AMELY), takes advantage of the homology between the two pseudoautosomal regions of the sex chromosomes and amplifies a segment of the XY-homologous amelogenin gene pair.[7] Although this method avoids the ambiguous nature of a negative result as a result of the failure of the PCR reaction rather than to the absence of Y material, the reliability of the test depends on the assumption that the tested individuals present normal karyotypes and intact sex chromosomes. Nevertheless, individuals presenting a discordance between their sex phenotype and their karyotype (i.e., sex-reversed 46, XX males or 46, XY females) are present at appreciable frequencies in the population and can constitute a source of error in sex assignment tests based only in the AMELY test. It is suggested, therefore, to perform other sex tests in parallel, such as amplification of the SRY gene, which will give complementary information to the AMELY test.[8]

IDENTIFICATION AND DISCRIMINATION OF MALE-SPECIFIC DNA

The analysis of Y-chromosome variation has been extensively used for forensic applications other than sex assignment. They can provide highly valuable information to resolve male–female DNA mixtures in, for example, sexual assault cases. In these cases, samples may present high amount of victims' cells, such as vaginal/anal washings, and sometimes no male autosomal profiles are identified. It has been shown that Y-chromosome microsatellite analysis becomes a useful alternative because the use of specific Y-primers can improve the chances of detecting small amounts of the perpetrator's DNA in a background of heterologous female DNA.[9] Another obvious application of the Y chromosome is in paternity testing, especially for deficiency cases where the alleged father is deceased. In this case, the absence of the biological father can be resolved by the analysis of any relative sharing the same male line. However, when comparing the Y-chromosome profile in a paternity test or between a suspect and a sample, inclusion remains an important caveat. Although Y-linked microsatellites can be confidently used for exclusion purposes, nonexclusion can be problematic because: (1) all the male relatives of a suspect/ alleged father share presumably the same Y haplotype and (2) the Y chromosome of a suspect/alleged father may be present at high frequencies in the population under study, considerably reducing the reliability of the test. In any case, the high level of polymorphism of microsatellite markers offers a significant degree of discrimination between individuals. For example, a study of a German population using a 9-locus microsatellite haplotype, demonstrated a discriminatory capacity between Y-chromosome haplotypes of 97%.[10] However, this may not be the case in other populations with different genetic backgrounds and different population histories, where, for example, male-specific migration processes, founder effects, or genetic drift may have led to overrepresentation of specific haplotypes in the population.

RETRACING ETHNIC AND GEOGRAPHIC ORIGINS

Another possible information of interest for forensic investigators is that of ethnic origin. Although the highest genetic variation is observed between individuals and not between populations, Y-chromosome polymorphisms may provide the best resolution to ascertain the approximate geographic origin of a suspect.[11] Some haplogroups tend to be restricted to European populations, while others are restricted to sub-Saharan African or East Asian populations.[2–4] Geographic clustering of Y-chromosome haplogroups or haplotypes can be even more restricted. A recent Y-chromosome survey of the British Isles has shown that haplotype frequencies can differ considerably even over relatively short geographic distances.[12] Lineages of recent origin, such as haplogroup R-SRY2627, may exhibit a very limited distribution. This lineage has been observed only among Europeans and is almost restricted to Basque and Catalan populations.[13] In this case, it can be highly informative to predict the origin of the population and for exclusion purposes. However, more populations need to be tested for additional markers in order to create highly discriminative databases that will increase the power of reliability when inferring the geographic origin of a sample.

PRESIDENTS AND TSARS: THE POWER OF Y CHROMOSOMES AND MITOCHONDRIAL DNAs (mtDNAs)

The usefulness of Y-chromosome data has been successfully used in some famous and controversial cases, such as U.S. President Thomas Jefferson's paternity of the children of one of his slaves. In 1802, he was accused of having fathered a child, Thomas Woodson, by Sally Hemings. Also, Sally's last son, Eston Hemings Jefferson, is thought to be the son of the president, although other scholars give more credence to the hypothesis that they were the sons of Jefferson's sister that fathered Eston. By analyzing several Y-polymorphisms in 14 members of the different male lines involved in the controversy (e.g., those of Thomas Woodson, Eston Hemings, President Jefferson, and his sister's sons, the Carr's line), the authors found that President Jefferson and Eston Hemings share the same haplotype.[14] In contrast, this haplotype was absent in Thomas Woodson and the Carr's lines, and its frequency in the general population is very low (0.1%). The authors concluded that President

Jefferson fathered his slave's last child, whereas he was not the father of Sally's first child. The possibility that any other president's male-line relative could have fathered Sally's last child cannot be excluded, highlighting therefore the difficulty of being conclusive in these studies, but no historical records support this hypothesis. This study nevertheless illustrates the usefulness of Y-chromosome data to disentangle complex hypothesis of alleged paternities and genealogical relationships.

In the absence of male-line comparison, a suitable partner of the Y chromosome is the mitochondrial DNA (mtDNA). This molecule can be highly informative in genealogical studies because, as the Y, it does not recombine and is transmitted through the maternal line without any modification other than naturally occurring mutations. Moreover, analyses of mtDNA in forensics can be even more powerful than Y-chromosome analyses because mtDNA is present at a higher copy number in cells and is more likely to survive prolonged periods than nuclear DNA. The most notorious example of its use in genealogical and forensic studies is the identification of the remains of Tsar Nicholas II and his family. By analyzing the amelogenin gene and autosomal STR variation in nine skeletons found in a grave in Yekaterinburg (Russia), the authors could confirm the presence of a family group, composed of two parents and three daughters and four additional unrelated bodies. To ascertain if the five related individuals corresponded to the Romanov family, mtDNA was analyzed in all bodies and in a living maternal relative of the Tsarina, Prince Philip the Duke of Edinburgh. Indeed, the mtDNA sequence of the putative Tsarina and her three daughters matched exactly that of the Duke of Edinburgh, supporting the hypothesis that their remains correspond to the Romanov family.[15] As to the putative body of the Tsar, his mtDNA sequence presented a very rare sequence heteroplasmy (presence of two alleles at a particular nucleotide position). By analyzing the body of the Tsar's brother and two living maternal relatives, the presence of this rare heteroplasmy was confirmed in the Tsar lineage, providing a powerful evidence supporting the identification of Tsar Nicholas II and his family.[16]

WHAT'S YOUR NAME? WHAT'S YOUR HAPLOTYPE?

In many societies, surnames are inherited patrilinearly, in the same manner as Y chromosomes. This parallel transmission can be used to tentatively associate surnames, or closely related human groups, and specific Y-chromosome haplotypes. For example, by using Y-chromosome microsatellites, it has been shown that half of the individuals with the English surname "Sykes" belong to a specific Y-chromosome haplotype that is not present in non-Sykes samples and, in general, in other UK samples.[17] The presence of other haplotypes among the remaining Sykes samples has been attributed to historical accumulation of nonpaternity. However, further studies have shown that this lineage is present in Baltic states, although it is virtually absent in the British Isles, Scandinavia, and Iceland. This study, although preliminary, may have important forensic and genealogical applications. Increasing the number of microsatellites analyzed in individuals may eventually reach such a level of resolution that surname-specific haplotypes could be observed, with obvious important applications.

The use of Y-chromosome variation has also given insights into genealogies dealing with Jewish identity. According to Jewish tradition, males of the Levi tribe, of which Moses was a member, were assigned special religious responsibilities, and male descendants of Aaron, his brother, were selected to serve as priests (Cohanim). Whereas in most cases Jewish identity is acquired by maternal descent, membership in the Cohen and Levi castes is determined by paternal descent. By studying a large group of Jewish individuals that include Israelites, Cohanim, and Levites, it was shown that whereas Israelites and Levites exhibit a heterogeneous set of Y chromosomes, the Cohanim are mainly characterized by a unique haplotype that is present at high frequencies in both Ashkenazic and Sephardic Cohanim.[18] This haplotype, known as the Cohen Modal haplotype, is thought to be a potential signature of Judaic origin.

CONCLUSION

The times when the Y chromosome was considered a chromosome of "low polymorphism" have passed away. Today, the Y chromosome phylogeny is robust and rich in variation that has been successfully used in population genetic and forensic studies. Nevertheless, the definition of Y-chromosome variation, through a well-defined battery of markers, in global samples of different human populations remains a task to be accomplished and represents an imperative prerequisite to establish population databases of Y-chromosome haplotypes to be used in forensics. In this context, the precise knowledge of Y-haplotype frequencies in a given population becomes very important when matching the DNA profile of a suspect with the corresponding population database, because this will give the probability of finding this haplotype by chance in the population. The creation of these high-resolution population databases will also increase the probability to deduce the geographic origin of a DNA sample and will shed light on the preliminary associations between certain surnames, or lineages, and Y-chromosome haplotypes with the consequent important forensic and legal applications.

REFERENCES

1. Lahn, B.T.; Pearson, N.M.; Jegalian, K. The human Y chromosome, in the light of evolution. Nat. Rev. Genet. **2001**, *2* (3), 207–216.
2. Underhill, P.A.; Shen, P.; Lin, A.A.; Jin, L.; Passarino, G.; Yang, W.H.; Kauffman, E.; Bonné-Tamir, B.; Bertranpetit, J.; Francalacci, P.; Ibrahim, M.; Jenkins, T.; Kidd, J.R.; Mehdi, S.Q.; Seielstad, M.T.; Wells, R.S.; Piazza, A.; Davis, R.W.; Feldman, M.W.; Cavalli-Sforza, L.L.; Oefner, P.J. Y chromosome sequence variation and the history of human populations. Nat. Genet. **2000**, *26* (3), 358–361.
3. Underhill, P.A.; Passarino, G.; Lin, A.A.; Shen, P.; Mirazon-Lahr, M.; Foley, R.A.; Oefner, P.J.; Cavalli-Sforza, L.L. The phylogeography of Y chromosome binary haplotypes and the origins of modern human populations. Ann. Hum. Genet. **2001**, *65* (1), 43–62.
4. Jobling, M.A.; Tyler-Smith, C. The human Y chromosome: an evolutionary marker comes of age. Nat. Rev., Genet. **2003**, *4* (8), 598–612.

5. Kayser, M.; Krawczak, M.; Excoffier, L.; Dieltjes, P.; Corach, D.; Pascali, V.; Gehrig, C.; Bernini, L.F.; Jespersen, J.; Bakker, E.; Roewer, L.; de Knijff, P. An extensive analysis of Y-chromosomal microsatellite haplotypes in globally dispersed human populations. Am. J. Hum. Genet. **2001**, *68* (4), 990–1018.

6. Kayser, M.; Sajantila, A. Mutations at Y-STR loci: Implications for paternity testing and forensic analysis. Forensic Sci. Int. **2001**, *118*, 116–121.

7. Sullivan, K.M.; Mannucci, A.; Kimpton, C.P.; Gill, P. A rapid and quantitative DNA sex test-fluorescence-based PCR analysis of X–Y homologous gene amelogenin. BioTechniques **1993**, *15*, 636.

8. Santos, F.; Pandya, A.; Tyler-Smith, C. Reliability of DNA-based sex tests. Nat. Genet. **1998**, *18*, 103.

9. Prinz, M.; Boll, K.; Baum, H.; Shaler, B. Multiplexing of Y-chromosome specific STRs and performance for mixed samples. Forensic Sci. Int. **1997**, *85*, 209–218.

10. Roewer, L.; Kayser, M.; Dieltjes, P.; Nagy, M.; Bakker, E.; Krawczak, M.; de Knijff, P. Analysis of molecular variance (AMOVA) of Y-chromosome-specific microsatellites in two closely related human populations. Hum. Mol. Genet. **1996**, *5* (7), 1029–1033.

11. Jobling, M.A. Y-chromosomal SNP haplotype diversity in forensic analysis. Forensic Sci. Int. **2001**, *118* (2–3), 158–162.

12. Capelli, C.; Redhead, N.; Abernethy, J.K.; Gratrix, F.; Wilson, J.F.; Moen, T.; Hervig, T.; Richards, M.; Stumpf, M.P.; Underhill, P.A.; Bradshaw, P.; Shaha, A.; Thomas, M.G.; Bradman, N.; Goldstein, D.B. A Y chromosome census of the British Isles. Curr. Biol. **2003**, *13* (11), 979–984.

13. Hurles, M.E.; Veitia, R.; Arroyo, E.; Armenteros, M.; Bertranpetit, J.; Pérez-Lezaun, A.; Bosch, E.; Shlumukova, M.; Cambon-Thomsen, A.; McElreavey, K.; Lopez De Munain, A.; Rohl, A.; Wilson, I.J.; Singh, L.; Pandya, A.; Santos, F.R.; Tyler-Smith, C.; Jobling, M.A. Recent male-mediated gene flow over a linguistic barrier in Iberia, suggested by analysis of a Y-chromosomal DNA polymorphism. Am. J. Hum. Genet. **1999**, *65* (5), 1437–1448.

14. Foster, E.A.; Jobling, M.A.; Taylor, P.G.; Donnelly, P.; de Knijff, P.; Mieremet, R.; Zerjal, T.; Tyler-Smith, C. Jefferson fathered slave's last child. Nature **1998**, *396*, 27–28.

15. Gill, P.; Ivanov, P.L.; Kimpton, C.; Piercy, R.; Benson, N.; Tully, G.; Evett, I.; Hagelberg, E.; Sullivan, K. Identification of the remains of the Romanov family by DNA analysis. Nat. Genet. **1994**, *6* (2), 130–135.

16. Ivanov, P.L.; Wadhams, M.J.; Roby, R.K.; Holland, M.M.; Weedn, V.W.; Parsons, T.J. Mitochondrial DNA sequence heteroplasmy in the Grand Duke of Russia Georgij Romanov establishes the authenticity of the remains of Tsar Nicholas II. Nat. Genet. **1996**, *12* (4), 417–420.

17. Sykes, B.; Irven, C. Surnames and the Y chromosome. Am. J. Hum. Genet. **2000**, *66* (4), 1417–1419.

18. Thomas, M.G.; Skorecki, K.; Ben-Ami, H.; Parfitt, T.; Bradman, N.; Goldstein, D.B. Origins of Old Testament priests. Nature **1998**, *394*, 138–140.

30 Forensic Identification
An Overview on Molecular Diagnostic Technology

Adrian Linacre, Yvonne E. Cruickshank
University of Strathclyde, Glasgow, U.K.

CONTENTS

INTRODUCTION

DNA profiling has greatly changed the way in which human identification is performed for the purpose of a forensic investigation. Since the first case in 1985 through to the most recent technological advances, DNA profiling has adapted techniques used in medical and diagnostic research. The use of microarrays and nanotechnology is now becoming part of diagnostic technology, so these same methods will be adapted to forensic science. This chapter outlines the current situation regarding forensic science and diagnostic techniques.

FORENSIC DNA PROFILING

Forensic science, being the application of science to law, has applied many scientific techniques used in other fields of the analytical sciences to address questions raised in a legal context. The application of DNA technology, and in particular the use of the polymerase chain reaction (PCR), has revolutionized the field of human identification. The first use of DNA in forensic science was to help solve a sexual assault in England. Professor Sir Alec Jeffreys coined the term "DNA fingerprinting"[1] when he was able to link two sexual assaults as committed by the same perpetrator and then to link the crime scene sample to the suspect, Colin Pitchfork. Previous examination of biological evidence used genetic polymorphisms producing different proteins; however, as the human genome is approximately 3 billion base pairs in size

DOI: 10.1201/9781003247432-30

and that less than 3% is gene related, greater polymorphism lay in the non-gene-related regions. It was always considered that the DNA from each individual was unique, with the exception of monozygotic twins. Professor Sir Alec Jeffreys noted that there were regions of DNA where the same sequence of DNA was repeated many times. The number by which this repeat motif was found along the chromosome of different people was found to differ. These regions, termed variable number tandem repeats (VNTRs, minisatellites), were exploited in DNA fingerprinting. The field of DNA profiling has now moved to the analysis of hypervariable short tandem repeats (STRs). Short tandem repeats are short regions of DNA, typically 2–8 DNA bases in length, that are repeated polymorphically along the chromosome. The first use of STRs was in 1994 when four separate DNA loci were amplified in one reaction.[2] Within 1 year the multiplex PCR had increased to six independent STR DNA loci.[3] Short tandem repeats are regions of DNA found close to genes, often within introns, in which a sequence of DNA is repeated. The repeat motif is frequently 4 bp in length with the number of repeats varying from 6 to over 30. The number of repeats is known to be highly variable, producing size variations at the loci. The flanking DNA on either side of the STR loci is known to be highly conserved allowing amplification of the locus by PCR using primers made to the flanking DNA. Primers have been designed that will work on all the major ethnic populations.

With increased knowledge of the PCR process the number of loci that could be successfully and routinely amplified increased to a point where currently 10 STR loci form the basis for the National DNA Database (NDNAD) in the United Kingdom. In the United States 13 loci form the Combined DNA Index System, or CODIS. Short tandem repeats are highly abundant in the human genome and are known to be polymorphic.[4,5]

While one STR locus provides little power of discrimination, it is common to analyze 10 STR loci in a single multiplex reaction. Each of the loci examined is independently inherited producing an overall power of discrimination of greater than a billion. This equates to close to identity and meets the current needs of the forensic community.

MITOCHONDRIAL DNA AND FORENSIC SCIENCE

The current situation in forensic science is that there are a number of autosomal STR loci that can be analyzed leading to possible links between a crime scene sample and that taken from a suspect. Mitochondrial DNA has been used in forensic science for human identification and for population studies.[6–8] Within each mitochondrion there is a circular loop of DNA being 16,569 bases in length in humans. There are no STR loci on the mitochondrial DNA, rather, DNA sequence polymorphisms are detected on either side of the point of replication. It is common practice to amplify this fragment of the mitochondrial genome and directly sequence the PCR product.

SINGLE NUCLEOTIDE POLYMORPHISMS

The sequence polymorphisms found within the mitochondrial DNA are termed single nucleotide polymorphisms (SNPs). SNPs are positions within the genome where there is a difference between people at one single base. It may be that at

Panel 1

Panel 2

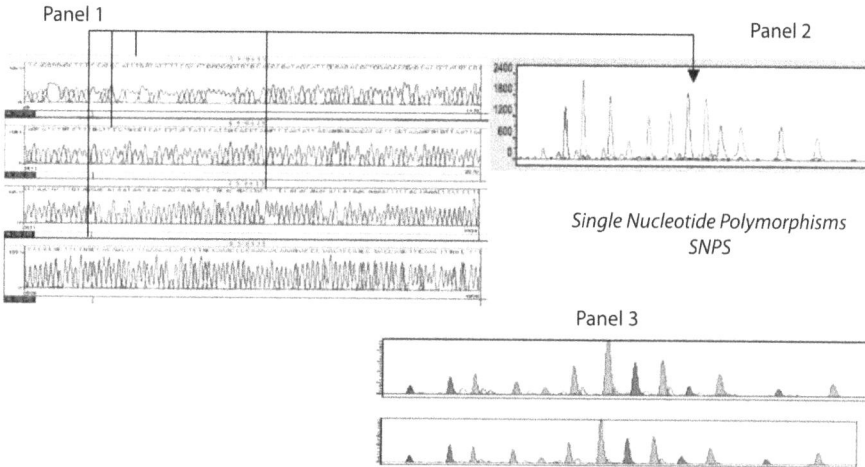

*Single Nucleotide Polymorphisms
SNPS*

Panel 3

FIGURE 30.1 Example of minisequencing of 12 SNP sites on the human mitochondrial genome. Panel 1 shows a DNA sequence trace from four samples of mitochondrial DNA and highlights single-base changes between one of the samples and the other three. Panel 2 shows the resulting minisequencing trace. Panel 3 shows a casework example where the top trace is from a hair found on clothing and the bottom trace is from an alleged victim of a crime. In this case there is a match between the DNA extracted from the hair and the reference sample from the victim.

Source: From work provided by Dr. Gillian Tully for publication purposes and based upon an illustration used in Ref. [10].

a particular position the nucleotide "A" is normally found, but in a minority of people the base "G" is present. In such an instance most people will have the base pair A/T with a minority being G/C at the same position. There are a number of SNP sites on the mitochondrial DNA within the loci examined. SNPs were first used in forensic science by the U.K. Forensic Science Service to examine mito-chondrial DNA from suspects and from crime scenes.[9] This method examined 12-point mutations known to be variable within the mitochondrial DNA by a technique termed minisequencing. Minisequencing of the mitochondrial DNA polymorphic sites allows rapid screening of the variable DNA bases. In forensic investigations this process is used to exclude all the samples in a case that do not match at the 12 points. An example of an SNP of mitochondrial DNA is shown in Figure 30.1.

Detection of SNP by minisequencing requires separation of DNA fragments by either gel or capillary electrophoresis. Much SNP testing for forensic and diagnostic purposes can now be performed using oligonucleotide microarrays.[10] Microarrays allow the single-base difference to be detected on a silicon chip. The two different versions, either A or G in the case example, are placed on the chip. If A is present then a DNA sequence with T will bind. If G is present a DNA sequence with C will bind. It is possible to detect SNP polymorphic sites on the mitochondrial DNA. An example is shown in Figure 30.2.

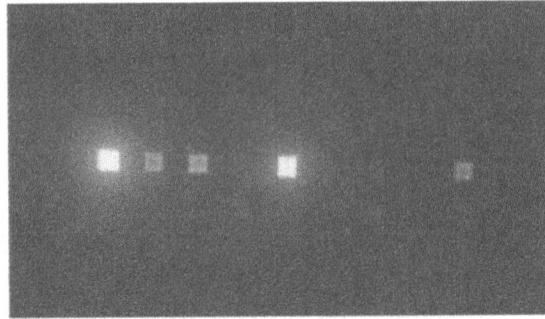

DNA SNP on the mitochondrial DNA

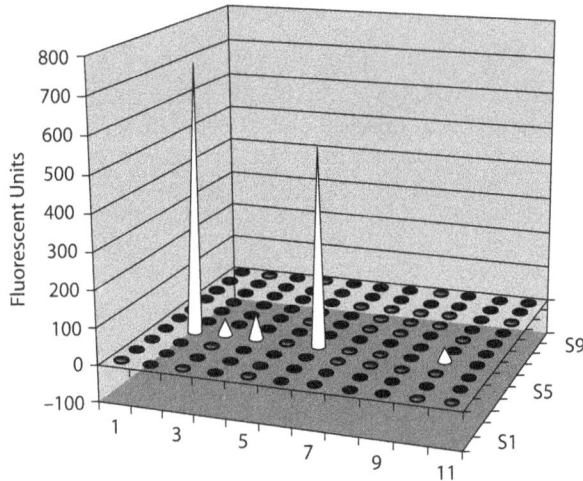

FIGURE 30.2 Analysis of two SNP loci on the human mitochondrial genome. The top panel shows the results of hybridization of a segment of DNA, mitochondrial DNA, to oligonucleotide sequences on a silicon chip. The position on the chip where the DNA binds to a complimentary sequence is detected by fluorescence. The bottom panel shows the amount of fluorescence at the two sites.

Source: **Research work by Dr. Barbara Llewellyn at the University of Strathclyde.**

SNP ANALYSIS IN CODING AND NONCODING DNA REGIONS

Within the chromosomal DNA, SNPs are highly abundant occurring on average 1 base in every 1000.[11,12] As the human genome is now known to be around 3 billion bases in size, there are therefore approximately 3 million SNPs within the human genome. To date, the International SNP Map Working Group has identified 1.42 million SNPs distributed throughout the human genome.[13] The SNP loci examined in forensic science are within the 97% of the human genome that is noncoding. SNPs that are present within coding regions of the DNA may have an effect on the protein transcribed by the gene and result in a genetic disorder.

As such, SNPs that cause a genetic disorder are of key interest to the medical community. These SNPs may be dominant or recessive and are responsible for a number of key genetic disorders. Additionally, SNPs within the regulator region of genes, such as in promoter sequences, may influence the susceptibility of individuals to particular disorders. The majority of SNPs are considered to have no known effect on the fitness of the organism; however, those SNP loci that affect the fitness of the organism have been the focus of interest to the medical community. The field of medical genetics has recently started to develop a number of diagnostic tests based upon single base-pair mutations in the human genome. A number of genetic diseases are the result of such single base-pair mutations, such as cystic fibrosis and beta thalassemia.[14,15] In parallel to the investment made in STRs by the forensic community, genetic diagnostics has focused on many such single point mutations. When the mutation becomes a common type, with the rarer allele at more than 1% of the frequency of the common allele, such loci are termed SNPs.[16]

Forensic science reacts to developments in other areas of science, frequently adapting techniques developed for nonforensic purposes. While DNA profiling is no exception, the analysis of STR loci has been highly successful. Tiny bloodstains can be examined and a DNA profile can be produced that has a match probability greater than 1 in a billion to a suspect. There has been much investment in the STR loci, particularly in the formation of DNA databases of felons. However, SNP analysis offers many advantages in the field of diagnostic analysis.

FORENSIC TOXICOLOGY

Forensic toxicology is defined as the study of the adverse effects of poisons on living organisms in relation to the law. Indeed, the most common poisons in today's society are drugs of abuse, controlled under different countries legislations, and the majority of forensic toxicology is concerned with the detection of such drugs within the blood or urine of a person. The analysis determines whether a controlled substance is present, and if so how much is present. In fatalities the quantity of a controlled substance must be known before a conclusion can be made as to the cause of death. To obtain the amount of drug present in the person at the time of death, a backtracking calculation is carried out. In such a calculation there are issues with the interpretation of the toxicology results. This is because one dose may be at a level that is therapeutic to one person but lethal to another. The interindividual differences in response to a drug are due to interindividual differences in drug metabolism, which in turn is a result of genetic polymorphisms in the production of drug receptors and drug transporters. A well-characterized genetic polymorphism occurs within the cytochrome–p450 system where there are subfamilies of isoenzymes including CYP2C9, CYP2D6, and CYP3A4.[17] Slow, medium, and fast metabolizers are the result of the production of different isoenzymes that is under a genetic control. SNPs to the gene system for the different metabolic systems would allow determination of whether the person had a slow metabolism, hence death resulting in a quantity of drug that would otherwise be therapeutic to a fast metabolizer.

CONCLUSION

One of the greatest advances in forensic science has been the introduction of DNA profiling. To date DNA profiling is performed using microsatellite amplicons (STRs) where hypervariability exists within the number of repeats of a short DNA sequence motif. Greater than 3 million samples have been analyzed by this method on the U.K. National DNA database alone, with currently an increasing number of samples on the U.S. CODIS and similar DNA databases throughout Europe. Due to the great investment in the science of DNA profiling using STR loci there would need to be a great advantage in a new technology. SNPs offer many advantages, particularly in the speed of throughput and option for automation. Approximately 50 SNP loci are required to have the same discrimination power as 10 STR loci, but 50 SNPs can be detected readily and therefore this level of DNA detection can be overcome. One major reason for microarrays not being more common in forensic science is that crime scene samples are frequently mixed, containing cellular material from more than one person. In such cases, a mix of a person with SNP of type A only and a person who is type G only could not be differentiated from a person who is A and G. Mixtures occur routinely in forensic science thereby adversely affecting the introduction of SNP typing in human identification. In forensic toxicology the samples are normally from one person only and it may be in application such as this that SNP typing will find a role.

REFERENCES

1. Jeffreys, A.J.; Wilson, V.; Thein, S.L. Individual-specific "fingerprints" of human DNA. Nature **1985**, *316*, 76–79.
2. Kimpton, C.P.; Gill, P.; Walton, P.A. Automated DNA profiling employing 'multiplex' amplification of short tandem repeat loci. PCR Methods **1993**, *3*, 13–22.
3. Sparkes, R.L.; Kimpton, C.P.; Gilbard, S. The validation of a 7-locus multiplex test for use in forensic casework. Int. J. Leg. Med. **1996**, *109*, 186–194.
4. Edwards, A.; Civitello, A.; Hammond, H.A. DNA typing and genetic-mapping with trimeric and tetrameric tandem repeats. Am. J. Hum. Genet. **1991**, *49*, 746–756.
5. Broman, K.W.; Murray, J.C.; Sheffield, V.C. Comprehensive human genetic maps: Individual and sex-specific variation in recombination. Am. J. Hum. Genet. **1998**, *63*, 861–869.
6. Sullivan, K.M.; Hopgood, R.; Gill, P. Identification of human remains by amplification and automated sequencing of mitochondrial DNA. Int. J. Leg. Med. **1992**, *105*, 83–86.
7. Ivanov, P.L.; Wadhams, M.J.; Roby, R.K. Mitochondrial DNA sequence heteroplasmy in the Grand Duke of Russia Georgij Romanov establishes the authenticity of the remains of Tsar Nicholas II. Nat. Genet. **1996**, *12*, 417–420.
8. Lutz, S.; Weisser, H.J.; Heizman, J.; Pollack, S. Location and frequency of polymorphic positions in the mtDNA control region of individuals from Germany: Errata. Int. J. Leg. Med. **1998**, *112*, 145–150.
9. Tully, G.; Sullivan, K.M.; Nixon, P. Rapid detection of mitochondrial sequence polymorphisms using multiplex solid-phase fluorescent minisequencing. Genomics **1995**, *34*, 107–113.
10. Linacre, A.; Graham, D. Role of molecular diagnostics in forensic science. Expert Rev. Mol. Diagn. **2002**, *2* (4), 346–353.

11. Cooper, D.N.; Smith, B.A.; Cook, H.J. An estimate of the unique DNA sequence heterozygosity in the human genome. Hum. Genet. **1985**, *69*, 201–205.

12. Wang, D.G.; Fan, J.B.; Siao, C.J. Large-scale identification, mapping, and genotyping of single-nucleotide polymorphisms in the human genome. Science **1998**, *280*, 1077–1082.

13. The International SNP Map Working Group. A map of the human genome sequence variation containing 1.42 million single nucleotide polymorphisms. Nature **2001**, *409*, 928–933.

14. Yamaguchi, A.; Nepote, J.A.; Kadivar, M. Allele specific PCR with microfluorometry: Application to the detection of del F508 mutation in cystic fibrosis. Clin. Chim. Acta **2002**, *316*, 147–154.

15. Chanock, S. Candidate genes and single nucleotide polymorphisms (SNPs) in the study of human disease. Dis. Markers **2001**, *17*, 89–98.

16. Risch, N.; Merikangas, K. The future of genetic studies of complex human disease. Science **1996**, *273*, 1516–1517.

17. Bertilsson, L.; Dahl, M.-L.; Dalen, P.; Al-Shurbaji, A. Molecular genetics of CYP2D6: Clinical relevance with focus on psychotropic drugs. Br. J. Pharmacol. **2002**, *53* (2), 111–122.

31 Denaturing Gradient Gel Electrophoresis (DGGE)

Luciana Cresta de Barros Dolinsky
Universidade Federal do Rio de Janeiro (UFRJ) and
Universidade do Grande Rio (UNIGRANRIO),
Rio de Janeiro, Brazil

CONTENTS

INTRODUCTION

Denaturing gradient gel electrophoresis (DGGE) is a DNA-based method that permits small mutation detection and single-nucleotide polymorphism (SNP) analysis. DGGE is the most sensitive mutation detection technique (detection rate close to 100%) and can be used for human hereditary disease diagnosis, large-scale mutation screening, and predictive genetic testing, including prenatal diagnosis and counseling. DGGE can also be used to analyze SNPs in hereditary complex disease traits. As DGGE is usually less expensive and less labor intensive and time-consuming than other point mutation detection techniques it is an excellent method to use in medical routine procedure.

DENATURING GRADIENT GEL ELECTROPHORESIS

DGGE is a DNA-based method that permits small mutation detection and SNP analysis with high-detection rate (close to 100%)[1] and is less labor intensive and time-consuming than other point mutation detection techniques. Despite this, it is easy to reproduce and can be used as a routine method.

TECHNICAL DESCRIPTION

The DGGE method is based on the electrophoretic mobility of a double-strand DNA molecule through linearly increasing concentrations of a denaturing agent (urea and formamide) on a polyacrylamide gel with ethidium bromide staining. As the

DOI: 10.1201/9781003247432-31

DNA fragment proceeds through the gradient gel, it will reach a position where the melting temperature (T_m) of its lowest melting domain equals the denaturing agent concentration, resulting in denaturation and consequent marked retardation of the DNA fragment mobility. As the T_m of a melting domain is dependent on its nucleotide sequence, even DNA fragments differing by a single nucleotide in their lowest domain will suffer branching and consequent retardation of their mobility at different positions along the DGGE gel, allowing DNA fragment separation.[2] The choice of the gel denaturant range is based upon the T_m of the fragment to be analyzed, and the electrophoretic runs can be performed at a constant temperature (58°C) that exceeds the T_m of an A–T-rich DNA fragment in the absence of denaturing agents.

DGGE cannot resolve fragments differing by nucleotide changes in the highest melting domain because of complete strand dissociation. This problem is overcome by introducing a GC clamp tail as short as 40 bp to serve as a high T_m domain and prevent complete dissociation of the DNA fragment. The GC clamp tail introduction increases the DGGE mutation detection percentage to close to 100%.[1] It is also possible to split DNA fragments into two segments to allow efficient mutation detection. The melting behavior of a DNA fragment can be simulated by computer software analysis.[3]

The DGGE method is conceptually similar to heteroduplex analysis (HA).[4,5] The PCR–DGGE combination is extremely efficient when applied to heterozygous nucleotide variants because of continuous denaturation and reannealing of single-strand molecules during PCR, allowing for the formation of heteroduplex and homoduplex molecules.[6] The presence of a single-nucleotide change within heteroduplexes decreases their melting domain temperature allowing separation from the homoduplexes and easy visual detection of the mutants (Figure 31.1). If working with an X-linked disease, it is necessary to mix male mutated DNA with a normal male DNA during PCR to ensure heteroduplex formation. To reduce workload, DNA fragments can be amplified in multiplex combinations. A further sequencing of the altered DNA fragment can determine the exact molecular alteration.

SENSITIVITY

Many different techniques such as single-strand conformation polymorphism (SSCP),[7] hydroxilamine and osmium tetroxide (HOT) chemical cleavage,[8] protein truncation test (PTT),[9] and HA[4,5] are available to identify single-nucleotide variants. However, in many cases, there is not a very high-mutation detection rate when we use each one of these techniques principally because of technical difficulties. For example, RNA-based techniques are difficult to use in a diagnostic setting because RNA expression is oftentimes extremely low.

Some laboratories are using haplotype analysis to assess risk status; however, this strategy may be inconclusive because in many cases there is intragenic recombination and high-mutation rate. This represents a great obstacle to genetic analysis of patients with small mutations and genetic counseling of their relatives.

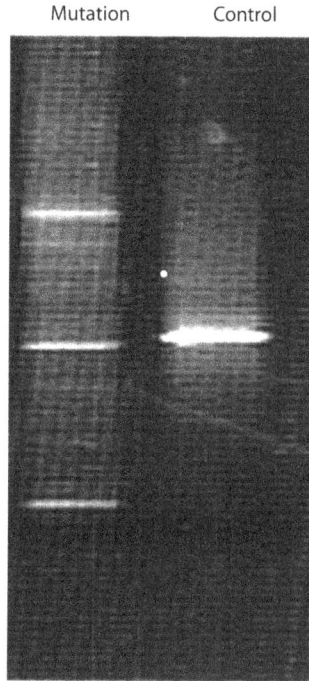

FIGURE 31.1 Small mutation detection in exon 17 of the dystrophin gene.

Recently, many diagnostic and research laboratories are using denaturing high-performance liquid chromatography (DHPLC)[10] to scan for single-nucleotide variations, but even with this method (best detection rate close to 92%) many families remain excluded from the analysis because the detection rate is not 100%. It is important to know that only mutation detection gives absolute certainty about diagnostics and counseling.

DGGE is the most sensitive electrophoretic method (mutation detection rate close to 100%)[1] and, despite this, does not require radioisotope use, is less labor intensive and time-consuming, is easy to reproduce, and is usually cheaper than the other routinely used methods for point mutation detection.[7]

CLINICAL APPLICATIONS

DGGE can be used for genetic diagnosis of hereditary human diseases caused by small mutations even when doing diagnosis for X-linked human diseases. It is also possible to analyze SNPs (genetic alterations where at least one allele frequency is higher than 1%) to associate with hereditary complex disease traits.

In many cases, mutation detection can be used in choosing adequate healthcare, as some mutation positions are associated with the worst prognostic. Knowing the exact mutation position makes it possible to establish the prognostic prior to the disease development course and, based on this, to decide which one is the best therapy. Knowing the polymorphism makes it possible to associate some polymorphisms with hereditary complex disease traits for each different population; for example, renin–angiotensin system polymorphisms are associated with high blood pressure in Japanese and African American.[11,12]

DGGE can also be used for predictive genetic testing, including prenatal diagnosis. Knowing a priori if a patient has inherited a familiar mutation with late expression, a polymorphism, or knowing during pregnancy if a baby has inherited a hereditary human disease can help health-care professionals in choosing an adequate strategy to care for the disease, to delay the disease development, or to establish the best life conditions during the disease course.

DGGE is also useful for counseling in an effort to avoid new cases of human hereditary diseases. In X-linked human diseases, for example, carrier couples can alternatively do in vitro fertilization with male embryo implantation to avoid new disease cases. Because mutation detection is close to 100% when using DGGE, we can do prenatal diagnosis and counseling knowing that we will give an answer to the family question and that not one family member will remain excluded from the analysis.

DGGE can also be used to perform large-scale population screening to improve genetic analysis. It is important for this kind of study as some mutations or polymorphisms have a high frequency in one population and a low frequency in another. Analyzing these frequency changes makes it possible to establish specific health-care rules for each different population, principally in hereditary complex disease traits where ethnic factors have an important role in inherited genes.[13]

DGGE is usually cheaper and less labor intensive and time-consuming than other routinely used point mutation detection techniques, does not require radioisotope use, is easy to reproduce, and, despite this, has a high-detection rate (close to 100%). Considering all these related qualities DGGE can be used as a medical routine procedure for practicing clinicians in hospitals and ambulatory settings.

CONCLUSION

We can conclude that DGGE is an excellent method for small mutation detection principally because it has a high-detection rate (close to 100%) and is less labor intensive and time-consuming than other point mutation detection techniques.

It can be useful for practicing clinicians in hospitals and ambulatory settings for diagnosis and predictive genetic testing, including prenatal diagnosis and counseling. It can also be used for large-scale mutation and SNP screening to improve genetic analysis and to establish adequate healthcare for each different population.

ACKNOWLEDGMENTS

I would like to acknowledge the diagnostic and research group from Leiden and Groningen Universities (The Netherlands), especially Drs. Johan T. Den Dunnen, Egbert Bakker, and Robert M.W. Hofstra for the important help during the beginning of this work. Thank you very much!

REFERENCES

1. Myers, R.M.; Fischer, S.G.; Lerman, L.S.; Maniatis, T. Nearly all single base substitutions in DNA fragments joined to GC-clamp can be detected by denaturing gradient gel electrophoresis. Nucleic Acids Res. **1985**, *13*, 3131–3145.
2. Myers, R.M.; Maniatis, T.; Lerman, L.S. Detection and Localization of Single Base Changes by Denaturing Gradient Gel Electrophoresis. In *Methods in Enzymology*; Wu, R., Ed.; Academic Press: New York, **1987**; Vol. 155, 501–527.
3. Lerman, L.S.; Silverstain, K. Computational Simulation of DNA Melting and Its Implication to Denaturing Gradient Gel Electrophoresis. In *Methods in Enzymology*; Wu, R., Ed.; Academic Press: New York, **1987**; Vol. 155, 482–501.
4. Soto, D.M.; Sukumar, S. Improved detection of mutations in the p53 gene in human tumors as single strand conformation polymorphism and double-strand heteroduplex DNA PCR. Methods Appl. **1992**, *2*, 96–98.
5. White, M.B.; Carvalho, M.; Derse, D.; O'Brien, S.J.; Dean, M. Detecting single base substitutions as heteroduplex polymorphisms. Genomics **1992**, *12*, 301–306.
6. Dolinsky, L.C.B. Denaturing Gradient Gel Electrophoresis (DGGE) for Mutation Detection in Duchenne Muscular Dystrophy (DMD). In *Neurogenetics: Methods and Protocols*, 1st Ed.; Potter, N.T., Ed.; Methods in Molecular Biology, Humana Press, Inc.: New Jersey, **2002**; Vol. 217, 165–175.
7. Orita, M.; Iwahana, H.; Kanazawa, H.; Hayashi, K.; Sekiya, T. Detection of polymorphisms of human DNA by gel electrophoresis as single-strand conformation polymorphism. Proc. Natl. Acad. Sci. U.S.A. **1989**, *86*, 2766–2770.
8. Cotton, R.G.H. Detection of single base changes in nucleic acids. Biochemistry **1989**, *263*, 1–10.
9. Roest, P.A.M.; Roberts, R.G.; Van der Tuijn, A.C.; Heikkop, J.C.; Van Ommen, G.J.B.; Den Dunnen, J.T. Protein truncation test (PTT) to rapidly screen the DMD gene for translation-terminating-mutations. Neuromuscul. Disord. **1993**, *3*, 391–394.
10. Bennet, R.R.; Den Dunnen, J.T.; Brien, K.F.; Darras, B.T.; Kunkel, L.M. Detection of mutations in the dystrophin gene via automated DHPLC screening and direct sequencing. BMC Genet. **2001**, *2*, 1–17.
11. Hata, A.; Namikawa, C.; Sasaki, M.; Soto, K.; Nakamura, T.; Tamura, K.; Lalowel, J.M. Angiotensinogen as a risk factor for essential hypertension in Japan. J. Clin. Invest. **1994**, *93*, 1285–1287.
12. Lifton, R.P.; Warnock, D.; Acton, R.T.; Harman, L.; Lalowel, J.M. High prevalence of hypertension associated with angiotensinogen variant T235 in African American. Clin. Res. **1993**, *41*, 260.
13. Niu, T.; Xu, X.; Rogus, J.; Zhou, Y.; Chen, C.; Yang, J.; Fang, Z.; Schmitz, C.; Zhao, J.; Roo, V.S.; Lindpainter, K. Angiotensinogen and hypertension in Chinese. J. Clin. Invest. **1998**, *101*, 88–194.

32 Heteroduplex Analysis (HA)

Philip L. Beales
Institute of Child Health (UCL), London, U.K.

CONTENTS

INTRODUCTION

The Human Genome Project provided a reestimate of the total number of genes within the human genome, and as specific links are made between genes and disease, the need for simple and cost-effective gene mutation detection becomes increasingly important. Nonetheless, the choice of test undertaken for any given screen presents a challenge, and in doing so, a number of considerations should be taken into account: (1) the level of throughput required; (2) the sensitivity and specificity permissible; (3) the resources available; (4) the speed with which a result is needed; and (5) cost.

Direct sequencing, for long considered the gold standard of mutation detection, is still costly and time-consuming; however, in terms of simplicity, heteroduplex analysis (HA) is unsurpassed in its application and analysis. This chapter will review the technique, its applications, and the latest developments that keep HA at the forefront of scanning technologies.

HETERODUPLEX ANALYSIS

More than 90% of disease-related mutations are caused by "microlesions" (single-base substitutions or small insertion/deletions [indels]) rather than large lesions (such as rearrangements or large indels) (Human Gene Mutation Database[1]). Given this

DOI: 10.1201/9781003247432-32

overrepresentation of microlesions, it is paramount that mutation detection methods are sensitive to single-base substitutions. The principle of HA relies on the formation of heteroduplex species formed when there is a complementary sequence mismatch in a double-stranded DNA fragment. The electrophoretic mobility of a heteroduplex is altered relative to a homoduplex and in general tends to retard migration of the fragment. Clearly, if an alteration is homozygous no heteroduplex species will be formed and therefore it will be necessary to mix wild-type and mutant fragments together prior to the denaturation and annealing steps to generate four species: a wild-type homoduplex, a mutant homoduplex, and two heteroduplexes. The occurrence of heteroduplex fragments was, in fact, first reported as a PCR artifact.[2]

Bhattacharyya and Lilley[3] predicted the formation of two forms of heteroduplex depending on the type of underlying mutation. A small insertion or deletion produces a "bulge" as a result of a relatively large stretch of unmatched bases in the duplex (Figure 32.1). Such "bulges" have been directly visualized on electron microscopy.[4] In contrast, a "bubble" will form at the site of a base substitution. The net effect on migration of a "bulge" heteroduplex is much greater relative to a homoduplex, owing to the bending of the DNA at the site of the bulge. In fact, the larger the sequence mismatch, the greater the bend and, consequently, the greater is the degree of retardation. In contrast, the "bubble"-type heteroduplex produces only subtle perturbations of structure, owing to less bending of the DNA and hence relatively minor migration defects. Nonetheless, the migration of these species can be further influenced by factors such as the length of fragment and even the nature and position of the mismatch.[5]

HETERODUPLEX ANALYSIS MATRICES

At its earliest inception, the sensitivity of HA was limited by the availability of suitable gel matrices. Slab-gel systems are limited to agarose and polyacrylamide. The improved MDE gel solution (Cambrex) is a polyacrylamide-like matrix with a high sensitivity for DNA conformational changes. The gel's unique structure allows DNA separation on the basis of both size and conformation, thus increasing the probability of detecting sequence differences from as low as 15% achieved in standard polyacrylamide gels to approximately 80%.

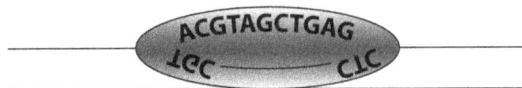

FIGURE 32.1 Types of heteroduplex. (a) "Bubble-type" heteroduplex forms in the presence of one or more base substitutions. (b) "Bulge-type" heteroduplex forms with larger lesion such as insertion or deletion.

A number of modifications to the matrix have been reported to improve the gel's ability to enhance fragment retardation. For example, the use of mild denaturants that exaggerate the effects of the mismatch forms the basis of conformation-sensitive gel electrophoresis (CSGE).[6–8] Ganguly et al.[9] have adapted the technique for use with fluorescently labeled fragments which can then be resolved using the ABI 377 (Applied Biosystems) DNA sequencer platform. They labeled PCR products with either 6-FAM, HEX, or NED prior to heteroduplex formation.

An array of matrices is now available for use with capillary electrophoresis such as cellulose derivatives and hydrophilic polymers substantially improving the resolution of HA.[10–12] One of the earliest reports to describe the adaptation of HA to capillary electrophoresis employed an entangled polymer matrix under nondenaturing conditions.[13] Rozycka et al.[14] adapted the technique for use with a single capillary fragment analyzer (ABI 310, Applied Biosystems) and were able to achieve high sensitivity in products of <350 bp. Such systems utilize replaceable linear polyacrylamides (LPAs) offering lower viscosity, high resolution, and rapid replacement of matrix thus permitting multiple reuse of the capillary. To attain high-throughput coupled with a greater resolution, Gao and Yeung[15] described temperature-gradient capillary electrophoresis. They exploited the lower melting temperature of heteroduplexes relative to homoduplexes and applied a temperature gradient to partially denature the DNA fragments during electrophoresis. By covering the entire range of melting temperatures of duplexes, they were able to achieve a high sensitivity for single-base substitutions set in a 96-well format. Our group, in an effort to raise the level of throughput without compromising sensitivity, designed a modification called multiplex capillary heteroduplex analysis (MCHA) using the 96-capillary MegaBACE1000 DNA Analyzer platform (Amersham Biosciences).[16] We described an additional level of throughput achieved by multiplexing up to six differently labeled (6-FAM, HEX, or TET) or sized products (up to 494 bp) in a single well. This coupled with 96-sample capability per run allows up to 576 samples in 1 h or as many as 3000 samples in a day. MCHA is extremely effective at detecting known alterations (specific) but has proven to be a useful scanning tool for the detection of novel mutations (Figure 32.2).

SENSITIVITY OF HA

The sensitivity of HA has consistently been reported to be high when utilized alone or together with another complementary technique such as single-strand conformation polymorphism (SSCP). Rossetti et al.[17] compared conventional horizontal gel HA with SSCP using known mutations from four different genes. HA detected 93% of mutations compared with 67% using SSCP; however, when combined, all mutations were revealed.

A comparison of HA, SSCP, and DGGE used for detecting apolipoprotein mutations in a large cohort of patients with hypercholesterolaemia revealed HA and DGGE to be equally as effective but superior to SSCP.[18]

Recently, Crepin et al.[19] designed an assay combining HA and an SSCP variation they called mutation detection gel analysis (MDGA) for the detection of multiple mutations in MEN1 using fluorescently labeled PCR fragments. Heteroduplexes

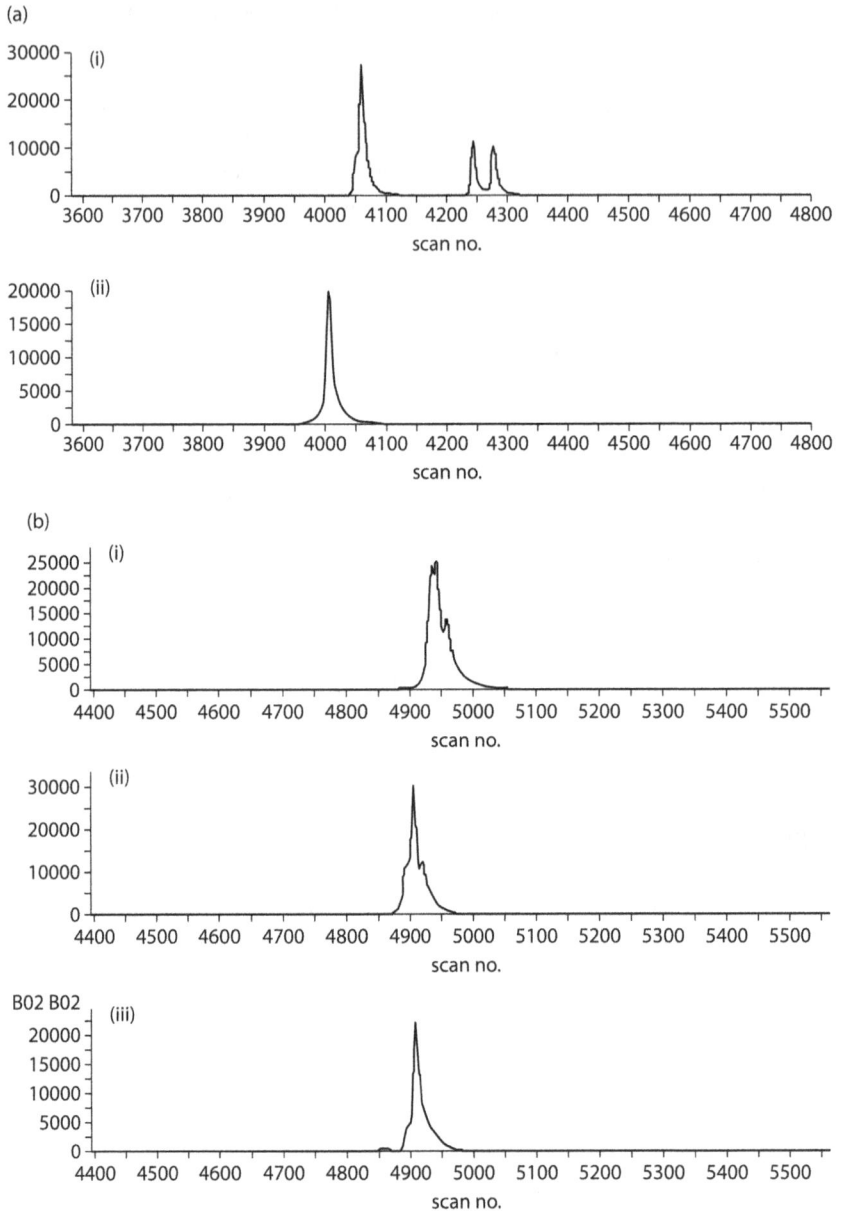

FIGURE 32.2 Peak traces of multiplex capillary heteroduplex analysis (MCHA) depicting profiles for different categories of mutations. (a) A 2 bp deletion in *BBS2×4*. (i) Homozygous patient sample mixed with control DNA giving four species: a mutant homoduplex, a wild-type homoduplex, and two heteroduplexes; (ii) control DNA homoduplex. (b) A G→C substitution in *BBS4×15*. (i) Patient sample; (ii) patient sample mixed with control DNA; (iii) control DNA homoduplex. (c) *BBS4×16* SNP. (i) Heterozygous sample mixed with control DNA; (ii) control DNA homoduplex. (d) A T→A substitution in *BBS2×4*. (i) Patient sample mixed with control DNA; (ii) control DNA homoduplex. *(Continued)*

Source: From Ref. [16].

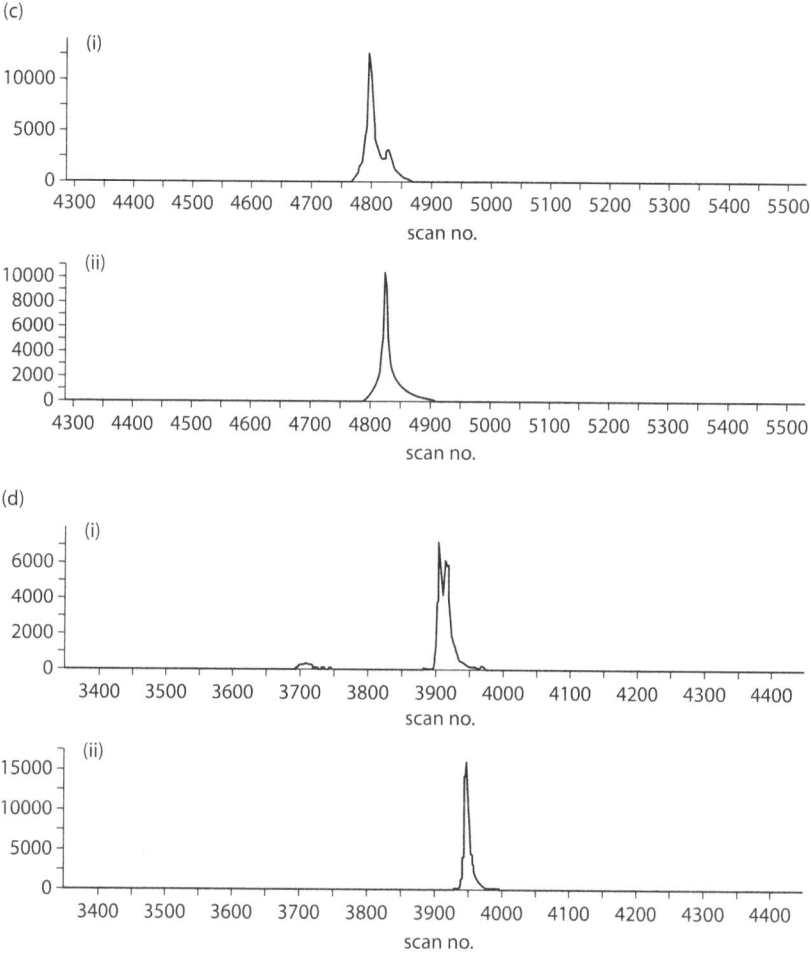

FIGURE 32.2   *(Continued)*

were resolved on a 0.5× MDE gel and for MDGA, single-stranded products on an Elliomut gel (Elliotek, Paris), both electrophoresed as a slab using an automated sequencer (ABI 377). Peaks were compared with controls and abnormal profiles subsequently sequenced to reveal the true nature of any underlying mutation. They demonstrated that each mutation cast a specific and reproducible pattern dependent on the base change. HA detected 88.5% of mutations and 100% when combined with MDGA. The authors questioned the specificity of the tests and suggested this be remedied by reducing the fragment length to less than 400 bp. They concluded that even if the positive predicted value of the combined test does not reach 100%, the negative predictive value does, a point of critical importance in genetic testing.

The study of Hoskins et al.[16] that tested 12 sequence changes (2 deletions and 10 substitutions) in four Bardet–Biedl syndrome genes concluded that HA when adapted for use on a 3% nondenaturing capillary injected matrix (LPA) gave outstanding sensitivity (100%) alone. Furthermore, they calculated the sample cost to be at least one-tenth that for direct sequencing.

FACTORS INFLUENCING SENSITIVITY

A number of factors are now known to influence the ability to detect single-base substitutions, which include the length of the fragment, the position of the mismatch within the fragment, the GC content, and the nature of the mismatch.

Fragment Length

Very short (<150 bp) or very long (>600 bp) fragments are considered unsuitable for HA. Rozycka et al.[14] showed a reduction in sensitivity as the fragment length increased above 350 bp. Hoskins et al.[16] using a capillary gel were able to detect mismatches in fragments of up to 500 bp with no deterioration in sensitivity.

Position of Mismatch

Mismatches within 50 bp of the product start and end are difficult to detect.[14] Therefore, attention to the design of amplification products is paramount.[16]

GC Content

The GC content of the fragment in question and the sequence flanking the change can influence the ability of HA to detect the mismatch. This is thought to be due to the three-dimensional conformation adopted by the double-stranded DNA that in turn is dependent on the GC content. The relationship of the GC-rich sequence in the region surrounding the mismatch will also influence detection rates especially if the mismatch is buried in the heteroduplex structure compared with a more superficial location.

Nature of the Mismatch

The presence of a mismatch, e.g., a C>T within a double-stranded fragment, will generate four species: a wild-type homoduplex (composed of C:G), a mutant homo-duplex (T:A), and two heteroduplexes, one containing a C:A mismatch and the other a T:G. Highsmith et al.[5] designed a DNA "toolbox," a set of molecules of varying length and GC content containing all possible mismatches at a particular location, to determine which factor had the greatest effect on the ability to detect a single-base substitution. They reported that the factor with the greatest effect was the nature of the nucleotide change; the order of detection being G:G/C:C>A:C/T:G>A:A/T:T. They concluded that this hierarchy of resolution was in agreement with the measured thermodynamic stability of oligonucleotides with different base pair mismatches.[20–22] It was presumed that the larger the distortion due to several unstacked bases, the greater the differential migration in the gel compared with fully complementary strands.

In an analysis of all mutations recorded in the Human Gene Mutation Database (HGMD), Krawczak et al.[23] reported that 60% were transitions. One reason for the predilection toward transitions in any set of alterations is the relatively high incidence of CG>TG and CG>CA substitutions caused by deamination of methylcytosine. Methylated cytosines most commonly occur in CpG dinucleotides and are susceptible to deamination favoring the formation of thymine (CG>TG). CG>CA substitutions occur if, following deamination of a methylcytosine in the antisense strand, there is a miscorrection of G>A in the sense strand. In an update to the study of Hoskins et al.[16] (unpublished) these types of mutation accounted for 21% of substitutions and 33% of transitions consistent with Krawczak et al.'s findings. The majority of the remaining transitions were A>G substitutions, with five of seven adenines preceded by another adenine. The local DNA sequence environment may therefore influence the sensitivity of techniques such as MCHA.

ADAPTATIONS

The use of denaturing HPLC systems to resolve heteroduplexes from homoduplexes has been reported for many gene mutations and has been discussed in another chapter. HA has been adapted for the detection of specific mutations in numerous conditions including von Willebrand's disease, PKU, hemochromatosis, and rifampicin resistance in tuberculosis.[24–27] As discussed above, test samples that are heterozygous for a mutation provide for a straightforward preparation of DNA heteroduplexes without the need for annealing to a reference sample, an additional step that can be time-consuming. For single-base substitutions the conformational change is often relatively small, but to enhance this difference for specific mutations a special fragment, termed a universal heteroduplex generator (UHG), has been developed.[28] This synthetic sequence is similar to genomic sequence but contains specific substitutions or indels at nucleotide positions corresponding to and contiguous with known mutation sites within the genomic DNA. A PCR-amplified test fragment is mixed with the UHG, denatured, and then slowly cooled to form a heteroduplex which is then electrophoresed on a suitable gel (Figure 32.3). If the test fragment contains a point mutation then the heteroduplex will contain both a bulge and bubble, and, consequently, will migrate much more slowly than a heteroduplex without the mutation or a homoduplex.

CONCLUSION

Despite continuing improvements to direct sequencing methodologies, SSCP and HA remain the most popular scanning techniques. HA, although similar in principle to SSCP, is considered to be the simpler and quicker of the two techniques. The relative sensitivities, although initially low, have now been substantially enhanced by the development of new gel matrices and, more recently, the adaptation to fluorescent capillary platforms (e.g., MCHA) enabling mutation detection rates close to 100%. The cost of HA compared with direct sequencing for scanning purposes is much more economical even when abnormal resulting fragments are subsequently sequenced to reveal the nature of the underlying mutation. Developments such as the UHG allow the technique to be tailored to specific mutations.

Already a number of reports have considered the use of microchip-based electrophoresis in resolving heteroduplexes. Footz et al.[29] described a simple but sensitive method for specific mutation detection for *BRCA1* and *BRCA2*. They utilized

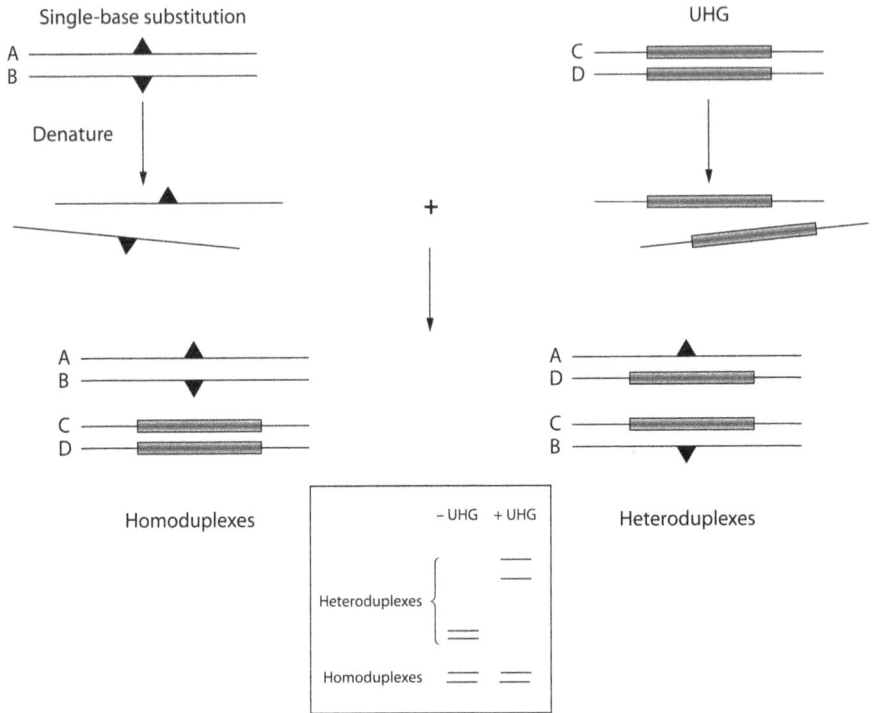

FIGURE 32.3 Schematic of the principle of the universal heteroduplex generator (UHG). The artificial mutation enhances the bubble effect of an existing mutation resulting in greater separation on gel electrophoresis.

a proprietary microfluidic chip (Micralyne, Edmonton, AB) consisting of a defined network of microchannels in glass, similar to conventional capillaries. The chip was filled with a commercially available sieving matrix (POP-6, Applied Biosystems) and loaded with 200–300-bp test PCR products. They experienced very high sensitivity rates for known mutations and reported rapid run times. Each run utilizes a very small quantity of PCR product (250 pL) and virtually no optimization is required. Clearly, similar efforts to increase throughput in a microchip-based system are on the horizon.

ACKNOWLEDGMENTS

I would like to thank Bethan Hoskins for her development of multiplex capillary heteroduplex analysis and for helpful comments in preparing this manuscript. This work is supported by the Wellcome Trust.

REFERENCES

1. HGMD. http://www.hgmd.org.
2. Nagamine, C.M.; Chan, K.; Lau, Y.F. A PCR artifact: Generation of heteroduplexes. Am. J. Hum. Genet. **1989**, *45* (2), 337–339.

3. Bhattacharyya, A.; Lilley, D.M. The contrasting structures of mismatched DNA sequences containing looped-out bases (bulges) and multiple mismatches (bubbles). Nucleic Acids Res. **1989**, *17* (17), 6821–6840.

4. Wang, Y.H.; Barker, P.; Griffith, J. Visualization of diagnostic heteroduplex DNAs from cystic fibrosis deletion heterozygotes provides an estimate of the kinking of DNA by bulged bases. J. Biol. Chem. **1992**, *267* (7), 4911–4915.

5. Highsmith, W.E. Jr.; Jin, Q.; Nataraj, A.J.; O'Connor, J.M.; Burland, V.D.; Baubonis, W.R.; Curtis, F.P.; Kusukawa, N.; Garner, M.M. Use of a DNA toolbox for the characterization of mutation scanning methods: I. Construction of the toolbox and evaluation of heteroduplex analysis. Electrophoresis **1999**, *20* (6), 1186–1194.

6. Korkko, J.; Annunen, S.; Pihlajamaa, T.; Prockop, D.J.; Ala-Kokko, L. Conformation sensitive gel electrophoresis for simple and accurate detection of mutations: Comparison with denaturing gradient gel electrophoresis and nucleotide sequencing. Proc. Natl. Acad. Sci. U.S.A. **1998**, *95* (4), 1681–1685.

7. Ganguly, A.; Prockop, D.J. Detection of mismatched bases in double stranded DNA by gel electrophoresis. Electrophoresis **1995**, *16* (10), 1830–1835.

8. Ganguly, A.; Rock, M.J.; Prockop, D.J. Conformation-sensitive gel electrophoresis for rapid detection of singlebase differences in double-stranded PCR products and DNA fragments: Evidence for solvent-induced bends in DNA heteroduplexes. Proc. Natl. Acad. Sci. U.S.A. **1993**, *90* (21), 10325–10329.

9. Ganguly, T.; Dhulipala, R.; Godmilow, L.; Ganguly, A. High throughput fluorescence-based conformation-sensitive gel electrophoresis (F-CSGE) identifies six unique BRCA2 mutations and an overall low incidence of BRCA2 mutations in high-risk BRCA1-negative breast cancer families. Hum. Genet. **1998**, *102* (5), 549–556.

10. Pariat, Y.F.; Berka, J.; Heiger, D.N.; Schmitt, T.; Vilenchik, M.; Cohen, A.S.; Foret, F.; Karger, B.L. Separation of DNA fragments by capillary electrophoresis using replaceable linear polyacrylamide matrices. J. Chromatogr. A **1993**, *652* (1), 57–66.

11. Barron, A.E.; Soane, D.S.; Blanch, H.W. Capillary electrophoresis of DNA in uncrosslinked polymer solutions. J. Chromatogr. A **1993**, *652* (1), 3–16.

12. Heller, C.; Pakleza, C.; Viovy, J.L. DNA separation with field inversion capillary electrophoresis. Electrophoresis **1995**, *16* (8), 1423–1428.

13. Cheng, J.; Kasuga, T.; Mitchelson, K.R.; Lightly, E.R.; Watson, N.D.; Martin, W.J.; Atkinson, D. Polymerase chain reaction heteroduplex polymorphism analysis by entangled solution capillary electrophoresis. J. Chromatogr. A **1994**, *677* (1), 169–177.

14. Rozycka, M.; Collins, N.; Stratton, M.R.; Wooster, R. Rapid detection of DNA sequence variants by conformation-sensitive capillary electrophoresis. Genomics **2000**, *70* (1), 34–40.

15. Gao, Q.; Yeung, E.S. High-throughput detection of unknown mutations by using multiplexed capillary electrophoresis with poly(vinylpyrrolidone) solution. Anal. Chem. **2000**, *72* (11), 2499–2506.

16. Hoskins, B.E.; Thorn, A.; Scambler, P.J.; Beales, P.L. Evaluation of multiplex capillary heteroduplex analysis: A rapid and sensitive mutation screening technique. Hum. Mutat. **2003**, *22* (2), 151–157.

17. Rossetti, S.; Corra, S.; Biasi, M.O.; Turco, A.E.; Pignatti, P.F. Comparison of heteroduplex and single-strand conformation analyses, followed by ethidium fluorescence visualization, for the detection of mutations in four human genes. Mol. Cell. Probes **1995**, *9* (3), 195–200.

18. Henderson, B.G.; Wenham, P.R.; Ashby, J.P.; Blundell, G. Detecting familial defective apolipoprotein B-100: Three molecular scanning methods compared. Clin. Chem. **1997**, *43* (9), 1630–1634.

19. Crepin, M.; Escande, F.; Pigny, P.; Buisine, M.P.; Calender, A.; Porchet, N.; Odou, M.F. Efficient mutation detection in MEN1 gene using a combination of single-strand conformation polymorphism (MDGA) and heteroduplex analysis. Electrophoresis **2003**, *24* (1–2), 26–33.

20. Aboul-ela, F.; Koh, D.; Tinoco, I. Jr.; Martin, F.H. Base–base mismatches. Thermodynamics of double helix formation for dCA3XA3G+dCT3YT3G (XY=A,C,G,T). Nucleic Acids Res. **1985**, *13* (13), 4811–4824.

21. Ke, S.H.; Wartell, R.M. Influence of nearest neighbor sequence on the stability of base pair mismatches in long DNA; determination by temperature-gradient gel electrophoresis. Nucleic Acids Res. **1993**, *21* (22), 5137–5143.

22. Werntges, H.; Steger, G.; Riesner, D.; Fritz, H.J. Mismatches in DNA double strands: Thermodynamic parameters and their correlation to repair efficiencies. Nucleic Acids Res. **1986**, *14* (9), 3773–3790.

23. Krawczak, M.; Ball, E.V.; Cooper, D.N. Neighboring-nucleotide effects on the rates of germ-line single-base-pair substitution in human genes. Am. J. Hum. Genet. **1998**, *63* (2), 474–488.

24. Wood, N.; Tyfield, L.; Bidwell, J. Rapid classification of phenylketonuria genotypes by analysis of heteroduplexes generated by PCR-amplifiable synthetic DNA. Hum. Mutat. **1993**, *2* (2), 131–137.

25. Wood, N.; Standen, G.R.; Murray, E.W.; Lillicrap, D.; Holmberg, L.; Peake, I.R.; Bidwell, J. Rapid genotype analysis in type 2B von Willebrand's disease using a universal heteroduplex generator. Br. J. Haematol. **1995**, *89*(1), 152–156.

26. Fruchon, S.; Bensaid, M.; Borot, N.; Roth, M.P.; Coppin, H. Use of denaturing HPLC and a heteroduplex generator to detect the HFE C282Y mutation associated with genetic hemochromatosis. Clin. Chem. **2003**, *49* (5), 822–824.

27. Thomas, G.A.; Williams, D.L.; Soper, S.A. Capillary electrophoresis-based heteroduplex analysis with a universal heteroduplex generator for detection of point mutations associated with rifampin resistance in tuberculosis. Clin. Chem. **2001**, *47* (7), 1195–1203.

28. Clay, T.M.; Culpan, D.; Howell, W.M.; Sage, D.A.; Bradley, B.A.; Bidwell, J.L. UHG crossmatching. A comparison with PCR-SSO typing in the selection of HLA-DPB1-compatible bone marrow donors. Transplantation **1994**, *58* (2), 200–207.

29. Footz, T.; Somerville, M.J.; Tomaszewski, R.; Sprysak, K.A.; Backhouse, C.J. Heteroduplex-based genotyping with microchip electrophoresis and dHPLC. Genet. Test **2003**, *7* (4), 283–293.

33 NEBcutter

A Program to Cleave DNA with Restriction Enzymes

Janos Posfai, R. J. Roberts, Tamas Vincze
New England Biolabs, Inc., Beverly, Massachusetts, U.S.A.

CONTENTS

INTRODUCTION

NEBcutter is a computer program offering services to research laboratories that employ nucleic acid technologies. The software provides bioinformatics tools for experiments where DNA molecules are manipulated with restriction enzymes. Type II restriction enzymes recognize specific four- to eight-nucleotide-long patterns in DNA and introduce cuts into the strands at predictable positions.[1,2] The specificity and precision of cleavage make these enzymes useful in a wide array of applications. Gene cloning, detection of point mutations, probing of repeat sequences, genome mapping, and characterization of genetic variations[3–5] are a few of these. The NEBcutter program supports such applications by providing routines for the systematic analysis of restriction site distributions within defined DNA sequences. The service is accessible over the web, through a link on http://www.neb.com.

The discovery of restriction enzymes and the accumulation of significant amounts of sequenced DNA prompted the development of the first restriction digest analysis computer programs.[6,7] Since then, hundreds of new restriction enzymes have been characterized. The authoritative restriction enzyme database, REBASE,[8] lists over 13,000 type II entries, with more than 500 different specificities. The amount of sequenced DNA has expanded equally rapidly. The GenBank[9] collection of DNA sequences now contains over 6.25 trillion bases, including complete genome sequences of dozens of species (human, mouse, fruit fly, yeast, bacteria, and so on).

DOI: 10.1201/9781003247432-33

Substantial changes in computer power and computing environments occurred during this time as well. Advanced algorithms,[10] network access, linked databases, and graphic presentations have become standard. In this new environment, NEBcutter provides tools for state-of-the-art restriction enzyme digest analysis.

METHODS

Analysis starts by specifying a target DNA sequence. Many popular vector and phage sequences (such as pBR322, lambda, and T7) are directly accessible for processing. Alternatively, DNA sequences can be pasted into an entry form, or uploaded from local user files. Sequences of GenBank entries can also be loaded simply by specifying their accession numbers. The program distinguishes between linear and circular sequences, and handles them according to their topology. Occasional ambiguities in the input sequence are permitted; these are resolved by putting special marks on hits that are conditional on the specific sequence. Analyzing inputs with runs of Ns (unspecified nucleotides) would be meaningless because the runs could be sites for any enzyme, and so such inputs should be avoided. Sizes of target sequences are currently limited to 300 kb.

The program determines the cut sites for a repertoire of restriction enzymes. This repertoire is selectable and may be New England Biolabs-marketed enzymes, all commercially available specificities or all specificities ever characterized. Users can add hypothetical enzymes with fictitious sites to the list manually. NEBcutter keeps its internal data structures up to date by connecting daily to REBASE and by fetching the latest information about restriction enzymes and specificities.

Results from a digest are presented in schematic figures, summary tables, and detailed textual pages. In the default setting, the site distribution of enzymes with single cuts is displayed (Figure 33.1). One-click options can bring up the distributions of double or triple cutters. One can also switch to a display of custom digests with particular enzyme(s). Cut sites sensitive[11] to the mammalian CpG methylation or to typical *Escherichia coli* methylation are highlighted on the output. Color-coded symbols distinguish cuts with blunt ends from cuts with 5' and 3' extensions. The user can zoom on selected regions, all the way to the actual sequence. Circular sequences need to be linearized (with a single click) before engaging this zoom option.

In many applications, locations of gene flanking cut sites are of particular interest. To find them, NEBcutter places arrows on the output to indicate where the genes of the input sequence are, assuming that GenBank-style coordinate annotations are available. When the input is a plain, unannotated sequence, NEBcutter displays open reading frames (ORFs) according either to its own internal ORF finder routine, or to manually entered and edited data. Clicking on an arrow brings the represented ORF into focus. On this page, sites that flank the ORF are indicated. ORF-specific silent mutagenesis analysis can also be run. This module computes silent mutations that would turn previously noncutting enzymes into cutters.

From the results of calculated hypothetical digests, the program is able to simulate how the fragments would separate in various gel electrophoresis experiments

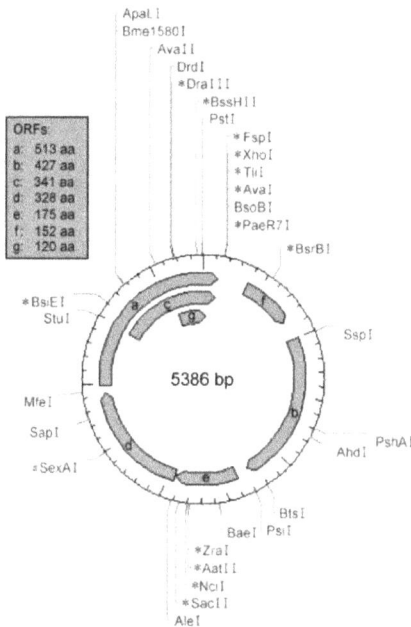

FIGURE 33.1 Restriction digest analysis results for phage PhiX174. Sites for enzymes with single cuts in the circular 3.4-kb sequence are displayed. Methylation-sensitive sites are marked with symbols. Color coding distinguishes blunt end cuts from ones that create overhangs. Gray arrows indicate the genes. Menu options for further actions are available at the bottom of the web page.

(Figure 33.2). A reference lane with selected marker fragments can be added to the view. Additional lanes will demonstrate the effects of possible DNA methylations.

With only a few clicks, NEBcutter will execute typical restriction digest analysis tasks. For more sophisticated tasks, sets of options are available. On-line context-specific help is accessible from every page. Pop-up windows show the basic information for every reported site. Direct links into REBASE are provided,

NEW ENGLAND BioLabs. NEBcutter

Custom Digest

Print | Close

Help | Comments

PhiX174 - digested with: MboII, MspI

Gel Type: **0.7% agarose** ⇕   Marker: **Lambda – HindIII Digest** ⇕   DNA Type: **Unmethylated** ⇕

L= 102 mm OK

marker unmeth. Cpg

#	Ends	Coordinates	Length (bp)
1	MboII-MboII	4756-492	1123
2	MboII-MboII	1654-2465	812
3	MspI-MboII	1104-1653	550
4	MboII-MboII	3740-4155	416
5	MboII-MboII	4380-4755	376
6	MspI-MboII	3368-3739	372
7	MspI-MspI	3020-3367	348
8	MboII-MspI	817-1103	287
9	MboII-MspI	493-729	237
10	MboII-MboII	4156-4379	224
11	MspI-MspI	2801-3019	219
12	MboII-MspI	2676-2800	125
13	MboII-MboII	2555-2672	118
14	MboII-MboII	2466-2554	89
15	MspI-MboII	730-816	87
16	MboII-MboII	2673-2675	3

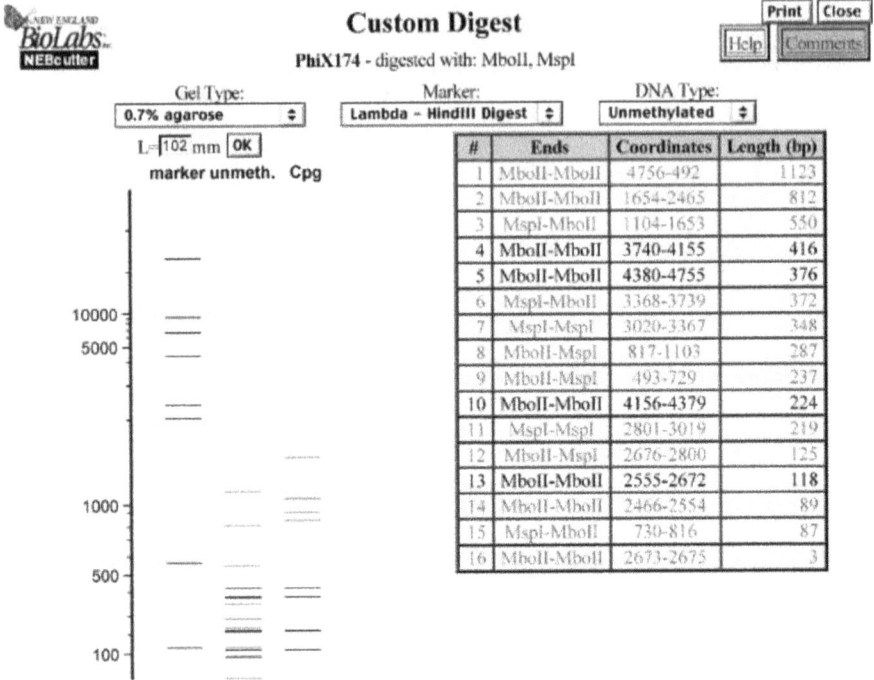

The virtual gel was generated by interpolating experimental data. See details.

FIGURE 33.2 Virtual gel for the custom double digest of PhiX174 with the restriction enzymes *Mbo*II and *Msp*I. A lambda *Hind*III digest marker lane is included to scale the gel. Separate lanes simulate runs for unmethylated and CpG-methylated targets. Fragments are ordered in the table for the unmethylated DNA experiment.

and detailed information about particular restriction enzymes is only a click away. Similarly, direct links connect to on-line enzyme ordering from New England Biolabs. Management tools enable users to save and later retrieve their own projects. User feedback is encouraged and facilitated via a built-in e-mail interface.

SOFTWARE AND ALGORITHMS

The program runs on a Sun Ultraspark-II server, under SunOS 5.7 operating system. NEBcutter is a suite of cooperating routines, where program modules execute different tasks. Calculations and queue management are implemented in C (gcc version 2.96) modules. Calls to functions of the GD library perform the visualization tasks. User interfaces are HyperText Markup Language (HTML) pages, generated dynamically by PHP scripts and tested extensively for compatibility with Netscape and Microsoft Explorer browsers. Connections are open for these standard browsers via an Apache web server.

The program localizes all restriction sites in a single pass over the target sequence with a finite state automaton[12] algorithm. With this algorithm, execution time increases only in a linear fashion as the input sequence length increases. In the neighborhoods of sequence ambiguities, the program switches to a brute force algorithm.

When identifying ORFs, the program considers maximal length segments from start to stop, assuming standard codon assignment and bacterial (nonsplicing) DNA sequences. Overlapping ORFs are pruned by a heuristic algorithm, which favors longer ORFs over shorter ones.

Gel run simulations are based on mobility data from our own laboratory experiments, or from the gel manufacturer's manual (www.elchrom.com/technic/manuals/man007.asp). Cubic spline curve interpolation[13,14] yields intervening data points.

Data Handling and Data Security

Suspended NEBcutter sessions can be resumed. Using the default settings, the program stores all project data (sequence, enzyme selection, display settings, and manual entry) on the server and deposits an associated session code on the user's computer. When the user later reconnects, these session codes are returned to the server, data from the user's earlier projects are retrieved, and results of those sessions are reconstructed. The identifier codes are randomly generated, unique numbers, and are stored in browser cookies. Each code is account-specific, so its use requires that the subsequent connection is made through the same log-in account. Another consequence is that people sharing an account also share each other's data. This data sharing can be avoided by using the "delete project" button on the main menu at the end of a session. Alternatively, users can block NEBcutter's save-and-resume feature by checking the appropriate menu option on the interface. Completely disabling a browser's use of cookies is another way of increasing data privacy.

The second method of saving and resuming NEBcutter sessions is activated through a menu option. With this option, project data are returned to the user and stored in a local data file, named by the user. To restart the project later, the user needs to upload this file. This second method is designed for reactivation of an analysis after an indefinite suspension of a session. The first method is intended for temporary interruption of a session. After 2 days of inactivity, server-stored session data files are deleted automatically.

By the time a session is reactivated, new restriction enzymes may have been added to the database, or the content of REBASE may have changed in other ways. Suspended sessions can still resume analysis with the original repertoire of enzymes, which makes the reconstruction of earlier results possible.

Concurrent users of NEBcutter do not notice each other, and they cannot access each other's data. For the purposes of troubleshooting, resource allocation, and optimization, NEBcutter administration routines monitor user activity, maintain log files, and compile usage statistics.

Graphics Output

NEBcutter gives users full access to the digest figures they create. The figures can be exported into graphic files in several standard formats. Resolution of the GIF raster format is selectable, and preview versions of vector graphics EPS format figures are adjustable. The graphic images are of high quality and are ready for inclusion in publications.

SUPPLEMENTARY MATERIALS

Results from genome-wide restriction enzyme digests have been precompiled and posted at http://www.neb.com/~posfai. For all prototype enzymes, site frequencies and densities, as well as average fragment lengths, appear here for mammalian, plant, and insect chromosomes, as well as for total genomic sequences. Site frequencies in one genome can be compared to site frequencies in other genomes, to frequencies predicted by first-order Markov models,[15] or to frequencies in randomized sequences. Both the Markov models and the randomized sequences are based on actual genome dinucleotide counts.

CONCLUSION

Restriction enzymes are invaluable tools of DNA engineering and analysis; they are used in many applications, both in research and in molecular diagnostic laboratories. The NEBcutter program provides tools for the design and interpretation of experiments that employ such enzymes. The software is freely available over the web, from the New England Biolabs website. The program is aimed at the bench scientist and has intuitive, easy-to-use interfaces. To assure that results are current, NEBcutter connects to REBASE directly and updates its internal restriction enzyme data structures daily. An important and distinguishing feature of our software is its ability to consider the effects of base methylation—the potential interference with cleavage.

ACKNOWLEDGMENTS

Gel mobility data are from the experiments of Shawn Stickel at New England Biolabs.

REFERENCES

1. Wilson, G.G.; Murray, N.E. Restriction modification systems. Annu. Rev. Genet. **1991**, *25*, 585–627.
2. Bhagwat, A.S. Restriction enzymes: Properties and use. Methods Enzymol. **1992**, *216*, 199–224.
3. Alberts, B.; Bray, D.; Lewis, J.; Raff, M.; Roberts, K.; Watson, J.D. *Molecular Biology of the Cell*, 3rd Ed.; Garland Publishing: New York, NY, **1994**.
4. Griffiths, A.J.F.; Miller, J.H.; Suzuki, D.T.; Lewontin, R.C.; Gelbart, W.M. *Introduction to Genetic Analysis*, 7th Ed.; W. H. Friedman and Co.: New York, NY, **1999**.
5. McClelland, M.; Nelson, M. Enhancement of the apparent cleavage specificities of restriction endonucleases: Applications to megabase mapping of chromosomes. Gene Amplif. Anal. **1987**, *5*, 228–257.
6. Blumenthal, R.M.; Rice, P.J.; Roberts, R.J. Computer programs for nucleic acid sequence manipulation. Nucleic Acids Res. **1984**, *10*, 91–101.
7. Bishop, M.J. Software for molecular biology: II. Restriction mapping and DNA sequencing programs. BioEssays **1984**, *1*, 75–77.

8. Roberts, R.J.; Vincze, T.; Posfai, J.; Macelis, D. REBASE: Restriction enzymes and methyltransferases. Nucleic Acids Res. **2003**, *31*, 418–420.

9. Benson, D.A.; Karsch-Mizrachi, I.; Lipman, D.J.; Ostell, J., Wheeler, D.L. GenBank. Nucleic Acids Res. **2003**, *31*, 23–27.

10. Gusfield, D. *Algorithms on Strings, Trees, and Sequences—Computer Science and Computational Biology*; Cambridge University Press: Cambridgeshire, UK, **1997**.

11. McClelland, M.; Nelson, M.; Raschke, E. Effect of site-specific modification on restriction endonucleases and DNA modification methyltransferases. Nucleic Acids Res. **1994**, *22*, 3640–3659.

12. Smith, R. A finite state machine algorithm for finding restriction sites and other pattern matching applications. Comput. Appl. Biosci. **1988**, *4*, 459–465.

13. Kozulic, B. Models of gel electrophoresis. Anal. Biochem. **1995**, *231*, 1–12.

14. Press, W.H.; Teukolsky, S.A.; Vetterling, W.T.; Flannery, B.P. *Numerical Recipes in C. The Art of Scientific Computing*, 2nd Ed.; Cambridge University Press: Cambridgeshire, UK, **1999**; 113–116.

15. Durbin, R.; Eddy, S.; Krogh, A.; Mitchinson, G. *Biological Sequence Analysis— Probabilistic Models of Proteins and Nucleic Acids*; Cambridge University Press: Cambridgeshire, UK, **1998**; 46–51.

34 Differential Display (DD) Analysis

Farid E. Ahmed
East Carolina University, Greenville, North Carolina, U.S.A.

CONTENTS

INTRODUCTION

Differential display (DD) technique was developed in the 1990s as a systematic method for studying eukaryotic gene expression. The updated method utilizes a combination of three frequently used molecular biology techniques: reverse transcription (RT) of messenger ribonucleic acid [Pol $(A)^+$ mRNA] followed by polymerase chain reaction (PCR) of the resulting complementary deoxyribonucleic acid (cDNA); polyacrylamide gel (PAG) electrophoresis; and cDNA cloning to visualize and compare gene expression patterns between two or more samples. RT uses one of three individual one-base fluorescently labeled anchored oligo-dT primers to amplify the 3′ terminal of the mRNA by PCR, instead of the original isotopic labeling method, in combination with one of the various 10–13 arbitrary (random) nucleotide primers. The resulting cDNAs are separated on a denaturing, and sometimes nondenaturing, PAG. A fluorescent scanner views the pattern of bands. The cDNA fragments of interest are retrieved from the gel, purified, reamplified, and either cloned or directly sequenced to identify the differently regulated genes. Finally, confirmation of the differential expression of resulting cDNAs can be carried out by a method such as Northern blotting, RNase protection, or quantitative real-time RT-PCR. Following confirmation, the cloned cDNA probes can be used either to screen a cDNA library for a full-length clone, or more easily to carryout rapid amplification of cDNA ends (5′-RACE).

PRINCIPLES AND GUIDELINES

The genome of higher eukaryotes contains close to 50,000 genes, of which between 10% and 15% are believed to be expressed at a given time in a cell to determine the regulatory mechanisms that control cellular processes controlling our lives

DOI: 10.1201/9781003247432-34

such as development and differentiation, homeostasis, response to insult, cell cycle regulation, aging, programmed cell death, and pathological changes such as cancer. Monitoring the pattern of gene expression under various physiological and pathological conditions is a critical step in understanding these diverse biological processes, and comprehending the mechanisms involved allows for needed interventions. Because of the large numbers of expressed genes, powerful tools are needed to characterize the overall pattern of gene expression.[1]

Older methods of identification of differentially expressed genes relied on differential or subtractive hybridization (SH), which although sensitive requires large amount of RNA, is error prone, nonsystematic, laborious, and time consuming, and results are not seen until the end of the process.[2]

The original DD protocol utilized the idea applied earlier for random amplification of fingerprinted genomic sequences. It was published in 1992 by prominent investigators at the Dana-Farber Cancer Institute in Boston, MA.[3] The principle of the method is to detect different types of gene expression patterns using three techniques: (1) RT of DNase I-treated total RNA (to remove any chromosomal DNA) using anchored primers, 12-mer long consisting of a stretch of 11 Ts plus one last non-T base to anchor primers to the pol $(A)^+$ tail of many RNAs. Use of total RNA is preferable to mRNA as it involves less preparatory steps and avoids background smearing[2]; (2) choosing 5' arbitrary 10 mers that hybridize to cDNA in a degenerate manner for setting lengths of cDNAs corresponding to mRNAs (tags) to be amplified by PCR. For a 5' primer of arbitrary base sequence, annealing position to cDNA should be randomly distributed from the pol $(A)^+$ tail. Therefore, the amplification provided from various mRNAs will differ in length; and (3) employing sequencing gels for isolation of cDNAs. The aim is to obtain a tag of a few hundred bases, long enough to uniquely identify mRNA, but short enough to electrophoretically separate by size. Pairs of primers are selected so that by probability each will amplify DNAs from 50 to 100 (average ~75) mRNAs, as this number can be adequately displayed on one lane of the gel.[3] This method was later on named DD-PCR[1] (Figure 34.1), to distinguish it from other DD methods discussed below.

Problems intrinsic to DD have been encountered such as high noise level due to smearing, misrepresentation of rare messages, bias toward high copy number mRNAs, not revealing differences due to mutational changes, incomplete cDNAs, contamination of purified PCR fragments by unrelated DNA sequences, and additional bands generated by arbitrary primers alone from palindromic sequences within an mRNA molecule; all led to high false-positive rates.[1–7] Therefore, many modifications have been periodically made to the original protocol to improve these intrinsic problems and to achieve specificity and efficiency as discussed below.

The original primer design was two-base anchored primer, but it resulted in suboptimal amplifications.[2] Three one-base anchored oligo-dT primers differing only at the last 3' non-T base are now used. This modification cuts down the number of RT reactions needed for each RNA sample and minimized the redundancy and underrepresentation of certain RNA species due to the degeneracy of the process. Substitution of decamer arbitrary primers with rationally designed 13-mers has increased the accuracy of priming.[5] Furthermore, even longer primers (25–29 mers)

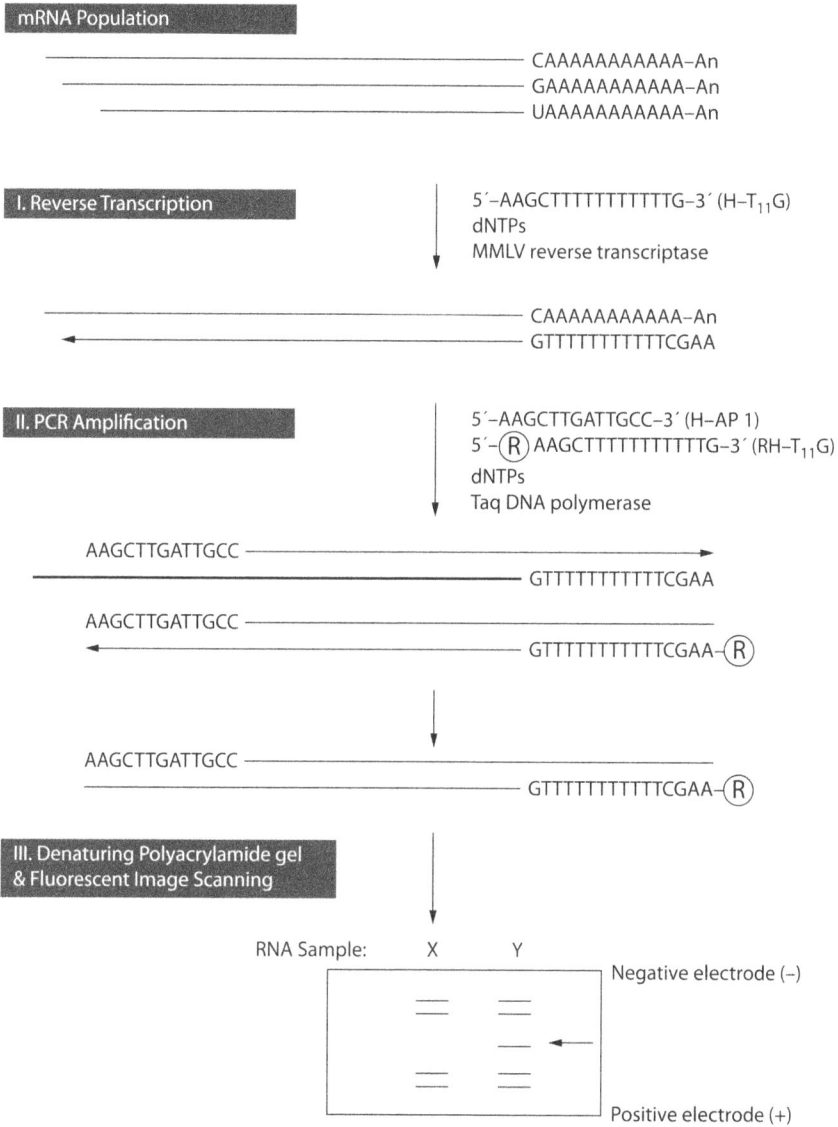

mRNA Population

CAAAAAAAAAAA–An
GAAAAAAAAAAA–An
UAAAAAAAAAAA–An

I. Reverse Transcription

5′–AAGCTTTTTTTTTTTG–3′ (H–T₁₁G)
dNTPs
MMLV reverse transcriptase

CAAAAAAAAAAA–An
GTTTTTTTTTTTCGAA

II. PCR Amplification

5′–AAGCTTGATTGCC–3′ (H–AP 1)
5′–(R) AAGCTTTTTTTTTTTG–3′ (RH–T₁₁G)
dNTPs
Taq DNA polymerase

AAGCTTGATTGCC
GTTTTTTTTTTTCGAA

AAGCTTGATTGCC
GTTTTTTTTTTTCGAA–(R)

AAGCTTGATTGCC
GTTTTTTTTTTTCGAA–(R)

III. Denaturing Polyacrylamide gel & Fluorescent Image Scanning

RNA Sample: X   Y

Negative electrode (–)

Positive electrode (+)

FIGURE 34.1 Schematic of fluorescent DD–PCR method. Total RNA is isolated from cells in culture or tissues. Then, I. A reverse transcription reaction is carried out using one of three specific one-based anchored oligo-dT primers (e.g., H-T₁₁G). II. PCR amplification is performed by using the corresponding fluorescent-labeled anchored oligo-dT primer (RH-T₁₁G; R, rhodamine labeling) in different combination with arbitrary 13-mer primer (H-AP1). III. The amplified PCR products are separated on a denaturing polyacrylamide gel, and the fluorescent-labeled cDNA fragments visualized by a fluorescence scanner. A differentially expressed cDNA fragment in the electropherogram is denoted by an arrow.

Source: Adapted from Ref. [7], with permission.

and optimal deoxynucleotide triphosphate (dNTP) concentrations were reported to allow arbitrary priming after an annealing temperature of 40°C in the first PCR cycle, followed by more stringent PCR annealing at 60°C to provide specificity. Additionally, use of hot-start PCR and other thermostable enzyme mixes suitable for long-length PCR resulted in highly reproducible and representative cDNA bands.[6] Moreover, the introduction of a restriction site at the 5′ end of both the anchored and arbitrary primers facilitated cloning of the cDNAs.[4]

Initially, isotope labeled oligos, α-[^{35}S] dATP, and later on [^{32}P]-end-labeled oligo-dT primer or α-[^{32}P] dATP, were used to autoradiographically detect amplified PCR products on sequenced gels.[2] Introduction of nonisotopic methods using digoxigenin and fluorescent dyes such as tetramethylrhodamine allowed coupling of the display with digital data analysis, resulting in increased throughput.[2,7] Use of nondenaturing 6% PAG (i.e., without urea) to reduce double bands into a single band was found to reduce band complexity and eliminated several bands of DNA molecules derived from different fragments that occupied the same position in the gel.[8] DD–PCR is only capable of determining the 3′ region of the gene, so full-length cDNA can only be achieved either by doing rapid amplification of cDNA ends (5′-RACE) or by probing a cDNA library.[1,7]

DD was reported to tolerate a broad range of annealing temperatures and elongation times. However, the major factors that impacted the reproducibility of the method were low concentration of dNTPs and random primers that made PCR amplifications susceptible to pipetting errors. A final concentration of dNTPs >2 µM and arbitrary primers to 0.2 µM improved the reproducibility of DD.[9] In one study utilizing cervical cancer cell line CaSki, it was found that—for most primer combinations—fourfold less cDNA and only 25 high-stringency PCR cycling produced reproducible complex band patterns with intensities that reflected two- to tenfold differences in expression levels (the most common levels of regulation).[10] To reduce or exclude the bands that are subject to statistical noise from consideration, it has been suggested to start with enough cDNA (i.e., about 30,000–100,000 molecules) to obtain 15 ng/mL of amplified product in 25–27 PCR cycles.[11] Although some publications reveal that as little as 1.1-fold amplification is detectable by DD–PCR, the threshold is not precisely known as an upper detection limit may be reached in this technique.[11]

If we consider that there are 15,000 genes expressed in a cell and each fingerprint contains ~70 mRNA species, then in an ideal situation if all mRNAs have the same probability of being displayed in each round, ~600 fingerprints (or 45,000 bands) are required to cover 95% of all mRNAs.[8] For a comprehensive analysis of all mRNA species in a given cell, statistical modeling that predicted at least 240 different DD primer combinations are needed.[3] An empirical determination of the comprehensiveness of DD-PCR in Chinese hamster ovary fibroblast HA-1 cell line that received (or not) hydrogen peroxide treatment used saturation DNA screening of 324 primer combinations. Results showed that a 100% comprehensive analysis by this technique is not possible regardless of the number of primer combinations. This may be due to a selective resistance to the identification of certain sequences by DD.[12]

Once the first few primer combinations have been tested, the results of the display should be examined keeping in mind that a redesign of the experiment should be made when >5% of the transcripts are differentially expressed, or if no differences exist.[11] It is instructive to keep in mind a few guidelines when performing DD-PCR: (1) drastically different conditions should not be compared; (2) the DD reaction should be repeated for each sample to control for false positives; and (3) multiple samples should be used for each DD experiment in order to provide internal controls.[13]

OTHER DD STRATEGIES AND IMPROVEMENTS TO THE ASSAY

Another strategy to provide RNA fingerprints employed an arbitrary primer instead of an anchored oligo-dT primer in the first step of RT, selecting those regions internal to the RNA that have six to eight base matches with the 3′ end of the primer. As in DD-PCR, this is followed by arbitrary priming of the resulting first strand of cDNA with the same or a different arbitrary primer, and then PCR amplification. This method is known as RAP-PCR.[14] This method samples anywhere in the RNA, including opening reading frames, and can be used on RNAs that are not polyadenylated (such as bacterial RNA). A limitation of this method too is that rare RNAs will be underrepresented.[1]

DD strategies designed to target specific sequences in bacteria, such as highly iterated palindromic (HIP) elements found in half of the genes in the genome of cyanobacteria, provided a convenient global expression strategy for identifying light intensity-regulated genes in bacteria using a limited number of primers.[15] Another area of wide application of DD is a study of scarcely sequenced and complex plant genes that contain large families of homologous genes.[16]

In order to improve the specificity of DD, attempts were made to replace the 10–13-base arbitrary primers—which hybridize nonspecifically to cDNA templates causing mismatches—with primer sites produced by restriction enzyme digests of double-stranded (ds) cDNAs; thus, allowing adapters for priming sites to be ligated into specific regions on the cDNA. These ds cDNAs are then amplified with either anchored oligo-dT primers combined with ligated primers, or with ligated primers alone, followed by cloning into a vector, amplifying, restriction digest again, sequencing, and identifying the produced fragments. Several of these restriction fragment length polymorphism (RFLP)-based DD strategies, which are variation on the same theme, have been devised, each with its name such as amplification of ds cDNA ends restriction fragments (ADDER), amplified differential gene expression (ADGE), gene calling, ordered DD (ODD), RFLP-coupled domain-directed DD (RC4D), and total gene expression analysis (TOGA).[1,7] However, RFLP-based strategies require more experimental steps such as second-strand cDNA synthesis, restriction digestion, and ligation of adapter primers over traditional DD, which can lead to error, and their precision over DD–PCR has not been substantiated in controlled experiments.[7]

APPLICATION OF DD TO BIOLOGICAL SYSTEMS

Because of the simplicity, specificity, and versatility of DD, it has been applied to countless applications more than any other genome expression profiling method in different fields of biomedical research as: studying gene expression in many phyla, polymorphism and gene silencing, identifying substrates for RNA-binding proteins, disease diagnosis and prognosis, cell cycle regulation, apoptosis, cancer research, cardiovascular diseases, neuroscience, endocrinology, immunology, and plant science.

A comprehensive compilation of DNA publications in various fields can be accessed through the website of GenHunter® Corporation (http://www.differentialdisplay.com), a company established by inventors of the DD-PCR.[3]

COMPARISON WITH OTHER EXPRESSION PROFILING TECHNOLOGIES

The number of methods employed for gene expression analyses is numerous; however, there are now only a few DD routinely used approaches such as PCR-based, SH, serial analysis of gene expression (SAGE), and microarrays. Additionally, there are variations on the theme that modify or combine features of these basic approaches.

A bottleneck in all variants of DD has been the confirmation of differential expression. Isolation and identification of clones, Northern blots, or RT-PCR analyses are time- and labor-intensive techniques. Therefore, several attempts have been made to combine DD with microarrays to facilitate parallel identification of frequently expressed tags simultaneously in many biological systems.[17,18] Furthermore, probes generated by DD from archived experiments can now be reamplified and used as sensitive probes for cDNA microarray studies, revealing more information than was yielded when originally resolved as fingerprints on PAG.[19] The use of fluorescent probes[20] combined with robotics for liquid dispensing and tube handling/rotation, digital analysis, and specialized computer programs for data acquisition, analysis, and storage increases method throughput and facilitates data interpretation.[1,7]

TABLE 34.1
Comparisons among Major Approaches to Genome Expression Profiling

Parameter	DD	SH	SAGE	Microarrays
Year published	1992	1985	1995	1995
Type of system	Open	Open	Open	Closed
Sensitivity	Moderate	High	High	Moderate
Specificity	High	High	High	Moderate
Quantification	Relative	Relative	Absolute	Relative
Previous knowledge of sequence	No	No	No	Yes
Detects novel genes	Yes	Yes	Yes	No
Labor intensity	Moderate	High	High	Moderate
Commercial services	Few	Very few	Few	Many
Cost	Moderate	Low	Moderate	High
Number of publications since inception[a]	3079	1197	386	3233

[a] Based on MEDLINE search via Ovid on 5 January 2004.

Another strategy combined DD with SH, rationalizing that removing the most commonly expressed mRNA and then displaying the remaining mRNAs might improve displaying.[21]

Table 34.1 lists the advantages and disadvantages of major approaches to gene expression studies. It can be seen that no one method is superior to any other in all parameters surveyed. Ultimately, the correct identification of the gene and its transcript in the hand of the experimenter will attest to the appropriateness of the method to its intended application.

CONCLUSION

DD is a systematic, sensitive, and convenient approach that allows many samples to be tested in parallel, and does not require any previous knowledge of mRNA or gene sequence, making it an open system. Initially, DD analysis suffered from a high rate of false positives (or noise) due to the use of short primers and low annealing PCR temperature, and from redundancy in display because short primers may anneal to different parts of the same transcript. However, progressive technical improvements as well as care in experimental design have reduced these shortcomings. Because of its simplicity, sensitivity, reproducibility, and yield to automation, which increases throughput and accuracy, DD, either by itself or in combination with another expression profiling method, has become one of the most widely used methods for studying differentially regulated genes in many biological systems.

REFERENCES

1. Ahmed, F.E. Molecular techniques for studying gene expression in carcinogenesis. J. Environ. Sci. Health **2002**, *C20*, 77–116.
2. Liang, P.; Bauer, D.; Averboukh, L.; Warthoe, P.; Rohrwild, M.; Muller, H.; Strauss, M.; Pardee, A.B. Analysis of altered gene expression by differential display. Methods Enzymol. **1995**, *254*, 304–321.
3. Liang, P.; Pardee, A.B. Differential display of eukaryotic messenger RNA by means of polymerase chain reaction. Science **1992**, *257*, 967–973.
4. Liang, P.; Zhu, W.; Zhang, X.; Guo, Z.; O'Connell, R.P.; Averbaukh, L.; Wang, F.; Pardee, A.B. Differential display analysis using one-based anchored oligo-dT primers. Nucleic Acids Res. **1994**, *22* (25), 5763–5764.
5. Liang, P. Factors ensuring successful differential display. Methods **1998**, *16*, 361–364.
6. Diachenko, L.B.; Ledesma, J.; Chenchik, A.A.; Siebert, P.D. Combining the technique of RNA fingerprinting and differential display to obtain differentially expressed mRNA. Biochem. Biophys. Res. Commun. **1996**, *219*, 824–828.
7. Liang, P. A decade of differential display. BioTechniques **2002**, *33* (2), 338–346.
8. Bauer, D.; Müller, H.; Reich, J.; Riedel, H.; Ahrenkiel, V.; Warthoe, P.; Strauss, M. Identification of differentially expressed mRNA species by an improved display technique (DDRT–PCR). Nucleic Acids Res. **1993**, *21* (18), 4272–4280.
9. Cho, Y.-J.; Prezioso, V.R.; Liang, P. Systematic analysis of intrinsic factors affecting differential display. BioTechniques **2002**, *32* (4), 762–766.
10. Ranamukhaarachchi, D.G.; Rajeevan, M.S.; Vernon, D.D.; Unger, E.R. Modifying differential display polymerase chain reaction to detect relative changes in gene expression profiles. Anal. Biochem. **2002**, *306*, 343–346.

11. Matz, M.V.; Lukyanov, S.A. Different strategies of differential display: areas of application. Nucleic Acids Res. **1998**, *26* (24), 5543–5572.
12. Crawford, D.R.; Kochheiser, J.C.; Schools, G.P.; Salamon, S.L.; Davies, K.J.A. Differential display: a critical analysis. Gene Expr. **2002**, *10*, 101–107.
13. Stein, J.; Liang, P. Differential display technology: a general guide. Cell. Mol. Life Sci. **2002**, *59*, 1235–1240.
14. Welsh, J.; Chada, K.; Dalal, S.S.; Cheng, R.; Ralph, D.; McClelland, M. Arbitrary primed PCR fingerprinting of RNA. Nucleic Acids Res. **1992**, *20* (19), 4965–4970.
15. Bhaya, D.; Vaulot, D.; Amin, P.; Takahashi, A.W.; Grossman, A.R. Isolation of regulated genes of the cyanobacterium *Synechocystis* spp. strain PCC 6803 by differential display. J. Bacteriol. **2000**, *182* (20), 5692–5699.
16. Breyne, P.; Zabeau, M. Genome-wide expression analysis of plant cell cycle modulated genes. Curr. Opin. Plant Biol. **2001**, *4*, 136–142.
17. Takeuch, H.; Fujiyaki, T.; Shirari, K.; Matsuo, Y.; Kamikouchi, A.; Iujinawa, Y.; Kato, A.; Tsujimoto, A.; Kubo, T. Identification of genes expressed preferentially in the honey bee mushroom body by combination of differential display and cDNA microarray. FEBS Lett. **2002**, *513*, 230–234.
18. Mandaokar, A.; Kunao, V.D.; Amway, M.; Broose, J. Microarray and differential display identify genes involved in jasmonate-dependent anther development. Plant. Mol. Biol. **2003**, *52*, 775–786.
19. Trenkle, T.; Welsh, J.; McClelland, M. Differential display probes for cDNA arrays. BioTechniques **1999**, *27* (3), 554–564.
20. Cho, Y.; Meade, J.; Walden, J.; Guo, Z.; Liang, P. Multicolor fluorescent differential display. BioTechniques **2001**, *30*, 562–572.
21. Fuchs, B.K.; Zhang, K.; Bolander, M.E.; Sarkar, G. Identification of differentially expressed genes by mutually subtracted RNA fingerprinting. Anal. Biochem. **2000**, *286*, 91–98.

35 Protein Truncation Test (PTT)

Sadanand Gite[1,2], Mark Lim[1],
Kenneth Rothschild[1,3]
[1]AmberGen Inc., Watertown, MA, U.S.A.
[2]First Light Diagnostics, Chelmsford, MA, U.S.A.
[3]Molecular Biophysics Laboratory, Photonics Center,
Department of Physics, Boston University,
Boston, MA, U.S.A.

CONTENTS

INTRODUCTION

The protein truncation test (PTT) is a mutation detection technique that specifically detects mutations in mRNA leading to the premature termination of protein translation. An increasing number of genes implicated in disease processes (primarily cancers) have been identified where the majority of the mutations result in premature termination of translation leading to an incomplete and nonfunctional protein product. These include the *APC* gene [1, 2], *BRCA*1/2 genes [3], *PKD*1 gene [4], *NF*1 gene [5], *TSC*1/2 genes [6], and *DMD* gene [7]. The PTT was first reported in 1993 [1, 8] and was used mainly in clinical research settings (http://www.genetest.org). However, the original form of the PTT had several limitations that included (1) slow readout because of the

DOI: 10.1201/9781003247432-35

use of electrophoresis followed by radioactive detection; (2) errors in the visual detection of mobility shifts on a gel (which depend on the level of the training of the technician); (3) safety issues involved with the use of radioactivity; and (4) the difficulty in automating Sodium Dodecyl Sulfate–Polyacrylamide Gel Electrophoresis (SDS-PAGE). Some other recent applications of PTT have been described in the literature [9–17]. Here, we describe some of the advances in the PTT, including the introduction of an enzyme-linked immunosorbent assay (ELISA)-based PTT (ELISA-PTT), which helped overcome these limitations [9, 10]. While rapid next-generation DNA sequencing is likely to replace PTT as the method of choice for detecting protein truncations, the advances in PTT described here may still find some niche applications in research and medical diagnosis.

TECHNICAL DESCRIPTION OF PTT

PTT PROTOCOL

The workflow for the standard-PTT and ELISA-PTT [9] is outlined in Figure 35.1. The first step involves the isolation of genomic DNA/mRNA and amplification of specific regions of coding sequences from the target gene using polymerase chain reaction/reverse transcriptase-polymerase chain reaction (PCR/RT-PCR). These PCR products are then used as templates for in vitro (cell-free) translation and the synthesized proteins are traditionally analyzed by SDS-PAGE followed by autoradiography. Shorter protein products of mutated alleles are distinguished from the full-length protein products of normal alleles due to mass/mobility differences. In the case of ELISA-PTT (Figure 35.2A), a reduced C- to N-terminal ratio (C/N) indicates the presence of mutations.

DESIGNING PRIMERS FOR PTT

Several factors are important when designing PCR primer pairs for PTT. Proper design of the forward primer is the most critical factor. This primer contains four specific regions that are essential for efficient in vitro translation of the PCR amplicon. The 5′-primer must have a promoter sequence at the 5′-end (generally for T7 polymerase) followed by a 5–7 bps spacer sequence, a eukaryotic translation initiation sequence (Kozak sequence), including the ATG start codon, and at the 3′-end, a region of the target gene sequence (~17–24 bps) in-frame with the ATG codon. In addition, for ELISA-PTT, the 5′-primer also contains sequences coding for the N-terminal detection tag (epitope) and/or binding tags. The 3′-primer must be complementary to the 3′-end of the target sequence and ELISA-PTT must also contain a sequence encoding the C-terminal detection tag [9]. To ensure that truncation mutations near the beginning or at the end of a fragment (i.e., at the 5′- or 3′-end) are not missed, flanking segments for a large region of coding sequence should have an overlap of 350–500 bps. When RNA-based PTT is used, primers from overlapping segments should be located in different exons to reduce the

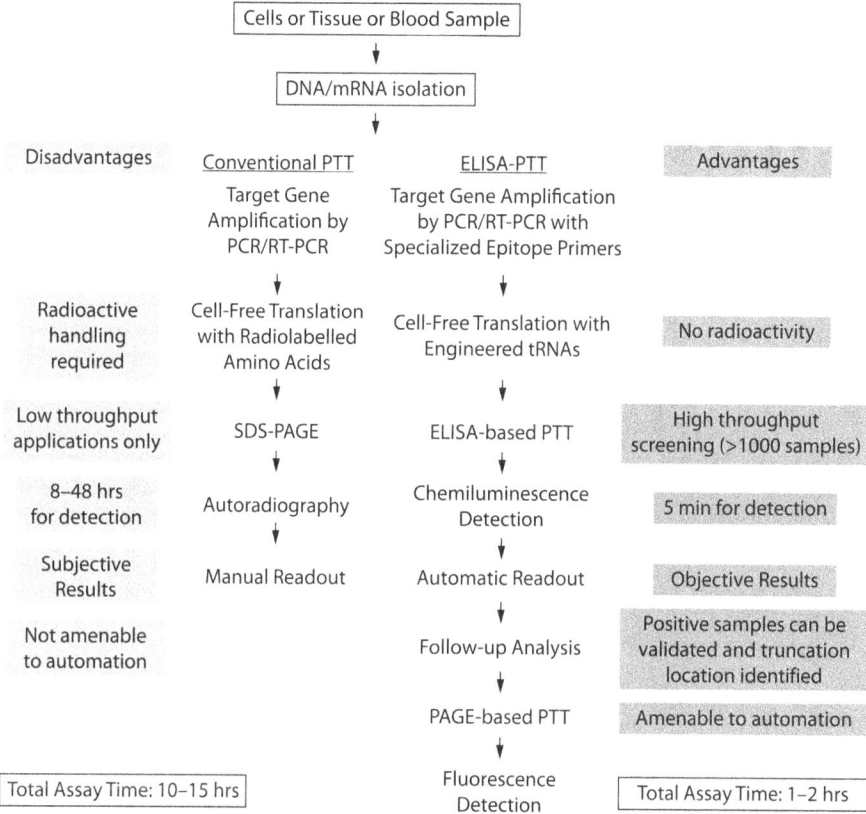

FIGURE 35.1 Various steps involved in typical protein truncation test.

possibility that a mutant allele having a deletion or splicing defect does not amplify with any of the primer sets.

TEMPLATE CONSIDERATIONS FOR PCR AMPLIFICATION

Although transcription/translation of PCR products up to 3 kb is not difficult, the best results are obtained using fragments of 1.3–2.0 kb size. If a gene contains several kilobases of coding sequence, it is necessary to divide it into multiple segments. Most published PTT reports use genomic DNA as the template to analyze large exons, e.g., exon 11 of *BRCA*1 and exon 15 of the *APC* genes [1, 3]. In addition, several attempts have been made to scan entire genes for truncating mutations using a combination of genomic DNA and mRNA (Table 35.1). In the case of RNA-based PTT, mRNA is first reverse transcribed to cDNA, which then functions as a template for amplification. The best source of RNA is cells in which the target gene is abundantly expressed. However, for practical reasons, RNA is generally isolated from freshly drawn peripheral blood lymphocytes.

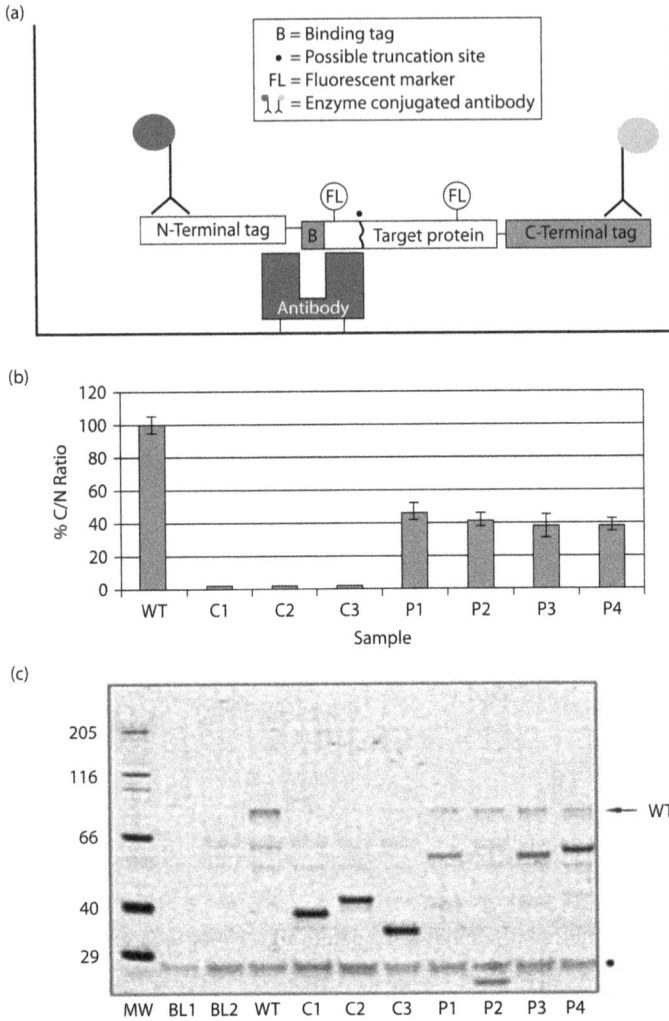

FIGURE 35.2 ELISA-PTT. (Panel A) Schematic representation of an ELISA-based protein truncation test (ELISA-PTT). Unlike conventional PTT, the test uses a solid-phase ELISA format. B, Binding tag (either biotin that is incorporated randomly along the sequence using mis-aminoacylated tRNAs or epitope tag incorporated at N-terminal) acts to immobilize the cell-free synthesized fragments on the wall surface. N- and C-terminal epitopes are detected by corresponding antibodies allowing for an estimate of the relative amount of truncated protein while simultaneously controlling for the level of translation. FluoroTags (FL) provide an independent confirmation of truncation and estimate of the fragment size by using fluorescence readout of gels. (Panel B) Detection of various truncation mutations in the APC gene using ELISA-PTT. (Panel C) Validation of ELISA-PTT result by gel-based fluorescence PTT. WT is wild-type, C1–C3 are mutant homozygous DNA samples from cell lines, and P1–P4 are the heterozygous DNA samples from patients pre-diagnosed with FAP. BL1 corresponds to a cell-free translation performed lacking both the added tRNAs and DNA. BL2 corresponds to a cell-free translation performed lacking only the added DNA. MW is molecular weight markers. The asterisk indicates the position of an auto-fluorescent protein band present in the cell-free translation extract [9].

TABLE 35.1
Top Ten Diseases Where PTT Is Applicable

No.	Disease	Gene	% Truncated Mutations	Ref.
1	Breast cancer	*BRCA1/2*	90	[3]
2	Colorectal cancer	*APC*	95	[28]
3	Familial adenomatous polyposis	*APC*	97	[1]
4	Hereditary nonpolyposis colon cancer	*MSH2/MLH1*	70–80	[29]
5	Neurofibromatosis	*NF1/2*	60–75	[5]
6	Polycystic kidney disease	*PKD1/2*	95	[4]
7	Duchenne muscular dystrophy	*DMD*	95	[7]
8	Cystic fibrosis	*CFTR*	50	[30]
9	Ataxia telangiectasia	*ATM*	90	[31]
10	Tuberous sclerosis	*TSC1/2*	75–98	[6]

PCR Amplification

High-fidelity polymerase such as PFU (Stratagene/Agilent) or Phusion (MJ research/Bio-Rad) should be used to reduce the possibility of PCR errors. In addition, PCR products should be analyzed by agarose gel electrophoresis prior to the transcription/translation. Direct analysis of the PCR products can indicate the presence of an abnormally sized amplicon suggesting a splicing error or genetic rearrangements such as deletions or duplications. However, purification of the PCR amplicon of the correct size is not necessary and should be avoided, because the aberrant-sized PCR products are often derived from mutant alleles.

In Vitro Transcription/Translation and Analysis of Translated Protein

The TNT® T7 Quick Coupled Transcription/Translation System (Promega, Madison, WI) and many other commercially available cell-free transcription/translation systems from Sigma Aldrich, Thermo Fisher, and New England Biolabs are convenient for PTT as both transcription and translation are performed simultaneously in a single tube using the PCR product as a template. There are several ways to detect the newly synthesized proteins. In the early days of PTT, one of the most common methods was the incorporation of radiolabeled amino acids (e.g., ^{35}S-methionine or ^{13}C-leucine). For separation of the translation products, appropriate SDS-PAGE conditions must be chosen for simultaneous detection of both full-length and smaller products. After electrophoresis, gels are dried and protein bands are detected using either X-ray film or a phosphorImager screen. In general, a wild-type (WT) sample will have a strong band at the expected size of the full-length translation product compared to bands corresponding to smaller proteins. Ideally, these weaker bands correspond to truncated protein fragments derived from mutant templates. However, weaker bands can also originate from internal weak translation initiation sites (AUG) and proteolytic degradation products. In some cases, these bands can obscure the analysis and/or detection of the truncated fragments derived from a mutant template.

In addition to radiolabeling, methods have been reported, which use non-isotopic detection. In one approach, biotin was incorporated during translation using biotin–lysine–tRNA and was detected by Western blotting [12]. Similarly, Kahmann and coworkers [18] used Western blotting to detect engineered epitope tags. Radioisotope labeling can also be eliminated by incorporating fluorescent labels using fluorescently labeled tRNAs [9, 19, 20]. Although these non-isotope-based methods have clear advantages compared to radioactive detection, they still suffer from the intrinsic throughput problems associated with electrophoresis. In order to overcome these limitations, ELISA-PTT has been developed, which provides higher throughput and lower cost because it circumvents electrophoresis [9, 10].

The basic ELISA-PTT approach is illustrated in Figure 35.2A. Specially designed primers, which incorporate N- and C-terminal epitopes, are used for amplification of target sequences. In addition to these epitopes, additional tags can be incorporated randomly along the protein chain for detection and capture purposes. This is accomplished using mis-aminoacylated tRNAs (e.g., biotin–lysine–tRNA or/and BODIPY–lysine–tRNA) which are added to the reaction mixture. A capture tag can also be incorporated into the protein by using a specially designed primer. After translation, the test proteins are captured in a single well of a microtiter plate, and the N- and C-terminal epitope tags are detected using appropriate antibodies. The signals obtained are used to compare the total amount of target protein captured (N-terminal signal) vs. the fraction that is full length (i.e., has a C-terminus).

The results of ELISA-PTT for the *APC* gene are shown in Figure 35.2B. In this experiment, DNA derived from normal controls, familial adenomatous polyposis (FAP) patients as well as cell lines with known mutations in the *APC* gene were analyzed. The C/N terminal ratio of WT was normalized to 100% and the values obtained for all other samples are expressed as a fraction of the WT. For cell line DNA (homozygous APC mutant), the C/N terminal ratio was close to 0%, as expected. The C/N terminal ratio for heterozygote samples derived from individuals with FAP ranged from 37% to 47% relative to the WT (Figure 35.2B). It is possible to validate the ELISA-PTT results and localize the mutation within the protein using the fluorescent labels incorporated into the in vitro translated proteins. Figure 35.2C shows results obtained from fluorescent imaging after SDS-PAGE using aliquots of the same translation mixtures used in Figure 35.2B. The WT sample produces a band of the expected molecular mass (~70 kDa), whereas the homozygous mutant cell line samples exhibit single bands at approximately 35, 40, and 32 kDa (C1–C3, respectively). Samples derived from FAP individuals heterozygous for a mutation exhibited two bands corresponding to both WT and mutant alleles (lanes P1–P4).

ELISA-PTT does not require electrophoresis for the primary screen but does require electrophoresis for sizing the proteins in order to locate and confirm the mutation in the amplicon. In contrast, Garvin et al. [21] reported a method to size in vitro synthesized proteins by mass spectrometry, thereby eliminating the need for electrophoresis. Experiments were carried out using a short sequence (21 bases) of the *BRCA*1 gene. After in vitro translation, the test peptide was purified using a FLAG-epitope and subjected to Matrix-assisted Laser Desorption/Ionization-Time of Flight (MALDI-TOF) mass spectroscopy. Truncating mutations in heterozygotes are easily detected, and importantly, single amino acid substitutions are also detectable because of the high resolution of mass spectroscopy. Although this process can be multiplexed and is amenable to automation, the small test sequence size reduces its effective throughput compared to gel-PTT or ELISA-PTT.

T³-ELISA-PTT: APPLICATION TO INHERITED BREAST CANCER

While the ELISA-PTT data in Figure 35.2B used a tRNA-mediated biotin label as the capture tag, as an alternative approach, the engineered tRNAs can be omitted and an entirely epitope tag-based system used (i. e. all tags incorporated using the 5′ and 3′ PCR primers). This totals three epitope tags used in this approach (N-terminal detection and capture tags in tandem, as well as a C-terminal detection tag), termed here T³-ELISA-PTT. Antibodies on the ELISA plate surface directed against the N-terminal capture tag are used to bind and isolate the in vitro translated protein, regardless of its truncation status (followed by detection of the N- and C-terminal tags similar to as before). This has the advantage that unlike the tRNA-based approach, 100% incorporation of the capture tag is achieved, and there is no competition between the unincorporated tRNAs and the in vitro translated protein for binding to the ELISA plate surface. Lim et al. developed and applied this approach to the detection of truncation mutations in the BRCA gene for inherited breast cancer [10]. Figure 35.3 shows a schematic representation of T³-ELISA-PTT.

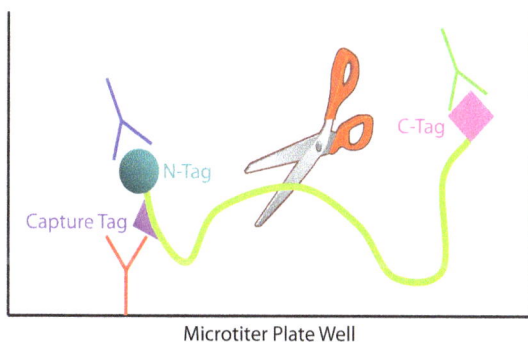

Microtiter Plate Well

FIGURE 35.3 T³-ELISA-PTT approach. This next-generation ELISA-PTT uses three different epitope tags, two in tandem at the N-terminus (one detection and one capture tag) and one at the C-terminus (detection tag), added using specially designed PCR primers. The chain truncation mutations are calculated using % C/N, the C- to N-terminal signal ratio of a given sample normalized against known WT controls.

Source: **Figure and legend adapted from Ref. [10].**

The following example results and discussion are adapted from Lim et al. [10], which reported the development and use of T³-ELISA-PTT for the detection of truncation mutations in the BRCA gene for inherited breast cancer.

The C-terminal signal arising from a full-length protein in a given sample is normalized against its N-terminal signal representing total nascent protein produced and captured on the ELISA plate. These C-to-N-terminal ratios (C/N ratios) of the test samples are calculated as a percentage of the ratio arising from known BRCA WT control samples (% C/N). Thus, in theory, BRCA WT samples would have a 100% C/N ratio, and heterozygous BRCA mutants would have a 50% ratio.

In practice, however, the % C/N ratios of the BRCA heterozygous mutants deviate from 50% in either direction in a segment-specific manner (for example, Figure 35.4A and B; red bars for BRCA truncations in various segments). This phenomenon can be explained by several possible factors, all of which essentially derive from the coexistence of two protein species (with potentially different folding patterns) in the heterozygous mutant samples (full-length and truncated) as opposed to a single protein species in the WT samples (full-length). Please refer to Lim et al. [10] for a more detailed discussion.

Figure 35.4C shows an example of the linearity of the T³-ELISA-PTT assay as a function of different amounts of translated protein input into the ELISA assay. The standard curves in Figure 35.4C are for the N- and C-terminal detection tags of the WT reference sample for BRCA2 exon 11 segment 1 (n = 7). This verified the excellent linearity of the assay over an approximate 10-fold range ($R^2 = 0.99$).

To evaluate the T³-ELISA-PTT for BRCA mutation analysis on clinical samples, Lim et al. [10] designed primer pairs to fully cover the large exons, 11 of both BRCA1 and BRCA2, dividing them into two and three overlapping segments, respectively; each segment was roughly 2 kb in size. One hundred clinical genomic DNA samples collected from blood were analyzed by utilizing the T³-ELISA-PTT (25 BRCA1 mutants, 25 BRCA2 mutants, and 50 normal controls). The mutation status of the samples had been previously determined by conventional gel-based PTT. T³-ELISA-PTT results are shown in Figure 35.4A and B. The range of mutations covered is detailed in Supplemental Table S1 in Additional File 1 of Lim et al. [10]. This sample cohort contained deletions as large as 40 bases, all of which were detected by the alternative T³-ELISA-PTT method.

Despite the aforementioned segment-specific % C/N ratios for heterozygous mutants in the T³-ELISA-PTT, a fixed method based on the variability within the WT cohort was still able to be employed for setting the diagnostic scoring cutoffs. T³-ELISA-PTT scoring cutoffs for a positive protein truncation were fixed at 3 standard deviations below the mean of the 50 normal (WT) samples (on a segment-by-segment basis) for an approximately 99% confidence interval. With these cutoffs (Figure 35.4A and B; green bar and green line), the sensitivity and specificity of the T³-ELISA-PTT (as compared with conventional PTT) were 100% and 96%, respectively. Note that the % C/N ratios for the only two false-positive calls (Figure 35.4A and B; black bars) were just slightly below the cutoff and were likely due to assay variance (approximately 1/100 false-positive calls are statistically expected with the aforementioned 99% confidence interval used).

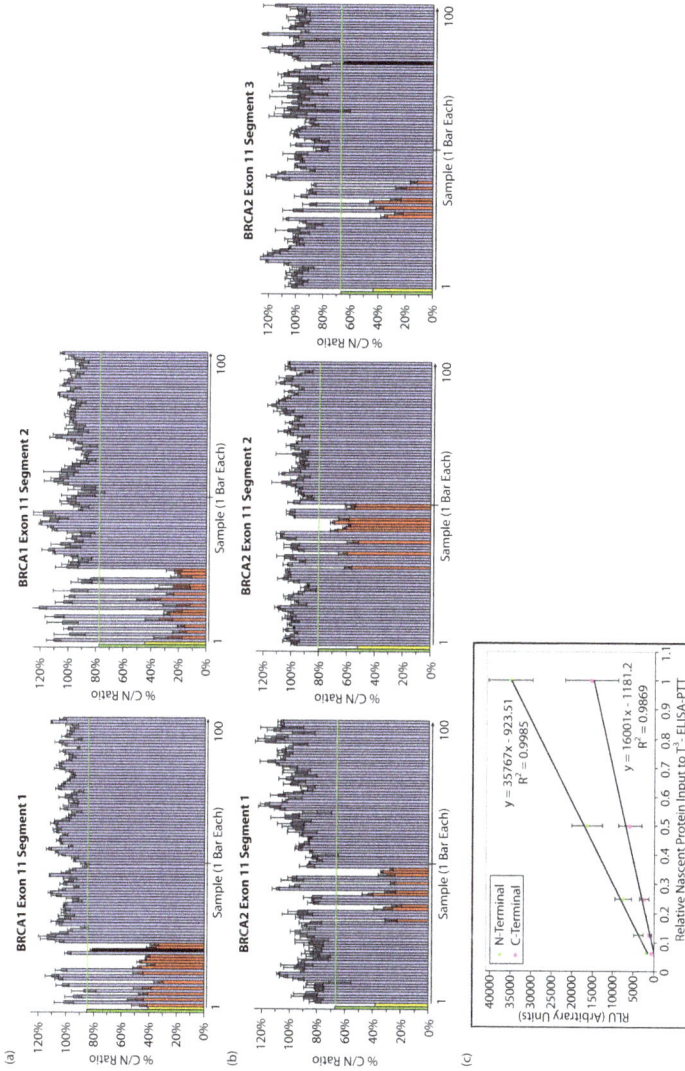

FIGURE 35.4 Clinical validation of T^3-ELISA-PTT on BRCA1/2 exons 11 for inherited breast cancer susceptibility [10]. One hundred blood-derived clinical genomic DNA samples were analyzed by T^3-ELISA-PTT. (Panel A) BRCA1 covered in two overlapping segments. (Panel B) BRCA2 covered in three overlapping segments. (Panel C) Standard curve of different protein inputs into the T^3-ELISA-PTT enzyme-linked immunosorbent assay (ELISA) using serial dilutions of known WT samples. A percentage of C/N is the C- to N-terminal signal ratio of a given sample normalized against known WT controls. The green bars and lines indicate the designated threshold (three standard deviations below the mean for the WT cohort), above which samples are scored as WT (blue bars) and below which samples are scored as mutant (red bars). Yellow bars indicate the predicted mutant % C/N based on the known WT standard curves (Panel C). Black bars denote false positives. The tick mark on the x-axis divides the 50 patients with known BRCA1/2 truncation mutations (left of tick mark) from the 50 WT patients (right of tick mark). BRCA1/BRCA2 = breast and ovarian cancer susceptibility gene 1/2; RLU = relative light unit.

GENERAL CONSIDERATIONS OF PTT

SENSITIVITY OF PTT

The PTT is most often used to detect truncating mutations in tumor suppressor genes from patients that are heterozygous for the mutation, using genomic DNA as starting material. The mutant allele represents 50% of the alleles present in the sample, and the PTT, like all other scanning methodologies, can easily detect mutations at the 50% level. There are other applications, however, where the mutant allele is present at much less than 50% (e.g., in a tumor biopsy or fecal material from colorectal cancer patients), and in these cases, sensitivity becomes an issue. In comparison with other mutation detection techniques, the sensitivity of gel-PTT is high and is capable of detecting mutant alleles at one part in five. Recently, Traverso et al. have demonstrated detection efficiencies of chain truncation mutations as low as 0.4% relative to WT [22]. This was achieved by first diluting genomic DNA samples so that no more than two to four DNA templates are present in each sample prior to PCR amplification. This is followed by the translation of the amplified DNA for approximately 144 samples and detection using radioactive gel-based PTT. As discussed previously [9], ELISA-PTT is ideal for such an application as radioactive gel-based detection is not suitable for automation.

ADVANTAGES AND DISADVANTAGES OF PTT

Despite the fact that chain-truncating mutations can be detected by conventional scanning methods such as DNA sequencing or single-strand conformation polymorphism (SSCP) analysis, PTT remains the method of choice for many laboratories. This is primarily due to its ability to accurately scan a large sequence (as large as 3 kb) in a single reaction. Once a chain truncation is identified and its position on a gel is known, a much smaller region of DNA can be scanned by sequencing to confirm and characterize the mutation. Although PTT detects only chain truncating mutations, these mutations are invariably associated with the disease, especially in the case of tumor suppressor genes because truncated proteins are almost always nonfunctional. In contrast, gene variants that result only in an amino acid substitution are often difficult for the clinician to interpret without additional information.

However, there are variety of limitations to the PTT approach. First, PTT is usually only appropriate for those genes that contain a high proportion of protein-truncating mutations. In addition, it is only effective for large exons as protein fragments encoded by smaller exons are not easily analyzed by SDS-PAGE. This limitation can be overcome by using mRNA as the starting material. However, there are handling and storage concerns with the use of mRNA due to nonsense-mediated mRNA decay [23] and the lack of mRNA expression in accessible tissue (peripheral blood). Also, PTT cannot detect mutations occurring outside the coding region, such as those that affect control of the gene expression.

A number of problems can arise during the PCR amplification step used in PTT. For example, if the mutated allele has a large insertion in the amplicon or if the mutation is a deletion that includes the primer binding site, the mutant allele will not amplify. Also, polymerase error in the first few cycles can lead to an artifact that can be mistaken for an authentic mutation. Artifacts can also be produced from in vitro translation and include false initiation from internal ATG codons (which is more severe in *Escherichia-coli*-derived reaction mixtures) and proteolytic degradation of the full-length protein by endogenous proteases present in the in vitro synthesis lysate. These and other factors can give rise to background bands in the WT control samples that can interfere with the detection of an overlapping mutant band.

ALTERNATIVE METHODS FOR DETECTING CHAIN TRUNCATIONS

Many genotyping methods exist which can detect chain truncating mutations as well as other types of mutations. DNA sequencing is considered the gold standard for genotyping as it can detect any mutation that can be amplified by PCR, including those that have never been previously described [23, 24]. Alternatively, hybridization-based approaches for detecting mutations such as TaqMan® (Applied Biosystems/ Thermo Fisher), Invader (Third Wave Technology/Hologic), and high-density chip arrays (Affymetrix/Thermo Fisher) are some of the high-throughput methods that are only capable of detecting previously identified mutations and will miss de novo mutations [25]. In one study [26], five different mutation detection methods to detect truncation mutations in *BRCA*1 were compared. These included four DNA-based methods: Two-dimensional gene scanning (TDGS), denaturing high-performance liquid chromatography (DHPLC), enzymatic mutation detection (EMD), SSCP, and PTT. Analysis of 21 samples showed that PTT correctly identified all 15 deleterious mutations. Not surprisingly, the DNA-based techniques did not detect a deletion of exon 22, and five truncating mutations were missed by SSCP. In another study [27], researchers compared the effectiveness of PTT and SSCP for detecting *BRCA*1 mutations and concluded that PTT is a superior screening test as the specificity of PTT was 100% and sensitivity was 82.6%, whereas for SSCP, the specificity was 99% but the sensitivity was only 60.9%. However, the rapid progress in the area of next-generation sequencing (NGS) and even single-molecule DNA sequencing in the last decade is significantly reducing the need for PTT, as well as the alternative methods described above. This is due to NGS's increased sensitivity, high throughput, reduced cost, and widespread availability of NGS instrumentation, including single-cell sequencing capabilities [32, 33].

CONCLUSION

Advances described here overcame some of the original limitations of the PTT. This includes the development of ELISA-PTT [9, 10], multicolor-PTT [19], and mass spectrometry-based PTT [21]. Research uses include detecting previously unidentified truncating mutations such as in tumor suppressor genes. In addition, by only identifying disease-causing truncation mutations, PTT analysis is not hampered by false-positive signals derived from phenotypically silent variants such as polymorphisms. Also, the size of the truncated product localizes the DNA mutation and thus enhances sequence analysis for confirmation. While the advent of rapid, low-cost next-generation DNA sequencing is likely to supplant PTT as the method of choice for the detection of chain-terminating mutations in both research and clinical settings, ELISA-PTT still provides a simple and inexpensive method accessible to most research labs since it requires only simple PCR and microtiter plate reader instrumentation.

ACKNOWLEDGMENTS

The development of ELISA-PTT was supported by SBIR Phase I and II grants (CA83396 and CA110403) from the NIH/NCI to AmberGen, Inc.

REFERENCES

1. Powell, S.M.; Petersen, G.M.; Krush, A.J.; Booker, S.; Jen, J.; Giardiello, F.M.; Hamilton, S.R.; Vogelstein, B.; Kinzler, K.W. Molecular diagnosis of familial adenomatous polyposis. N. Engl. J. Med. **1993**, *329*(27), 1982–1987.
2. van der Luijt, R.; Khan, P.M.; Vasen, H.; van Leeuwen, C.; Tops, C.; Roest, P.; den Dunnen, J.; Fodde, R. Rapid detection of translation-terminating mutations at the adenomatous polyposis coli (APC) gene by direct protein truncation test. Genomics **1994**, *20*(1), 1–4.
3. Garvin, A.M. A complete protein truncation test for BRCA1 and BRCA2. Eur. J. Hum. Genet. **1998**, *6*(3), 226–234.
4. Rossetti, S.; Strmecki, L.; Gamble, V.; Burton, S.; Sneddon, V.; Peral, B.; Roy, S.; Bakkaloglu, A.; Komel, R.; Winearls, C.G.; Harris, P.C. Mutation analysis of the entire PKD1 gene: Genetic and diagnostic implications. Am. J. Hum. Genet. **2001**, *68*(1), 46–63.
5. Messiaen, L.M.; Callens, T.; Mortier, G.; Beysen, D.; Vandenbroucke, I.; Van Roy, N.; Speleman, F.; Paepe, A.D. Exhaustive mutation analysis of the NF1 gene allows identification of 95% of mutations and reveals a high frequency of unusual splicing defects. Hum. Mutat. **2000**, *15*(6), 541–555.
6. Mayer, K. Application of the protein truncation test (PTT) for the detection of tuberculosis sclerosis complex type 1 and 2 (TSC1 and TSC2) mutations. Methods Mol. Biol. **2003**, *217*, 329–344.
7. Roest, P.A.; Roberts, R.G.; van der Tuijn, A.C.; Heikoop, J.C.; van Ommen, G.J.; den Dunnen, J.T. Protein truncation test (PTT) to rapidly screen the DMD gene for translation terminating mutations. Neuromuscul. Disord. **1993**, *3*(5–6), 391–394.
8. Roest, P.A.; Roberts, R.G.; Sugino, S.; van Ommen, G.J.; den Dunnen, J.T. Protein truncation test (PTT) for rapid detection of translation-terminating mutations. Hum. Mol. Genet. **1993**, *2*(10), 1719–1721.
9. Gite, S.; Lim, M.; Carlson, R.; Olejnik, J.; Zehnbauer, B.; Rothschild, K.J. A high-throughput nonisotopic protein truncation test. Nat. Biotechnol. **2003** Feb, *21*(2), 194–197.
10. Lim, M.J.; Foster, G.J.; Gite, S.; Ostendorff, H.; Narod, S.; Rothschild, K.J. An ELISA-based high throughput protein truncation test for inherited breast cancer. Breast Cancer Res. **2010**, *12*(5), R78.
11. Vossen, R.; den Dunnen, J.T. Protein truncation test. Curr. Protoc. Hum. Genet. **2004**, *9*, 11

12. Kirchgesser, M.; Albers, A.; Vossen, R.; den Dunnen, J.; van Ommen, G.J.; Gebert, J.; Dupont, C.; Herfarth, C.; von Knebel-Doeberitz, M.; Schmitz-Agheguian, G.; Kirchgesser, M. et al. Optimized non-radioactive protein truncation test for mutation analysis of the adenomatous polyposis coli (APC) gene. Clin. Chem. Lab. Med. **1998** Aug, *36*(8), 567–570.

13. Petrakis, E.C.; Trantakis, I.A.; Kalogianni, D.P.; Christopoulos, T.K.; Petrakis, E.C. et al. Screening for unknown mutations by a bioluminescent protein truncation test with homogeneous detection. J. Am. Chem. Soc. **2010** Apr, *132*(14), 5091–5095.

14. Hauss, O.; Müller, O. The protein truncation test in mutation detection and molecular diagnosis. Methods Mol. Biol. **2007**, *375*, 151–164.

15. Du, L.; Lai, C.H.; Concannon, P.; Gatti, R.A. Rapid screen for truncating ATM mutations by PTT-ELISA. Mutat. Res. **2008** Apr, *640*(1–2), 139–144.

16. Real, S.M.; Marzese, D.M.; Gomez, L.C.; Mayorga, L.S.; Roqué, M. Development of a premature stop codon-detection method based on a bacterial two-hybrid system. BMC Biotechnol. **2006** Sep, *6*, 38.

17. Hardy, C.A. The protein truncation test. In *PCR Mutation Detection Protocols. Methods in Molecular Biology*; Theophilus B.D.M.; Rapley R., Eds.; Humana Press: Totowa, NJ, **2002**; Vol. *187*. 10.1385/1-59259-273-2:087

18. Kahmann, S.; Herter, P.; Kuhnen, C.; Muller, K.M.; Muhr, G.; Martin, D.; Soddemann, M.; Muller, O. A nonradioactive protein truncation test for the sensitive detection of all stop and frameshift mutations. Hum. Mutat. **2002**, *19*(2), 165–172.

19. Traverso, G.; Diehl, F.; Hurst, R.; Shuber, A.; Whitney, D.; Johnson, C.; Levin, B.; Kinzler, K.W.; Vogelstein, B. Multicolor in vitro translation. Nat. Biotechnol. **2003**, *21*(9), 1093–1097.

20. Gite, S.; Mamaev, S.; Olejnik, J.; Rothschild, K. Ultrasensitive fluorescence-based detection of nascent proteins in gels. Anal. Biochem. **2000**, *279*(2), 218–225.

21. Garvin, A.M.; Parker, K.C.; Haff, L. MALDI-TOF based mutation detection using tagged in vitro synthesized peptides. Nat. Biotechnol. **2000**, *18*(1), 95–97.

22. Traverso, G.; Shuber, A.; Levin, B.; Johnson, C.; Olsson, L.; Schoetz, D.J. Jr.; Hamilton, S.R.; Boynton, K.; Kinzler, K.W.; Vogelstein, B. Detection of APC mutations in fecal DNA from patients with colorectal tumors. N. Engl. J. Med. **2002**, *346*(5), 311–320.

23. Bateman, J.F.; Freddi, S.; Lamande, S.R.; Byers, P.; Nasioulas, S.; Douglas, J.; Otway, R.; Kohonen-Corish, M.; Edkins, E.; Forrest, S. Reliable and sensitive detection of premature termination mutations using a protein truncation test designed to overcome problems of nonsense-mediated mRNA instability. Hum. Mutat. **1999**, *13*(4), 311–317.

24. Schuurman, R.; Demeter, L.; Reichelderfer, P.; Tijnagel, J.; de Groot, T.; Boucher, C. Worldwide evaluation of DNA sequencing approaches for identification of drug resistance mutations in the human immunodeficiency virus type 1 reverse transcriptase. J. Clin. Microbiol. **1999**, *37*(7), 2291–2296.

25. Gavert, N.; Yaron, Y.; Naiman, T.; Bercovich, D.; Rozen, P.; Shomrat, R.; Legum, C.; Orr-Urtreger, A. Molecular analysis of the APC gene in 71 Israeli families: 17 Novel mutations. Hum. Mutat. **2002**, *19*(6), 664.

26. Andrulis, I.L.; Anton-Culver, H.; Beck, J.; Bove, B.; Boyd, J.; Buys, S.; Godwin, A.K.; Hopper, J.L.; Li, F.; Neuhausen, S.L.; Ozcelik, H.; Peel, D.; Santella, R.M.; Southey, M.C.; van Orsouw, N.J.; Venter, D.J.; Vijg, J.; Whittemore, A.S. Comparison of DNA- and RNA-based methods for detection of truncating BRCA1 mutations. Hum. Mutat. **2002**, *20*(1), 65–73.

27. Geisler, J.P.; Hatterman-Zogg, M.A.; Rathe, J.A.; Lallas, T.A.; Kirby, P.; Buller, R.E. Ovarian cancer BRCA1 mutation detection: Protein truncation test (PTT) outperforms single strand conformation polymorphism analysis (SSCP). Hum. Mutat. **2001**, *18*(4), 337–344.

28. Fearon, E.R.; Vogelstein, B. A genetic model for colorectal tumorigenesis. Cell **1990**, *61*(5), 759–767.

29. Kohonen-Corish, M.; Ross, V.L.; Doe, W.F.; Kool, D.A.; Edkins, E.; Faragher, I.; Wijnen, J.; Khan, P.M.; Macrae, F.; St. John, D.J. RNA-based mutation screening in hereditary nonpolyposis colorectal cancer. Am. J. Hum. Genet. **1996**, *59*(4), 818–824.

30. Romey, M.C.; Tuffery, S.; Desgeorges, M.; Bienvenu, T.; Demaille, J.; Claustres, M. Transcript analysis of CFTR frameshift mutations in lymphocytes using the reverse transcription-polymerase chain reaction technique and the protein truncation test. Hum. Genet. **1996**, *98*(3), 328–332.

31. Broeks, A.; de Klein, A.; Floore, A.N.; Muijtjens, M.; Kleijer, W.J.; Jaspers, N.G.; van't Veer, L.J. ATM germ-line mutations in classical ataxia–telangiectasia patients in the Dutch population. Hum. Mutat. **1998**, *12*(5), 330–337.

32. Yohe, S.; Thyagarajan, B. Review of clinical next-generation sequencing. Arch. Pathol. Lab. Med. **2017** Nov, *141*(11), 1544–1557.

33. Pfisterer, E.; Bräunig, J.; Brattås, P.; Heidenblad, M.; Karlsson, G.; Fioretos, T. Single-cell sequencing in translational cancer research and challenges to meet clinical diagnostic needs. *Genes Chromosomes Cancer* **2021** Jul, *60*(7), 504–524.

Index

Note: Locators in *italics* represent figures and **bold** indicate tables in the text.

For Product Safety Concerns and Information please contact our EU
representative GPSR@taylorandfrancis.com
Taylor & Francis Verlag GmbH, Kaufingerstraße 24, 80331 München, Germany

www.ingramcontent.com/pod-product-compliance
Lightning Source LLC
Chambersburg PA
CBHW060758220326
41598CB00022B/2476

* 9 7 8 1 0 3 2 1 6 1 9 0 7 *